Smart Penetration of Electric Vehicles in Distribution Grids

Smart Penetration of Electric Vehicles in Distribution Grids

Linni Jian
Department of Energy and Power Engineering, City University of Hong Kong (Dongguan), Dongguan, China

Hang Yu
School of Naval Architecture, Ocean and Energy Power Engineering, Wuhan University of Technology, Wuhan, China

ELSEVIER

Elsevier
Radarweg 29, PO Box 211, 1000 AE Amsterdam, Netherlands
125 London Wall, London EC2Y 5AS, United Kingdom
50 Hampshire Street, 5th Floor, Cambridge, MA 02139, United States

ISBN: 978-0-443-27706-1

For information on all Elsevier publications visit our website at
https://www.elsevier.com/books-and-journals

Publisher: Priscilla Braglia
Acquisitions Editor: Dan Egan
Editorial Project Manager: Somoshri Banerji
Production Project Manager: Karthikeyan Durairaj
Cover Designer: Arumugam Kothandan

Typeset by TNQ Tech

Contents

3. Industrial standards and infrastructure technologies

4. Exploiting electric vehicles' benefits within household nanogrids

Preface

The electrification of transportation is reshaping the planning, operation, and control of modern power distribution systems. Electric vehicles (EVs) provide an effective pathway to reduce fossil fuel consumption and carbon emissions in the transportation sector. However, the rapid growth of EVs also introduces new challenges to distribution grids, especially when large-scale charging demand is imposed on low-voltage networks in an uncoordinated manner. Random and concentrated charging behaviors may increase peak loads, aggravate transformer and feeder congestion, cause voltage deviations, deteriorate power quality, and require costly upgrades of the existing grid infrastructures.

At the same time, EVs should not be regarded merely as additional electrical loads. With proper charging coordination, bidirectional power conversion, communication, and energy management, they can act as flexible and distributed energy resources. Vehicle-to-grid (V2G) technology enables EVs to exchange power with distribution grids and provide multiple grid-support functions, such as peak shaving, valley filling, renewable energy accommodation, voltage regulation, and auxiliary services. Therefore, the large-scale penetration of EVs is both a challenge and an opportunity for future active distribution grids.

This book focuses on the smart penetration of EVs within distribution grids. It aims to provide a systematic discussion of EV–grid integration from theoretical foundations to engineering infrastructures and from general scheduling schemes to practical application scenarios. The main concern of the book is how EVs can be effectively integrated into distribution grids through coordinated charging, V2G operation, advanced power architectures, hierarchical energy management, and cyber–physical coordination. Particular attention is given to active distribution grids, household nanogrids, hybrid AC/DC microgrids, nearly zero–energy buildings, old residential communities, and the internet of smart charging points.

The book is written to offer researchers, engineers, graduate students, and practitioners in EVs, power systems, smart grids, microgrids, charging infrastructures, and building energy systems a comprehensive reference. It is also expected to support further research and engineering practices for low-carbon transportation, flexible distribution grid operation, and renewable energy–oriented energy management.

The book is divided into seven chapters.

Chapter 1 introduces the background, motivation, and objectives of the book. It discusses the development of EVs, the characteristics of distribution grids, the need for carbon emission reduction on the distribution side, and the sustainable thinking required for large-scale EV deployment. The chapter also presents the overall structure of the book.

Chapter 2 presents the general schemes and mathematical foundations of EV–grid interaction. The chapter introduces plug-in EVs, uncoordinated charging behaviors, and their negative impacts on power grids. It then discusses coordinated V2G operation, including unidirectional and bidirectional power interaction modes. Centralized and decentralized scheduling schemes are compared, and the mathematical foundations of optimal scheduling are explained, including the objective functions, scheduling constraints, solving algorithms, and computational performance evaluation.

Chapter 3 reviews industrial standards and infrastructure technologies for EV integration in active distribution grids. The chapter introduces nanogrids, microgrids, and clusters as important architectures in active distribution systems. It also discusses charging standards, grid integration standards, on-board and off-board charging circuits, and typical architectures for the integration of EVs into distribution grids. AC-coupled, DC-coupled, hybrid AC/DC-coupled, and advanced interlinking converter–based architectures are assessed and compared in terms of charging demand compatibility, power quality, V2G availability, scalability, control complexity, and technology maturity.

Chapter 4 investigates how EVs can be used to exploit operational benefits within household nanogrids. A DC nanogrid architecture incorporating photovoltaic generation, battery energy storage, local loads, and EVs is presented. A pseudo-hierarchical energy management architecture is introduced to achieve both autonomous local power coordination and regulated power dispatching. The chapter analyzes multiple operation scenarios in peak, shoulder, and valley periods and verifies the effectiveness of the proposed strategy through simulation and experimental studies.

Chapter 5 discusses the role of EV–grid coordination in facilitating nearly zero–energy buildings. A hybrid AC/DC microgrid architecture for building applications is presented, together with a hierarchical management structure. The chapter focuses on local control strategies, power flow dispatching strategies, and the coordinated operation between AC and DC subgrids. It shows how EVs, renewable energy resources, and energy storage systems can be integrated to improve local renewable energy consumption, enhance operational flexibility, and support the energy performance of nearly zero–energy buildings.

Chapter 6 addresses the difficult-to-charge issue in old residential communities. Many existing residential distribution networks were not originally designed to accommodate large-scale EV charging, and conventional grid reinforcement may be restricted by high cost and limited physical space. This chapter presents a shared fast charging scheme and a hybrid AC/DC microgrid reconstruction method for old residential communities. Hierarchical energy management and coordinated operation strategies are developed to improve EV accommodation capacity, realize peak shaving and valley filling, and enhance the operation performance of existing low-voltage distribution grids.

Chapter 7 introduces the internet of smart charging points for V2G operation. The chapter discusses the physical scheme, communication network, distributed computing structure, and privacy-preserving characteristics of smart charging point systems. A cyber–physical comodeling method is presented to integrate energy flow and information flow. The chapter further investigates optimal energy dispatching for internal and external V2G regulation and analyzes a case study involving photovoltaic integration. The proposed framework highlights the potential of edge computing and smart charging infrastructure in enabling scalable, efficient, and secure EV–grid coordination.

The authors hope that this book will help readers understand the technical principles, infrastructure requirements, control methods, and practical implementation pathways for the smart penetration of EVs into distribution grids. As EVs, renewable energy resources, and active distribution networks continue to develop, their coordinated operation will become an essential component of future low-carbon energy systems.

Linni Jian
Dongguan, China
Hang Yu
Wuhan, China
April 2026

Acknowledgments

This book summarizes our research on the smart interaction between electric vehicles and distribution grids from 2018 to 2026. The enabled V2G technology would bring multiple benefits to distribution network systems at different levels and to electric vehicle owners themselves. It is impossible to name all the students and colleagues from academia and industry who have contributed to this edition of the book by asking intelligent questions or making constructive comments—thanks to all of you. We are especially grateful to Prof. C. C. Chan for his great help in inspiring and guiding our research ideas. We also thank Prof. Yanchong Zheng, Prof. Yitong Shang, and Prof. Xiang Lei who made many valuable suggestions for improvement. Special thanks are due to Dr. Jiahao Zhong for his careful examination of the entire book. We also wish to thank Dr. Songyan Niu, Dr. Haibiao Cheng, Mr. Cheng Zhang, Ms. Man Liu, and Ms. Yumeng Zhang. Finally, we appreciate the assistance and help from the staff at Elsevier.

Linni Jian
Hang Yu

Chapter 1

Introduction

1.1 Electric vehicles

Electric vehicles (EVs) have moved from a peripheral technological option to a central pathway in the decarbonization of road transport. This shift is driven by the combined pressures of climate mitigation, energy transition, urban air-quality management, and advances in electrochemical energy storage [1]. The transport sector remains a major source of greenhouse gas emissions worldwide, and road transport contributes the largest share within that sector [2]. In this context, the electrification of vehicles is increasingly regarded not merely as a substitution of propulsion technology, but as a structural transformation of the transport-energy nexus. When supplied by low-carbon electricity, EVs can substantially reduce the emissions intensity of road mobility [3]. At the same time, their widespread deployment introduces new operational interactions with the power system, especially at the distribution level. This development is significant because road transport has historically been one of the most difficult sectors to decarbonize. Unlike stationary energy uses, transport demand is highly dispersed, strongly behavior-dependent, and closely tied to daily routines, spatial accessibility, and infrastructure availability. Conventional decarbonization options in this sector have often been constrained by fuel logistics, vehicle cost, and the limited controllability of combustion-based technologies. Electrification changes this situation by connecting mobility demand to an energy carrier that can be generated from diverse primary resources, transmitted through networked infrastructure, and increasingly optimized through digital monitoring and control. In this sense, EVs are important not only because they are cleaner vehicles, but because they make road transport more compatible with the broader evolution of modern energy systems.

From a technical perspective, the term electric vehicle is often used in a broad sense to describe road vehicles that employ electric drivetrains either fully or partially. In common engineering and policy discourse, EVs encompass battery electric vehicles (BEVs), plug-in hybrid electric vehicles (PHEVs), and hybrid electric vehicles (HEVs) [4]. BEVs are propelled solely by electric motors and store onboard energy in rechargeable batteries that must

Smart Penetration of Electric Vehicles in Distribution Grids. https://doi.org/10.1016/B978-0-443-27706-1.00008-8

be charged from an external electricity source. PHEVs combine an electric drivetrain with an internal combustion engine and can also be charged by plugging into the grid. They typically operate in electric mode for a limited distance before switching to fuel-based propulsion. HEVs likewise combine electrical and combustion-based propulsion, but their batteries are charged internally, such as through regenerative braking, rather than by external connection to the grid. In a stricter grid-integration context, however, the most relevant categories are BEVs and PHEVs, because these vehicles can draw power directly from external charging infrastructure and thus participate in managed charging or vehicle-to-grid (V2G) interactions. This distinction is not merely taxonomic. It has direct implications for infrastructure planning, charging demand, and power system operation. HEVs improve fuel efficiency but do not materially alter electricity demand patterns in the distribution network. By contrast, BEVs and PHEVs function as mobile electrical loads whose charging behavior is shaped by user routines, battery capacity, charging power, and access to residential, workplace, or public charging facilities. For this reason, many studies and industrial reports use the term plug-in electric vehicles (PEVs) when referring specifically to BEVs and PHEVs. In this book, EVs refer specifically to BEVs and PHEVs, because they constitute the physical interface through which the transport sector becomes coupled with the electricity sector. This scope definition is methodologically important. The broader universe of electrified transport may include rail systems, electric two-wheelers, hybridized drivetrains, and other specialized mobility devices, yet not all of these interact with the power grid in the same manner. The focus on plug-in vehicles reflects the fact that externally chargeable batteries create a direct and recurring link between transport activity and electricity demand. Once a vehicle must be connected to the grid for energy replenishment, its operation becomes relevant not only to transport planning, but also to load forecasting, infrastructure sizing, tariff design, and network management. Moreover, the key technical attributes of plug-in EVs are not limited to the battery itself. Electric motors, onboard chargers, battery management systems, and power-electronic converters jointly determine vehicle efficiency, charging compatibility, operational safety, and the feasibility of smart control. Battery capacity influences range and charging duration; charging power affects both user convenience and instantaneous demand on the grid; and the power-electronic interface makes it possible, at least in principle, to modulate charging behavior in response to network conditions or price signals. These properties distinguish plug-in EVs from conventional end-use transport devices and make them especially relevant in the context of distribution-grid analysis.

At the same time, it is increasingly important to recognize that electric mobility is not limited to standard passenger cars. In many developing or infrastructure-constrained settings, a distinct category of small and low-speed electric vehicles has emerged as a practical complement to mainstream EV development. These vehicles generally include compact four-wheeled battery-powered vehicles, electric tricycles, and other short-range personal mobility platforms that are positioned between electric bicycles and conventional electric cars in terms of size, speed, cost, and technical complexity. Although they often fall outside the mainstream classification frameworks established for highway-capable passenger EVs, they nevertheless represent a meaningful form of transport electrification, particularly for short-distance travel, first- and last-mile mobility, rural access, community circulation, and low-income household transport needs. Recent discussion has highlighted that such low-speed EVs deserve greater attention in global sustainability discourse because they provide a clean and affordable mobility option for user groups that are often underserved by the conventional EV transition, including older adults, rural residents, and lower-income populations [5]. Compared with full-sized electric cars, these vehicles typically rely on smaller battery packs, lower-voltage electrical architectures, and simpler manufacturing processes, which reduce both their purchase cost and their material intensity. In practical terms, this means that electrified mobility can be extended to a much broader segment of society, rather than remaining concentrated in markets where consumers can afford large, sophisticated, and resource-intensive electric vehicles.

The rapid expansion of EV deployment over the past decade has been enabled by several reinforcing developments. First, battery technologies have progressed in energy density, safety, durability, and manufacturability, while costs have declined sufficiently to improve the competitiveness of electrified transport [6]. Second, policy support has accelerated market penetration through purchase incentives, fuel economy standards, zero-emission vehicle mandates, public procurement, and charging infrastructure programs [7]. Third, consumer acceptance has increased as vehicle range, model diversity, digital functionality, and charging convenience have improved [8]. By the end of 2024, the global stock of EVs had reached almost 58 million worldwide, indicating the rapid expansion of electric mobility in recent years [9]. Despite this momentum, the significance of EVs cannot be understood solely through market diffusion statistics. Their broader importance lies in the fact that they alter both the energy carrier and the control architecture of road transport. Conventional vehicles rely on liquid fuels supplied through a largely one-way logistics chain. EVs, by contrast, depend on electricity, a highly flexible and increasingly low-carbon energy carrier that is delivered through networked infrastructure and can, in principle, be managed in time and space. This attribute makes EVs fundamentally different from traditional automobiles. An EV is at once a transport asset and an electrical device. It provides mobility services to the user, yet it is also a schedulable load connected to the grid through power-electronic interfaces. Under appropriate technical and regulatory conditions, it may further be regarded as a distributed storage resource capable of offering flexibility services [10]. Accordingly, EV deployment should not be viewed simply as fleet substitution, it should be understood as the introduction of millions of controllable electrical endpoints into the energy system. This interpretation has several important implications. First, the impact of EV growth depends not only on the number of vehicles sold, but also on when they are charged, at what power level, and in which parts of the network they are connected. Second, the electrification of transport links consumer mobility choices to system-level electricity planning in a way that has few precedents in conventional automotive history. Third, EV adoption creates a new class of distributed electrical assets whose aggregate behavior may either intensify network stress or contribute to system flexibility, depending on how charging is organized. It is therefore insufficient to view EV diffusion purely through the lens of market penetration or emissions accounting. A more complete perspective recognizes that electrified transport introduces a new layer of coupling between infrastructure systems that were previously more loosely connected. Vehicle ownership, travel behavior, parking duration, charging access, and electricity pricing all become interdependent variables. This is precisely why EVs have emerged as a key subject not only in transport engineering, but also in power systems, energy economics, urban planning, and sustainability governance.

From the standpoint of sustainability, EVs possess several widely recognized advantages [11]. The most immediate is the elimination of tailpipe emissions in the case of BEVs, which contributes to improved urban air quality and reduced local exposure to pollutants such as nitrogen oxides and particulate matter [12]. In addition, electric drivetrains are generally more energy-efficient than internal combustion engines because they convert a larger fraction of input energy into wheel propulsion. EVs also offer regenerative braking, lower maintenance requirements for BEVs due to fewer moving parts, and the possibility of synergistic integration with renewable electricity. These features explain why electric mobility has been widely identified as a key route toward transport decarbonization. Nonetheless, the magnitude of environmental benefit depends on the carbon intensity of the electricity used for charging, the lifecycle profile of battery manufacturing, and the operational context in which the vehicles are deployed. Therefore, EVs are not inherently or automatically sustainable in all settings. Their net contribution depends on how they are embedded in larger technological and institutional systems. This point deserves emphasis because discussions of EVs are often polarized between technological optimism and lifecycle skepticism. In practice, both perspectives capture only part of the picture. EVs can substantially reduce operational emissions, especially in regions with cleaner power systems, but they also shift part of the environmental burden upstream to electricity generation, battery materials extraction, manufacturing, and end-of-life treatment. Their sustainability performance is therefore systemic rather than intrinsic. It depends on the composition of the power mix, the design of industrial supply chains, the durability of batteries, and the effectiveness of recycling and second-life strategies. In addition, EV deployment raises several practical challenges that must be acknowledged alongside its advantages. These include relatively high upfront vehicle cost in many markets, uneven access to convenient charging, uncertainty in user charging behavior, and the potential mismatch between rapid vehicle adoption and slower infrastructure reinforcement. Concerns related to battery degradation, resource dependency, and charger interoperability also remain relevant. Including these issues at the outset is important because the transition to electric mobility is not simply a matter of replacing one vehicle technology with another; it is a process of coordinated technical, economic, and institutional adaptation. A balanced understanding of EVs therefore requires attention not only to their promise, but also to the conditions under which that promise can be realized.

Another important feature of EVs is the temporal flexibility of charging. Unlike refueling a conventional car, charging is often compatible with extended parking durations at home, at work, or in public locations [13]. This means that, in many cases, the timing of electricity consumption can be shifted without compromising the primary mobility function of the vehicle. Such flexibility creates opportunities for coordinated charging and V2G operation. In V2G operation, bidirectional power exchange allows EV batteries to discharge electricity back to buildings or power networks, thereby supporting peak shaving, reserve provision, and local energy balancing [14]. Through these mechanisms, EVs can evolve from passive electricity consumers into active participants in distributed energy management. The technical feasibility of such services has been widely recognized, although their practical implementation depends on communication systems, charging standards, user incentives, battery management, and regulatory support. The importance of this flexibility cannot be overstated. Most private vehicles remain parked for the vast majority of the day, and this idle time creates a charging window that is absent in most other forms of electrical demand. A washing machine, lighting system, or industrial motor typically consumes electricity at the moment when the service is required, whereas a vehicle often requires mobility at one time but energy replenishment at another. This temporal decoupling is what makes EV charging potentially schedulable. However, flexibility is not automatic. If users charge immediately upon returning home during evening peak periods, EV adoption may aggravate local demand spikes and increase stress on transformers and feeders. If charging is shifted toward valley periods or aligned with periods of high renewable generation, the same vehicles may help flatten load curves and improve energy utilization. Hence, EVs can be either problematic or beneficial from a grid perspective, depending on the design of charging strategies, market incentives, and control mechanisms. This duality is one of the central themes of smart EV integration. The relationship between EVs and distributed renewable energy is also increasingly important. As solar photovoltaics and other distributed energy resources become more prevalent, EVs may serve as flexible demand that absorbs otherwise underutilized generation, particularly at the distribution level. In more advanced configurations, coordinated charging can increase local renewable self-consumption, reduce curtailment, and improve the operational efficiency of microgrids, nanogrids, and building-integrated energy systems. Thus, the value of EV flexibility extends beyond mobility alone; it also lies in its capacity to support a more decentralized and responsive electricity system.

For the purposes of this book, it is therefore useful to conceptualize EVs through a dual lens. On the one hand, they are end-use transport technologies intended to replace internal combustion vehicles and reduce dependence on petroleum fuels. On the other hand, they are flexible electrochemical assets that interact directly with charging infrastructure and, through it, with the distribution grid. This duality is essential. If EVs are considered only as cleaner vehicles, analysis remains confined to transport technology and consumer adoption. If they are considered also as energy-system components, a broader set of questions emerges: how charging should be coordinated, how local networks should be planned, how renewable generation can be better utilized, and how vehicle batteries can support rather than burden distribution systems. These questions form the intellectual bridge between electric mobility and smart distribution grids. This dual perspective also helps explain why EV research has become increasingly interdisciplinary. The deployment of electric mobility involves not only vehicle engineering, but also charging infrastructure, user behavior, tariff design, communication architecture, regulatory governance, and distribution-network operation. No single disciplinary lens is sufficient to capture the full implications of EV penetration. For a book concerned with smart distribution grids, this broader framing is especially necessary, because the relevant issues extend far beyond the vehicle itself and into the organization of local energy systems. In this sense, EVs should be understood as both a challenge and an opportunity for the distribution side of the power sector. They introduce new and potentially volatile electricity demand, but they also provide a basis for flexibility, coordination, and renewable integration that conventional vehicles never possessed. Whether EVs ultimately become a burden or a benefit to distribution grids depends on the technical architectures, operational strategies, and policy frameworks through which they are integrated. This is the central premise from which the subsequent chapters of this book proceed.

1.2 Distribution grids

Distribution grids (DGs) form the physical and operational interface between bulk power systems and end users. They are responsible for delivering electricity from higher-voltage transmission systems or primary substations to residential, commercial, and light industrial customers through medium- and low-voltage networks. In conventional power systems, DGs were primarily designed for unidirectional power delivery, that is, from centralized generation sources to dispersed loads. Their planning and operation were therefore largely based on relatively predictable demand patterns and passive end-use behavior. However, this traditional paradigm has been fundamentally challenged by the rapid proliferation of distributed energy resources, power-electronic loads, and increasingly flexible demand [15]. Among these emerging factors, the large-scale integration of EVs has become one of the most influential drivers of change at the distribution level [16]. The importance of DGs in the context of transport electrification lies in the fact that EV charging demand is ultimately imposed on local networks rather than on the power system in the abstract. Although EV adoption is often discussed at the national or system level, its direct technical impacts are usually first manifested in distribution feeders, service transformers, and end-user connection points. In other words, the distribution level is the first layer of the power system that experiences the aggregated consequences of charging behavior. When a growing number of EVs are connected in homes, workplaces, public parking facilities, or commercial charging stations, their electricity demand becomes embedded in the existing local load structure. As a result, the operational compatibility between EVs and DGs becomes a central issue for both system reliability and future infrastructure planning. This consideration is also relevant to the growing presence of small and low-speed EVs discussed in Section 1.1. Although such vehicles usually have smaller battery capacities and lower charging power than standard passenger BEVs, their impacts on DGs should not be overlooked. In many cases, these vehicles are charged through ordinary household outlets or other low-voltage access points, which makes their electricity demand even more directly embedded in residential networks. Individually, their charging loads may appear modest. However, once adopted at scale in specific neighborhoods, rural communities, or peri-urban areas, they can still reshape local load profiles, alter transformer utilization, and increase the importance of low-voltage network management. Therefore, small and low-speed EVs do not change the basic logic of EV–DG interaction, but rather extend it to a broader and more diverse range of electrified mobility scenarios.

Compared with transmission systems, DGs are generally more sensitive to localized fluctuations in demand and more constrained in terms of equipment capacity and voltage regulation. Their assets, including feeders, cables, and transformers, are typically designed according to historical consumption characteristics rather than high-density electrified transport loads. This makes them particularly vulnerable to clustered charging behavior. For example, when many EV users return home and begin charging during evening hours, the resulting simultaneity can significantly increase peak demand in residential networks [17]. Such load concentration may accelerate transformer aging, increase thermal stress on feeders, intensify network losses, and worsen voltage deviation. Similar concerns arise in commercial districts and public charging hubs, especially where fast charging introduces large instantaneous power demand. Therefore, the challenge posed by EVs is not simply the amount of additional electricity consumed, but the temporal and spatial concentration of that consumption within already constrained local infrastructures. This sensitivity is rooted in several structural characteristics of DGs. First, distribution equipment is usually rated with smaller thermal margins than transmission assets, making it less tolerant of sudden or persistent load increases. Second, voltage regulation in distribution systems is more vulnerable to local demand variation because of longer radial feeders, higher R/X ratios, and the proximity of loads to the point of delivery. Third, many distribution assets were installed under assumptions of conventional household and commercial demand growth, rather than under scenarios involving rapid electrification of transport, heating, and distributed generation. As a result, even when overall energy consumption remains manageable, localized coincidence in EV charging can produce disproportionately large technical consequences. The network impacts associated with EV charging are also multi-dimensional. Beyond peak demand growth, they may include increased phase imbalance in low-voltage networks, deterioration of power quality due to widespread power-electronic interfacing, altered reactive power conditions, and more frequent operation of voltage regulation equipment. In areas with substantial fast-charging deployment, short-duration but high-magnitude power draws may create particularly challenging operating conditions. Therefore, the technical question is not simply whether DGs can accommodate more energy, but whether they can accommodate new patterns of demand that are more concentrated, less predictable, and more electronically mediated than traditional end-use consumption.

In this regard, DGs should not be understood merely as passive carriers of electricity. As discussed in Section 1.1, EVs are not only transport assets but also electrical loads with growing flexibility potential. This characteristic is particularly important at the distribution level, where multiple energy resources, flexible loads, control systems, and market mechanisms increasingly interact. The integration of EVs reinforces this transition. On the one hand, EVs behave as new electrical loads whose charging patterns affect distribution-level power quality, asset utilization, and operational security. On the other hand, if coordinated appropriately, they may also provide flexibility that supports DG operation. This dual role gives EVs a special significance in distribution systems. They can aggravate local stress under uncoordinated charging, yet they can also help mitigate variability and improve operational efficiency under coordinated control. Hence, DGs are not simply the location where EV-related problems emerge. They are also the principal domain in which EV-related solutions can be designed and implemented. This transition from passive accommodation to active coordination is fundamental. In traditional distribution engineering, the grid was expected to supply whatever demand emerged at the customer side, with only limited real-time interaction between network operation and end-use behavior. The emergence of EVs weakens this assumption. Because charging demand is often temporally shiftable, at least within certain user-defined boundaries, the operation of DGs increasingly depends not only on infrastructure adequacy but also on the quality of charging management. In other words, the relevant question is no longer only how much network capacity exists, but also how intelligently that capacity is used. This shift has important operational implications. If EV charging is left unmanaged, distribution utilities may face accelerated reinforcement needs and reduced asset efficiency. If, however, charging is coordinated through time-of-use tariffs, direct load control, aggregator-based dispatch, or local energy management platforms, the same EV population may be integrated with lower cost and greater system benefit. Therefore, flexibility is not simply a technological attribute of EV batteries; it is an operational resource whose value depends on the institutional and control framework through which it is activated.

From a planning perspective, the rise of EVs requires a reassessment of how DGs are designed, operated, and upgraded. Traditional load forecasting methods, which rely heavily on stable historical consumption patterns, may be insufficient when EV adoption accelerates and charging behavior evolves rapidly. Infrastructure adequacy can no longer be judged solely on the basis of existing household or commercial demand. It must also reflect future electrified mobility patterns. This creates the need for more granular forecasting, probabilistic analysis, and scenario-based planning at the feeder and transformer level. In addition, charging infrastructure deployment itself must be coordinated with grid capability. Expanding charging access without considering local hosting capacity may shift the burden to already stressed network assets, whereas strategic siting and intelligent management can reduce reinforcement requirements and improve service quality. Planning under EV penetration therefore becomes a problem of uncertainty as much as a problem of scale. Future charging demand depends on uncertain adoption trajectories, heterogeneous user routines, charger power levels, and evolving policy incentives. This uncertainty makes deterministic planning increasingly inadequate, especially in rapidly changing urban and peri-urban environments. As a result, DG planning must move toward more adaptive methods that combine scenario construction, probabilistic load representation, sensitivity analysis, and staged reinforcement strategies. At the same time, the concept of hosting capacity becomes especially relevant. A distribution network may be able to connect additional EV load in principle, but not necessarily at every location, at every time, or at every charging power level without violating thermal or voltage limits. Hence, EV planning is closely linked to the locational capability of DGs. This is why strategic charger placement, coordinated permitting, and grid-aware infrastructure development are often as important as hardware upgrades themselves. In many cases, better information and better coordination can defer or reduce the need for costly network reinforcement.

The strategic importance of DGs is further amplified by the broader transformation of power systems toward decentralization, digitalization, and decarbonization [18]. With more distributed energy resources (DERs) connected at the distribution level, power flows are no longer strictly one-way, and local networks are increasingly expected to support active management rather than passive delivery. Smart meters, advanced inverters, local controllers, communication networks, and data-driven energy management platforms are gradually changing the operational logic of distribution systems [19–21]. In this context, EVs should be regarded as part of a wider ecosystem of flexible and interactive distribution resources. Their charging demand can be coordinated with Photovoltaic (PV) generation, stationary storage, building loads, and tariff structures. Their batteries may also contribute, under suitable conditions, to local balancing and flexibility provision. This perspective naturally leads from conventional DGs toward the concept of active distribution systems. In an active framework, distributed resources are monitored, scheduled, and coordinated to improve efficiency, resilience, and sustainability. Such a framework is highly relevant to EV integration, because the value of EV flexibility can only be captured when distribution-level infrastructure is equipped with appropriate sensing, communication, and control capabilities. The subsequent chapters of this book will examine these issues from different viewpoints, including charging schemes, industrial standards, microgrid architectures, coordinated management in buildings and residential communities, and networked smart charging infrastructures. All these themes share a common premise: EV integration is fundamentally a distribution-level problem and, at the same time, a distribution-level opportunity.

1.3 Carbon emission reduction in distribution side

Carbon emission reduction is one of the most important motivations for the large-scale deployment of EVs [22]. However, from the perspective of power systems, the low-carbon value of EVs should not be understood in an overly simplified manner. In public discourse, EVs are often treated as inherently low-carbon because they replace direct fossil-fuel combustion in road transport. Yet, from an energy-systems perspective, this conclusion is only conditionally valid. Electrification changes the location and form of emissions rather than eliminating them automatically. As noted in Section 1.1, EVs can reduce petroleum consumption and eliminate tailpipe emissions in the case of BEVs, but their overall contribution to decarbonization depends on how the required electricity is generated and how charging demand is integrated into the grid. The Intergovernmental Panel on Climate Change (IPCC) has emphasized that increasing transport electrification affects indirect emissions, particularly where electricity systems remain carbon-intensive [23]. Therefore, for this book, the question is not only whether transport energy use can be electrified, but also how such electrification can contribute to carbon reduction at the distribution side. This shift in perspective is essential because, once electricity becomes the energy carrier of mobility, the carbon outcome of transport is increasingly influenced by the operational characteristics of local power systems rather than by vehicle technology alone.

The distribution side is especially important because it is the point at which electrified transport, end-use demand, distributed generation, and local network operation converge. In conventional transport systems, carbon emissions are directly associated with fuel combustion in vehicles. After electrification, a substantial part of the emission issue is shifted upstream and becomes linked to the temporal and spatial characteristics of electricity consumption. This means that the emissions performance of 1 kW-hour consumed by an EV is not necessarily identical across time and location. The carbon intensity of charging electricity may vary with local generation mix, renewable availability, system load conditions, and even feeder-level operating circumstances. As a result, two identical EVs may deliver different carbon outcomes if they are charged under different distribution-side conditions. In practical terms, this means that the carbon performance of EVs is influenced not only by vehicle technology, but also by when and where charging occurs, whether local renewable generation is available, and whether distribution-level operation can accommodate charging demand efficiently. As a result, carbon reduction at the distribution side is not merely an accounting issue but also an operational and planning issue. It is an accounting issue because emissions must be traced through electricity supply pathways, but it is also an operational issue because charging schedules, network constraints, and local resource coordination directly affect how much low-carbon electricity can actually be used by EVs in practice.

From this perspective, EV integration can support emission reduction through several mechanisms. First, coordinated charging can improve the utilization of low-carbon electricity by shifting charging demand toward periods with higher renewable output or lower marginal generation emissions [24]. The IEA has consistently noted that the full emission-reduction potential of EVs depends on the decarbonization of electricity generation and on their effective integration into power systems. At the distribution level, this implies that EV charging can be aligned with local PV generation, off-peak low-carbon electricity supply, or other favorable operating conditions. This mechanism is particularly important under increasing renewable penetration. In many distribution systems, renewable generation does not always coincide naturally with local demand patterns. EV charging, because of its temporal flexibility, can act as a controllable sink for cleaner electricity during favorable periods. In this sense, coordinated charging is not only a demand-management tool, but also a carbon-management tool. Second, coordinated charging can alleviate peak demand and reduce the need for carbon-intensive peak generation. If EV charging is concentrated in already stressed periods, the additional load may increase reliance on marginal generation sources with higher emissions. By contrast, if charging demand is temporally managed, the distribution system can operate in a more balanced and carbon-efficient manner.

This point deserves emphasis because average grid carbon intensity does not fully capture the emissions consequences of charging behavior. In many systems, the marginal generation unit dispatched during peak periods may be more carbon-intensive than the average supply mix. Therefore, unmanaged charging can weaken the decarbonization benefit of EVs even when the overall power system is becoming cleaner. Smart charging helps reduce this mismatch by moving electricity demand away from periods in which additional generation is both operationally stressful and environmentally unfavorable. Third, the flexibility of EVs creates opportunities to enhance the accommodation of DERs connected at the distribution level. As discussed in Section 1.2, DGs are increasingly characterized by the coexistence of variable renewable resources, flexible loads, and digital control. In this environment, EVs can act as adjustable demand that helps absorb otherwise curtailed renewable electricity. This role is increasingly relevant in active distribution systems where rooftop PV, local storage, and controllable loads coexist within the same local network. Without sufficient flexible demand, part of the renewable electricity generated at the distribution side may be exported under constrained conditions or curtailed altogether. EVs provide a way to convert such variable renewable output into useful transport energy, thereby improving both local energy efficiency and carbon performance. In more advanced configurations, V2G may further strengthen this role by allowing EV batteries to return electricity to local grids or buildings when needed. Although the primary objective of V2G is often discussed in terms of flexibility and peak support, it may also contribute indirectly to carbon reduction by improving the operational value of renewable electricity and reducing the need for less efficient dispatch patterns. Put differently, the carbon benefit of V2G does not arise simply from bidirectional power flow itself, but from the possibility of using stored electricity more strategically in time. If charging and discharging actions are coordinated in accordance with renewable generation conditions and local demand characteristics, EV batteries can support a more carbon-efficient pattern of electricity use at the distribution side. At the same time, these mechanisms should not be idealized. The carbon-reduction value of coordinated charging and V2G depends on enabling conditions such as communication infrastructure, pricing signals, charging standards, user participation, and the availability of sufficiently clean electricity during chargeable periods. Without such conditions, EV electrification may still reduce direct vehicle emissions but fail to achieve its full system-level decarbonization potential. This is precisely why the distribution side matters: it is the practical layer at which technological potential is either realized or weakened by operational reality.

In summary, the distribution side is a critical arena for translating transport electrification into actual carbon reduction. It is here that charging demand can be coordinated with renewable generation, local network operation, and flexible energy management. It is also here that the limits of uncoordinated electrification become visible. Therefore, the contribution of EVs to decarbonization should be evaluated not only in terms of vehicle technology, but also in terms of distribution-level integration and operation.

1.4 Sustainable thinking on electric vehicles

The discussion in Sections 1.1–1.3 suggests that EVs should not be evaluated solely as a substitute for internal combustion vehicles. Their significance is broader and more structural. On the one hand, EVs represent an important pathway for transport electrification and carbon mitigation. On the other hand, their large-scale deployment reshapes electricity demand, charging infrastructure requirements, and the operational conditions of DGs. For this reason, a sustainable understanding of EVs must go beyond the vehicle itself and extend to the wider technical, economic, and social systems within which EVs are deployed. The IEA has explicitly framed EV integration as both a grid challenge and a flexibility opportunity, rather than a simple increase in electricity consumption. In this book, sustainable thinking on EVs refers to a system-oriented perspective in which environmental performance, grid compatibility, economic viability, and user acceptability are considered together. Environmental sustainability remains a central objective, since the electrification of road transport can reduce dependence on petroleum fuels and lower emissions when supported by cleaner electricity. Yet, as noted in Section 1.3, these benefits are conditional rather than automatic. The carbon outcome of EV deployment depends not only on vehicle technology, but also on electricity supply characteristics, charging behavior, and the effectiveness of system integration. The IPCC has emphasized the importance of indirect emissions associated with electricity use, while the IEA has repeatedly shown that the benefits of electric mobility depend on both power-sector decarbonization and charging integration.

At the same time, sustainability in the EV context must include the operational sustainability of DGs. As discussed in Section 1.2, the distribution level is where EV charging demand becomes physically visible and technically consequential. An EV fleet may be environmentally desirable, but if it is integrated through uncoordinated charging, it can produce local peak amplification, voltage deviation, transformer overloading, and inefficient asset utilization. A sustainable EV pathway therefore requires that electrified mobility be compatible with the secure and efficient operation of local power systems. This shifts the focus from simple EV adoption to intelligent EV integration. In this sense, sustainability is not merely a question of how many EVs are deployed, but also of how they are charged, controlled, and incorporated into network operation.

Economic sustainability is equally important. The large-scale deployment of EVs requires investment in charging infrastructure, distribution network reinforcement, communication systems, and management platforms. If this transition is pursued without adequate coordination, costs may rise unnecessarily for utilities, charging service providers, and end users. By contrast, coordinated charging and V2G can improve asset utilization, reduce avoidable peak-driven upgrades, and create additional system value from EV flexibility. This does not mean that all EVs should immediately participate in advanced grid services. Rather, it means that sustainable EV development should seek cost-effective integration pathways in which infrastructure expansion, operational control, and user incentives evolve in a mutually supportive manner.

A further dimension is user-oriented sustainability. EV systems ultimately depend on user participation, because charging behavior is embedded in everyday mobility routines [25]. Technical optimization alone cannot ensure successful integration if charging is inconvenient, inaccessible, or inconsistent with user expectations. Residential charging access, workplace charging availability, public charging coverage, tariff design, digital interfaces, and perceptions of battery degradation all influence whether users are willing to engage with coordinated charging or V2G programs. Sustainable thinking therefore requires attention not only to engineering feasibility, but also to behavioral realism and social practicality.

From a broader perspective, sustainable thinking on EVs also implies a transition from isolated technology replacement to integrated energy-system design. EVs should be considered alongside DERs, PV generation, Battery Energy Storage System (BESS), buildings, tariff mechanisms, and digital control architectures. Their role is not limited to consuming electricity but can also make them flexible and interactive components of low-carbon distribution systems under suitable conditions. This is especially important in emerging active distribution environments, where the value of EVs depends on coordination with other distributed resources rather than standalone deployment. In such a framework, EV sustainability is achieved not through vehicle electrification alone, but through the coordinated design of infrastructure, operation, and incentives across the transport-power interface.

In summary, sustainable thinking on EVs requires a shift from a narrow vehicle-centered view to a broader system-centered view. EVs should be assessed not only by their ability to replace fuel consumption, but also by their environmental effects, grid impacts, economic implications, and user compatibility. Such a perspective is essential for this book, because the central concern here is not simply the diffusion of EVs, but their smart and sustainable penetration within DGs.

1.5 Aim and objectives of the book

The rapid penetration of EVs is reshaping not only the transportation sector but also the architecture, operation, and planning logic of distribution grids. As discussed in the preceding sections, EVs should not be understood merely as low-emission transport technologies. They also function as flexible electrical loads and, under suitable conditions, as distributed storage resources capable of interacting actively with power systems. This duality creates both opportunities and challenges. On the one hand, EVs can support carbon reduction, renewable energy utilization, and grid flexibility. On the other hand, uncoordinated charging may aggravate local peak demand, intensify voltage and thermal constraints, and increase the complexity of distribution-grid management. In this context, the central aim of this book is to develop a systematic understanding of how EVs can penetrate distribution grids in a smart, efficient, and sustainable manner.

More specifically, this book aims to bridge three layers of inquiry that are often treated separately in existing literature. The first is the conceptual and methodological layer, which concerns the basic interaction modes between EVs and power grids, the principles of coordinated charging and V2G operation, and the mathematical foundations needed for scheduling and optimization. The second is the infrastructural and technological layer, which concerns charging standards, converter topologies, active distribution-grid architectures, and the engineering conditions under which EV-grid coordination becomes feasible. The third is the application-oriented layer, which concerns how these concepts and technologies can be implemented in representative scenarios such as household nanogrids, nearly-zero-energy buildings, old residential communities, and internet-enabled smart charging networks. Taken together, these layers reflect the broader purpose of the book: to move from general principles to practical architectures and from theoretical coordination to real deployment contexts.

Against this background, the book pursues several specific objectives. The first objective is to clarify the technical mechanisms through which EVs interact with distribution grids, including uncoordinated charging, coordinated charging, and bidirectional V2G operation, as well as centralized and decentralized scheduling approaches. This objective is necessary because the benefits and risks of EV integration cannot be evaluated without a clear understanding of the relevant power interaction modes and control methodologies. The second objective is to establish the mathematical and analytical basis for EV scheduling problems, including objective functions, operational constraints, solution algorithms, and model evaluation. This provides the methodological foundation for both academic analysis and engineering implementation. The third objective is to examine the hardware and architectural conditions required for effective EV-grid integration. This includes charging infrastructures, grid-connection standards, converter technologies, and typical AC-, DC-, and hybrid-coupled architectures in active distribution grids. Since EV performance in grid interaction is inseparable from the underlying power architecture, the book treats infrastructure not as a secondary issue but as a core determinant of operational feasibility and system performance. The fourth objective is to investigate how EV coordination can be exploited in differentiated local energy systems. Rather than discussing EV-grid interaction only at an abstract level, the book places strong emphasis on representative application scenarios, including residential nanogrids, building-scale hybrid microgrids, constrained old communities, and cyber-physical charging-point networks. This objective reflects the view that the practical value of EV integration depends on its adaptation to specific social, electrical, and spatial contexts. The fifth objective is to identify pathways through which EVs can contribute simultaneously to grid operation improvement and carbon-emission reduction. In this book, EVs are not treated only as additional loads to be accommodated, but also as potential agents for peak shaving, valley filling, voltage support, renewable-energy absorption, local balancing, and energy-management enhancement. Accordingly, the book seeks to show how properly designed architectures and strategies can transform EV penetration from a distribution-grid burden into a distribution-grid opportunity. Finally, the book aims to provide an integrated reference for researchers, engineers, and decision-makers who are concerned with the future evolution of EV-integrated distribution systems. By combining review-based chapters with scenario-driven technical studies, the book offers both a broad conceptual map and a set of concrete solution pathways for the smart penetration of EVs within distribution grids.

1.6 Structure of the book

This book is organized into seven chapters, progressing from introductory concepts and general methodologies to infrastructure technologies and then to representative application scenarios.

Chapter 1 introduces the background and motivation of the book. It explains the significance of EVs in the decarbonization of road transport, clarifies the role of distribution grids in the context of transport electrification, and discusses carbon reduction and sustainability issues associated with EV deployment. On this basis, the chapter defines the aims of the book and positions the subsequent discussion within the broader transition toward smart and active distribution systems.

Chapter 2 presents the general schemes and mathematical foundations of EV-grid integration. It begins with the basic characteristics of plug-in electric vehicles and then distinguishes between uncoordinated charging and coordinated V2G operation. Centralized and decentralized scheduling schemes are introduced and compared, followed by a structured discussion of objective functions, scheduling constraints, solving algorithms, and computational performance evaluation. This chapter establishes the conceptual and analytical foundation for understanding how EV charging and discharging can be modeled and optimized.

Chapter 3 examines industrial standards and infrastructure technologies for EV integration in active distribution grids. It reviews the architectures of nanogrids, microgrids, and clusters, and then discusses charging standards, grid-connection standards, and the technical roles of EVs as loads, energy resources, and storage units. The chapter further introduces on-board and off-board charging circuits, as well as typical AC-coupled, DC-coupled, hybrid AC/DC-coupled, and advanced interlinking-converter-based integration architectures. These architectures are assessed from the perspectives of charging demand compatibility, power quality, V2G availability, scalability, control complexity, and technology maturity.

Chapter 4 focuses on the exploitation of EV benefits within household nanogrids. It introduces a DC nanogrid architecture and a pseudo-hierarchical energy management framework built around smart charging points. The chapter develops the corresponding system configuration and modelling methods, including EV battery, battery energy storage system, photovoltaic generation, and load models. It then presents operation strategies at both the local management level and the power dispatching level, followed by simulation analysis and experimental verification under multiple operating scenarios. This chapter demonstrates how EVs can support autonomous coordination and regulated power dispatching in compact residential energy systems.

Chapter 5 addresses nearly-zero-energy buildings through EV-grid coordination. It starts from the energy and carbon characteristics of buildings and then introduces a novel hybrid AC/DC microgrid architecture tailored to nearly/net zero energy buildings (nZEB) applications. A hierarchical management structure and corresponding local and dispatching control strategies are developed to coordinate power exchanges among distributed generators, loads, storage devices, and EVs. The operation performance of the proposed architecture is then analyzed under different power-flow scenarios, including power transfer between AC and DC subgrids and islanded operation. This chapter shows how EV integration can enhance local renewable-energy utilization and support the realization of building-scale low-carbon energy systems.

Chapter 6 turns to the difficult-to-charge problem in old residential communities. It first analyzes the practical barriers in such areas, including limited charging parking spaces, insufficient grid capacity, and power quality concerns. To address these issues, the chapter proposes a shared fast-charging scheme and an easy-to-implement hybrid AC/DC microgrid reconstruction strategy based on a smart interlinking unit. A hierarchical coordinated operation strategy is then introduced, involving upper-level V2G dispatching and lower-level coordinated control. Through case studies oriented toward peak shaving, valley filling, and price incentives, the chapter demonstrates how existing infrastructure can be reused and upgraded to improve EV accommodation and grid operation performance in constrained urban communities.

Chapter 7 explores the internet of smart charging points for V2G operation. It presents a cyber-physical architecture in which smart charging points equipped with local computing and storage capabilities enable scalable, privacy-preserving, and efficient V2G coordination. The chapter discusses the physical system scheme, communication network, distributed computing framework, and user privacy preservation, and then develops a cyber-physical co-modeling approach that integrates both energy and information flows. On this basis, optimal internal and external V2G dispatching strategies are introduced, with particular attention to photovoltaic integration, load flattening, voltage regulation, and computational efficiency. This final chapter extends the discussion of EV-grid interaction into the domain of edge computing and networked smart charging infrastructures.

Taken together, Chapters 2 and 3 provide the methodological and technological foundations of the book, while Chapters 4 to 7 demonstrate how those foundations can be translated into concrete application scenarios. Through this structure, the book seeks to provide a coherent pathway from the basic principles of EV-grid interaction to practical solutions for smart penetration of EVs within modern distribution grids.

References

[1] Banol Arias N, Hashemi S, Andersen PB, Traeholt C, Romero R. Distribution system services provided by electric vehicles: Recent status, challenges, and future prospects. IEEE Trans Intell Transport Syst 2019;20(12):4277–96.

[2] Kim Y-G, Lim H, Lee J. Decarbonizing road transport in Korea: Role of electric vehicle transition policies. Transport Res Transport Environ 2024;128:104084.

[3] Guo X, Zhang D, Li M, Chang L, Wang M, Zhang L. Decarbonizing transportation through electric vehicles: A life cycle perspective across China, Europe, and the USA. J Environ Manag 2026;401:128932.

[4] Onat NC, Kucukvar M. A systematic review on sustainability assessment of electric vehicles: Knowledge gaps and future perspectives. Environ Impact Assess Rev 2022;97:106867.

[5] Jian L, Chen Y, Chan C. Affordable mobility for all: Why we need smaller, cheaper electric vehicles. Nature 2026;651:585–7.

[6] Wang C-Y, et al. Fast charging of energy-dense lithium-ion batteries. Nature 2022;611(7936):485–90.

[7] Liu Y, Zhao X, Lu D, Li X. Impact of policy incentives on the adoption of electric vehicle in China. Transport Res Pol Pract 2023;176:103801.

[8] Ramachandaramurthy VK, Ajmal AM, Kasinathan P, Tan KM, Yong JY, Vinoth R. Social acceptance and preference of EV users—A review. IEEE Access 2023;11:11956–72.

[9] Global EV Outlook 2025 – Analysis. IEA. Accessed: Feb. 10, 2026. Available: https://www.iea.org/reports/global-ev-outlook-2025.

[10] Zhong J, Lei X, Shao Z, Jian L. A reliable evaluation metric for electrical load forecasts in V2G scheduling considering statistical features of EV charging. IEEE Trans Smart Grid 2024;15(5):4917–31.

[11] Farzaneh F, Jung S. Lifecycle carbon footprint comparison between internal combustion engine versus electric transit vehicle: A case study in the U.S. J Clean Prod 2023;390:136111.

[12] Requia WJ, Mohamed M, Higgins CD, Arain A, Ferguson M. How clean are electric vehicles? evidence-based review of the effects of electric mobility on air pollutants, greenhouse gas emissions and human health. Atmos Environ 2018;185:64–77.
[13] Zhang Y, et al. Analysis of electric vehicle charging behaviour in existing regional public and workplace charging infrastructure: A case study in the North-East UK. Transp Engg, vol 2025;19:100309.
[14] Zhong J, Lei X, Shao Z, Jian L. Value-oriented data-driven approach for electrical load forecasting apt to facilitate vehicle-to-grid scheduling. IEEE Trans Ind Inf 2025;21(7):5235–44.
[15] Razavi S-E, et al. Impact of distributed generation on protection and voltage regulation of distribution systems: A review. Renew Sustain Energy Rev 2019;105:157–67.
[16] Gonzalez Venegas F, Petit M, Perez Y. Plug-in behavior of electric vehicles users: Insights from a large-scale trial and impacts for grid integration studies. eTransportation 2021;10:100131.
[17] Rafique S, Nizami MSH, Irshad UB, Hossain MJ, Mukhopadhyay SC. EV scheduling framework for peak demand management in LV residential networks. IEEE Syst J 2022;16(1):1520–8.
[18] Shaukat N, et al. Decentralized, democratized, and decarbonized future electric power distribution grids: A survey on the paradigm shift from the conventional power system to micro grid structures. IEEE Access 2023;11:60957–87.
[19] Steffel S, Dinkel A. Absorbing the rays: Advanced inverters help integrate PV into electric utility distribution systems. IEEE Power Energy Mag 2013;11(2):45–54.
[20] Zhang K, Shi W, Zhu H, DallAnese E, Basar T. Dynamic power distribution system management with a locally connected communication network. IEEE J Sel Top Signal Process 2018;12(4):673–87.
[21] Thirugnanam K, Moursi MSE, Khadkikar V, Zeineldin HH, Al Hosani M. Energy management of grid interconnected multi-microgrids based on P2P energy exchange: A data driven approach. IEEE Trans Power Syst 2021;36(2):1546–62.
[22] Mangipinto A, Lombardi F, Sanvito FD, Pavičević M, Quoilin S, Colombo E. Impact of mass-scale deployment of electric vehicles and benefits of smart charging across all European countries. Appl Energy 2022;312:118676.
[23] Chapter 10: Transport. Accessed: Apr. 11, 2026. Available: https://www.ipcc.ch/report/ar6/wg3/chapter/chapter-10/?utm_source=chatgpt.com.
[24] Colmenar-Santos A, Muñoz-Gómez A-M, Rosales-Asensio E, López-Rey Á. Electric vehicle charging strategy to support renewable energy sources in Europe 2050 low-carbon scenario. Energy 2019;183:61–74.
[25] Zhang C, Kitamura H, Goto M. Behavioral insights into vehicle-to-grid (V2G) adoption: Beyond economic perspectives. Renew Sustain Energy Rev 2026;228:116603.

Chapter 2

General schemes and mathematical foundations

2.1 Introduction

With the increasing concern on fossil fuel shortages and environmental deteriorations, the electrification of transportation sector has attracted more and more attentions around the world [1]. EVs have been regarded as a sustainable alternative to reduce our dependence on traditional fuel sources as well as reducing our carbon footprint. Compared with conventional internal combustion engine vehicles, EVs also provide an important pathway for coupling transportation demand with increasingly cleaner electricity supply. This feature makes them particularly relevant in the context of energy transition, because the decarbonization of transport can proceed in parallel with the decarbonization of power generation. At the initial stage, the development of EVs remains very slow owing to the high manufacturing costs and the severe shortage of charging facilities [2,3]. Many policies and regulations have been implemented across the world to conquer these barriers to the large scale adoption of EVs so that they might become the future vehicles. According to International Energy Agency in its report entitled "Global EV Outlook 2018", the global EV population is projected to reach nearly 130 million by 2030 [4]. Recent market figures have further confirmed the rapid growth of electric mobility. This means that EVs are no longer a small or emerging option, but are becoming an increasingly important part of modern power systems.

It is foreseeable that the large adoptions of EVs will bring a huge challenge to the current power system. The unmanaged loads resulting from EV charging would cause a significant negative impact on girds such as the power quality degradation and the transformer overloading. This problem is especially prominent at the distribution level, where the charging behavior of EV users is often spatially concentrated and temporally synchronized. In residential areas, for example, a large number of vehicles may start charging shortly after returning home in the evening, thereby coinciding with the conventional peak demand period. Such charging simultaneity can aggravate feeder congestion, increase voltage deviation, and accelerate the thermal stress of transformers

Smart Penetration of Electric Vehicles in Distribution Grids. https://doi.org/10.1016/B978-0-443-27706-1.00004-0

and other network components. In this situation, it is inevitable to reinforce grid utilities. However, it involves a great deal of extra investments in upgrading grid utilities and deploying DGs [5]. Correspondingly, the V2G technology can provide an effective and economic solution to integrate the additional charging loads into girds friendly. Although substantial research efforts have demonstrated the potential to mitigate the adverse impact of EV charging on power grids by smart charging and/or discharging, the intelligent power interaction strategy between EVs and grids still remains an open research hotspot.

The core challenge lies in transforming a potential grid liability—uncoordinated charging demand—into a valuable grid asset. This transformation is not merely a technical problem but a complex socio-techno-economic one, requiring sophisticated coordination mechanisms. The V2G technology can achieve the bidirectional energy exchange between EVs and grids. Thus, it can store the surplus power during lower-demand periods and feeding them back to grids during higher-demand periods to achieve the balance of the supply and demand of electricity. In this situation, EVs are not only the pure electric loads, but also served as the distributed energy storage system. This dual role is one of the most distinctive features of EV-grid integration. On the one hand, EVs increase electricity demand and may burden existing infrastructures. On the other hand, if properly coordinated, the aggregated battery capacity of EV fleets can provide considerable flexibility for balancing variable renewable generation and fluctuating customer demand. In consequence, the V2G mode can utilize the EV fleet to provide spinning reserve, frequency regulation and so forth for grids [6]. Moreover, EV users could benefit from rewards based on the provision of grid services [7]. However, a significant technical barrier to the widespread adoption of V2G is its impact on vehicle batteries, primarily due to the additional degradation caused by frequent discharging [8]. Battery degradation is not only a cost issue but also a behavioral issue, because the willingness of EV users to participate in grid services depends strongly on whether the perceived benefits can compensate for battery wear and the possible inconvenience in charging availability. In spite of this, the cost-benefit analysis for EVs participating in ancillary services to grids suggests that it is worth trying the V2G operation from the economic perspective, especially with the advancement of battery technology and the rapid increase of the renewable energy [9,10]. Thus, a nuanced understanding of battery degradation models and their integration into scheduling algorithms is paramount for designing economically viable V2G schemes.

The incorporation of EVs with grids is regarded as the efficient integration of information and energy. The key about this issue is how to build an intelligent communication link to achieve the coordinated energy exchange between EVs and grids for the win-win ecosystem. As an effort to integrate EVs into power system, the existing research efforts devote to either the centralized or decentralized methods. In centralized methods, the charging/discharging processes of EVs are directly controlled to improve the operation efficiency of grids. In decentralized methods, the charging/discharging behaviors of EV users are coordinated with an incentive strategy to follow the expectations of grids. These two paradigms reflect different philosophies of control and coordination. Centralized schemes usually pursue system-level optimality by gathering information and issuing unified decisions, whereas decentralized schemes often emphasize scalability, user autonomy, and lower communication requirements. Each of them has its own strengths and limitations under different grid environments and market structures. Within the grid communication and control architecture, various objectives are adopted to obtain and implement the optimal V2G strategy, such as minimizing power losses, minimizing load variances, maximizing load factors, minimizing generation costs and so on [11,12]. These objectives reflect the fact that EV scheduling is not governed by a single criterion, but by a combination of technical, economic, and sometimes environmental considerations.

However, the existence of well-defined objectives does not automatically imply that the corresponding optimization problems are easy to solve. In practice, EV-grid scheduling often involves high-dimensional decision variables, nonlinear relationships, stochastic charging behavior, and multiple layers of operational constraints. For example, the arrival and departure times of EVs are uncertain, the state of charge differs from one vehicle to another, and network constraints may vary across feeders or time periods. Once these factors are incorporated into scheduling, the mathematical problem quickly becomes more complicated than a conventional load-control task. Although much of the literature focuses on the effectiveness of optimization models for EV integration, relatively limited attention has been devoted to the solution approaches themselves and to their feasibility in obtaining satisfactory charging strategies, especially for large-scale EV populations. This issue is particularly important because an optimization model that is theoretically sound may still be difficult to implement if its solution process is too computationally expensive, too data-intensive, or insufficiently robust under realistic uncertainty. In real engineering applications, computational tractability is not a secondary concern. A scheduling method that cannot produce timely decisions under practical data and hardware limitations may lose much of its operational value, regardless of its theoretical optimality. Accordingly, the mathematical foundation of EV scheduling deserves systematic discussion, not merely as a supporting detail, but as a central component of V2G implementation.

Against this background, this chapter provides a general introduction to the major schemes and mathematical foundations underlying EV-grid integration. It first revisits the basic characteristics of plug-in electric vehicles and then discusses uncoordinated charging and its impacts on power grids. The chapter next introduces coordinated vehicle-to-grid operation, including both unidirectional and bidirectional forms, and compares their main characteristics. On this basis, two mainstream scheduling schemes—centralized and decentralized—are presented and examined. The chapter then turns to the mathematical foundation of EV scheduling, including objective functions, scheduling constraints, solution algorithms, and computational performance evaluation. Through this structure, the chapter aims to establish the conceptual and analytical basis for the later chapters of the book, where infrastructure technologies, active distribution architectures, and representative application scenarios are discussed in greater detail.

In this sense, the present chapter serves as a methodological bridge between the general introductory discussion in Chapter 1 and the more application-oriented analyses in the subsequent chapters. Its role is not limited to summarizing earlier literature, but rather to organize the main interaction modes, scheduling logics, and optimization foundations into a coherent framework suitable for book-length discussion. Such a framework is necessary because the smart penetration of EVs within distribution grids depends not only on the existence of charging technologies, but also on a rigorous understanding of how charging and discharging behaviors can be modeled, coordinated, and evaluated under realistic system conditions.

2.2 Plug-in electric vehicles

According to the power sources for propulsion, EVs generally fall into three categories: HEV, BEV and FCEV. HEVs combine the conventional ICE system with the electric propulsion system to pursue higher fuel economy. The more efficient and eco-friendly alternative of HEV is the PHEV that has a grid-chargeable battery pack. It can use either the onboard gasoline or the electricity stored in the battery pack for propulsion. Moreover, BEVs run on electricity only whereas FCEVs are driven by hydrogen fuel. Overall, only PHEVs and BEVs require to be charged from power girds for energy supply. Therefore, they belong to PEVs.

In the global PEV market, BEV has drawn more and more attention from the automotive manufacturers in recent years. As shown in Table 2.1 [13], there are 13 kinds of BEV products accounting for 65% of the top 20 best-selling PEV products in the global market. It can be observed that BEVs are almost equipped with large capacity battery packs for the longer driving range. This substantial battery capacity is not merely a vehicle attribute but a defining factor for their potential role as distributed energy resources. And that is exactly what makes the BEV manufacturing cost high, especially in the primary stage of BEV development. PHEVs are served as a temporary solution until more advanced EV technologies become mature in the future [14]. It can be seen that there are only seven kinds of PHEV products with a small portion of the total. In general, the capacity of the battery packs equipped on the PHEVs is much less than BEVs, so the all-electric range of PHEV is rather short.

TABLE 2.1 Top 20 best-selling PEV products in global market in first half of 2018.

Ranking	Brand	Manufacturing country	Model	Capacity (kWh)	All-electric range (km)	Classification
1	Nissan	Japan	Leaf	30.0	172.0[a]	BEV
2	BAIC	China	EC series	20.3/20.5	156.0[b]/162.0[b]	BEV
3	Tesla	America	Model 3	80.0	345.0[a]	BEV
4	Toyota	Japan	Prius prime PHEV	8.8	35.0[a]	PHEV
5	BYD	China	Qin PHEV	23.0	100.0[b]	PHEV
6	Tesla	America	Model S	85.0	550.0[a]	BEV
7	Tesla	America	Model X	100.0	552.0[a]	BEV
8	BYD	China	Song PHEV	16.9	80.0[b]	PHEV
9	Renault	France	Zoe	22.0	240.0[a]	BEV
10	JAC	China	iEV E/S	29.2/41.5	255.0[b]/310.0[b]	BEV
11	Mitsubishi	Japan	Outlander PHEV	12.0	52.0[a]	PHEV
12	BYD	China	e5	43.0	305.0[b]	BEV
13	BMW	Germany	i3	33.0	183.0[a]	BEV
14	Roewe	China	i6 PHEV	8.0	60.0[b]	PHEV
15	JMC	China	E200	29.2	252.0[b]	BEV
16	Chevrolet	America	Bolt	60.0	383.0[a]	BEV

Continued

TABLE 2.1 Top 20 best-selling PEV products in global market in first half of 2018.—cont'd

Ranking	Brand	Manufacturing country	Model	Capacity (kWh)	All-electric range (km)	Classification
17	Roewe	China	eRX5 PHEV	12.0	60.0[b]	PHEV
18	Zhidou	China	D2 EV	18.0	155.0[b]	BEV
19	BMW	Germany	530e	9.2	45.0[a]	PHEV
20	BAIC	China	EX series	48	318.0[b]	BEV

PEVs, both BEVs and PHEVs, are regarded as an effective way to reduce the emissions of harmful gases and relieve the dependence on fossil fuels. BEVs are known to entirely rely on the electricity stored in the onboard battery packs for propulsion. Consequently, the range anxiety is regarded to be one of the major barriers to the large-scale expansion of BEVs. However, this is not an issue for PHEVs because PHEVs can adopt their ICE as their power system when the onboard battery is exhausted. In spite of this, the all-electric range of PHEVs is also very limited. Hence, PHEVs cannot move away from a reliance on fossil fuels completely in most scenarios. In the future, it is very likely to replace the ICEVs with the BEVs with the advancement of new battery technology that has higher power and energy density in the future.

Notes to Table 2.1:

1) All-electric range is the driving range of a vehicle using only power from its electric battery pack to traverse a given driving cycle.
2) The superscript *a* indicates the all-electric range data released by the United States Environmental Protection Agency.
3) The superscript *b* indicates the all-electric range data released by Ministry of Industry and Information Technology of the People's Republic of China.

2.3 Uncoordinated charging

2.3.1 Charging power modes

According to the charging power level of the chargers, the battery charging mode can be classified into slow and fast charging mode. Their basic characteristics for PEV charging are summarized in Table 2.2. This classification is pivotal, as the chosen mode directly influences charging duration, infrastructure requirements, and, most importantly, the magnitude and timing of the load imposed on the grid. These charging modes can be applied in different fields in accordance with their features.

TABLE 2.2 Basic characteristics of slow and fast charging.

Type Feature	Slow charging mode	Fast charging mode
Advantages	• easy for charging posts installation • home charging	• relatively short replenishing time
Drawbacks	• rather long replenishing time	• hard for charging posts installation • remarkable negative impacts on grids

The slow charging mode employs the AC, and it is made in different standards. In China, the home power system offers 220 V of AC, with 16 or 32 A for PEV slow charging; therefore, the charging power is either 3.52 kW or 7.04 kW. However, the charging power in America varies from 1.44 to 19.2 kW. It is definitely convenient to recharge PEVs by directly using the home power systems through the slow charger. Consequently, the slow charging points are the popular method for charging PEVs at residential garage or roadside parking. The problem is that, given a PEV with a 24 kWh battery, it will take about 17 hours to charge it from empty to full with a charging power of 1.44 kWh. This requires the PEV users have enough free time for the charging tasks. Apart from overnight charging at home, the slow charging mode is also applicable to the circumstances in which they find the charging time in the workplace and public. This mode adopts a small rate of charging power; therefore, it consumes a rather long charging serving time.

The fast charging mode uses the DC instead of AC. The fast chargers provide power at up to 50 kW or more to achieve a relatively short charging time. Tesla built 120 kW supercharger stations for fast charging. The supercharger can charge Model S with a 90 kWh battery from empty to 80% in about 40 minutes. These fast charging options are more high-tech than a simple household outlet due to extremely high currents and voltages. So the fast chargers should be equipped with a tethered cable rather than the ordinary household cable. The fast charging stations aim to serve those PEVs, which need to be recharged in a short amount of time. Although the less consuming time is attained for PEV charging in this situation, it may trigger extreme surges in demand and threaten the stability and security of the electric grid.

2.3.2 Negative impacts on power grids

Uncoordinated charging would bring about significant negative impacts on the distribution grid due to the inflexible nature of its charging loads. Uncoordinated charging refers to the charging mode in which PEVs start charging immediately when plugged into the grid, and stop charging only when they are fully charged or disconnected [15]. In this situation, PEVs are only considered as a novel kind of electric load for grids. This type of charging load is projected to be huge with the rapid growth of PEVs. Due to the random temporal and spatial characteristics of PEV charging, it is extremely difficult to accurately predict and control these charging loads. If left uncontrolled, such charging behavior may result in adverse influences on power quality and grid infrastructures. In other words, the problem caused by uncoordinated charging is not merely the increase in electricity consumption itself, but the fact that this additional demand is often concentrated in particular places and periods, thereby placing considerable stress on local power systems.

The impacts of PEV charging on power quality have been investigated in substantial works [10,15,17,18]. A massive deployment of PEVs will definitely lead to the remarkable change on the power flow of distribution networks. It has been demonstrated that PEV concentration at a certain location at a given time would cause huge surges in power demand, thereby increasing the line losses and dropping the voltage. What is more, it may result in the increased risk of grid failures due to the deterioration arising from power network congestion. According to the mandatory EN50160 standard [18], voltage deviations up to 10% in low voltage grids, for 95% of the time, are acceptable for the stability of power grid. Clement-Nyns et al. conducted a research based on the radial network of the IEEE 34-node test feeder, and they found that the voltage deviations are close to 10% for a PEV penetration of 30% during the early evening periods [10]. In addition, the incremental energy losses are evaluated for different penetrations of PEVs by a comprehensive approach [14]. The results show that the energy losses will increase up to 40% during off-peak hours for a scenario with 60% of PEVs. This finding indicates that the negative impacts of PEV charging are not confined to conventional peak-load periods. Once the penetration level becomes sufficiently high, even time intervals that were previously regarded as relatively light-load periods may experience considerable operational stress. Moreover, this kind of nonlinear loads caused by PEV charging may generate the severe harmonic pollution on the power system [16]. According to the research based on a typical 19 bus residential network, the harmonic analysis reveals that the total harmonic distortions of voltage levels will reach 45% when employing uncoordinated charging [17]. Such high levels of distortion can interfere with the operation of sensitive equipment and further exacerbate energy losses.

On the other hand, the large and inflexible charging loads triggered by the large-scale adoption of PEVs may also pose an additional threat on distribution transformers. At the beginning of substations construction, distribution transformers are planned and designed without the consideration of the large charging demands from PEVs. Predictably, it would increase the hot spot temperature of the transformer and accelerate the transformer aging once the heavy PEV loads are clustered in a certain residential distribution grid [18]. Moreover, it would lead to the increased risk of transformer failure [19]-[20]. Taylor et al. showed that the transformer percent aging tends to increase linearly with PHEV penetration [21]. Ghazal et al. conducted a study in the context of the National Household Travel Survey, and found that uncontrolled charging has a litter effect on the transformer life time with Level 1 (namely, 1.44 kW) whereas it will result in very high aging rates and possibly even transformer failure at Level 2 (namely, 7.2 kW) [22]. Additionally, the uncoordinated charging behaviors would increase the burden on lines' loading [15], [23]-[24]. This issue is particularly important in residential areas, where transformers and feeders are usually designed according to historical household demand rather than the simultaneous charging of multiple vehicles. Therefore, as PEV penetration continues to rise, uncoordinated charging may gradually turn existing network assets into operational bottlenecks, thereby increasing the need for either infrastructure reinforcement or more intelligent charging management strategies.

2.4 Coordinated vehicle-to-grid operation

Developing a smart charging/discharging strategy is an economic, reliable and efficient way to mitigate the negative impacts mentioned above. The key about this issue is how to build an intelligent communication link for the integration of energy and information between PEVs and grids. With smart grid technologies, PEV is not only regarded as a kind of electric load, but also served as a distributed energy storage system for power grid. In this context, PEVs are able to store the surplus power during the load-valley periods and feed power back to the grids during the load-peak periods for achieving the balance between supply and demand of electricity. Consequently, it is technically feasible to accommodate a certain penetration of PEVs for energy supply by using the coordinated charging/discharging approach.

The premise for PEV coordinated charging/discharging is to build an intelligent information platform to achieve the optimal energy management between PEVs and the grids. Fig. 2.1 depicts the corresponding relevance on the information and energy flow between the grid and power utilizations for PEV integration. It can be seen that the power grid is composed of generation system, transmission system, distribution system and etc. In such a conceptual framework, the system operators are important to integrate the actions of the

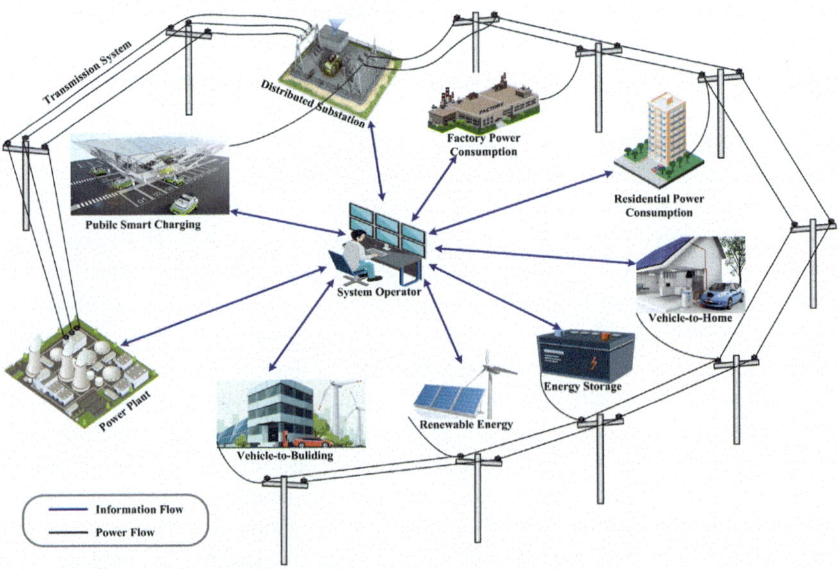

FIGURE 2.1 Implementation framework for PEV coordinated charging/discharging.

power consumers and develop the rational strategy in generation, transmission, and distribution for the supply and demand coordination of electricity. This role evolves from a one-way supplier to a two-way system manager, requiring advanced forecasting, optimization, and control capabilities. Among all the power utilizations, the power loads can fall into two types: inflexible and flexible loads. In general, the inflexible loads are not easy to be shifted such as household appliances, some commercial users and so forth. On the other hand, the flexible loads are demand-responsive and generally driven by the electricity market or system operators through the interaction with power system. In the aspect of load flexibility, PEVs can be regarded as either inflexible or flexible loads, depending entirely on their charging behaviors. Therefore, how to make the additional PEV loads flexible is the focal point for system operators especially when renewable energy resources are involved. The inherent storage capability of PEV batteries makes them an ideal flexible resource to compensate for the intermittency of renewables such as solar and wind power.

As stated above, the system operators should design the appropriate electricity allocation strategy on the basis of acquiring all the useful information to schedule the charging loads. The two-way communication between the system operators and PEVs or PEV aggregators should be available to achieve the information exchange. Herein, the PEV aggregators are served as the third party to exploit the business opportunities in electricity markets by managing a group of PEVs. In general, when PEVs or PEV aggregators report their buy/sell bids to the system operators, the system operators will evaluate whether their bidding strategies are validated or not, then negotiate with PEVs or PEV aggregators until a bidding agreement is achieved.

The power interaction relationship between PEVs and grids could be uncoordinated or coordinated depending on the adopted charging strategy of PEVs. The uncoordinated charging mode of PEVs operates regardless of the supply and demand status of distribution networks, and is more likely to jeopardize the grid stability and reliability when a large number of PEVs are involved. The coordinated V2G mode is developed for integrating more PEVs into the existing power system. The current works for developing a smart charging/discharging system for PEVs can generally fall into unidirectional and bidirectional V2G. The unidirectional V2G mode is the so-called coordinated charging mode, and the bidirectional V2G mode is the so-called coordinated charging/discharging mode. In the uncoordinated charging mode and the unidirectional V2G mode, the energy only flows from the grid to PEVs (namely, Grid to Vehicle) whereas the energy not only flows from the grid to PEVs but also flows from PEVs to the grid in the bidirectional V2G mode. The power flow direction of three operating modes and their alternative terms are provided in Table 2.3.

The concept of Vehicle-to-Home (V2H) is a small version of V2G technology, which allows a PEV to supply homes with power generated from its battery [25]. With the help of V2H technology, PEVs can be severed as a

TABLE 2.3 Power flow direction of three operating modes and their alternative terms.

Operating mode	Power flow direction	Alternative term
Uncoordinated charging	From grid to PEV	• Dumb charging • Uncontrolled charging • Conventional charging
Unidirectional V2G	From grid to PEV	• Coordinated charging • Regulated charging • Smart charging
Bidirectional V2G	From grid to PEV From PEV to grid	• Coordinated charging/ discharging • Regulated charging/ discharging • Smart charging/discharging

battery storage system to feed any electrical appliance where the PEV is parked. In spite of this, they don't have the capability to operate as an off-line uninterruptible power supply (UPS). Therefore, using PEVs as an off-line UPS is explored to support the smart grid and the smart home, and it is has been proved in Ref. [26]. Besides, the V2G technology can utilize PEVs to achieve the reactive power compensation so as to provide the grid with an economical in operation mode [27]. Additionally, the Vehicle-to-Vehicle (V2V) concept is developed for a local PEV community that can charge or discharge PEV battery among them [28]. It can be seen that these concepts enable PEVs act as distributed sources for the power grid within the V2G technology.

The uncoordinated charging mode refers to the conventional charging mode in a traditional power grid that PEVs start charging immediately once they are plugged in. In this case, the arbitrary plugged-in time is entirely determined by PEV users, and there are no PEV aggregators to regulate the PEV charging profile. Therefore, the charging processes of PEVs are completely dominated by their users as their own will. It has been investigated that a majority of the uncoordinated charging events occur in the home-charging situation, in which PEV users charge their vehicles as soon as they arrive home during early evening hours [29]; however, the load level of power gird is still rather high at this moment. Needless to say, the frequent PEV charging activities during this period would significantly stress the distribution system.

There are serval aspects of PEVs with the regard to the impact on distribution networks, such as vehicle pattern, vehicle penetration, vehicle load demand, plugged-in/plugged-out time, vehicle charging location, vehicle charging pattern and so forth [30]. It doesn't make significant impacts on distribution grids at low PEV penetrations [31]. Nonetheless, it is likely to result in the collapses of power grid if a high penetration level of PEVs is intensely recharged during load-peak periods. As a result, a coordinated power interaction relation between PEVs and grids should be established to tackle the charging problem of large-scale PEVs.

2.4.1 Unidirectional vehicle-to-grid

Unidirectional V2G indicates the technical capability to modulate the charging power of PEVs to better correspond to the grid requirements with the assistance of communication technology. To that end, a smart infrastructure capable of supporting two-way information exchange between PEVs and grids is indispensable. Nowadays, a majority of PEVs have been equipped with the on-board unidirectional chargers. It is more likely to implement unidirectional V2G in the short term in comparison to bidirectional V2G. From a practical perspective, this operating mode is often regarded as the most feasible entry point for smart EV-grid integration, because it does not require the more demanding hardware conditions associated with reverse power flow. In this sense, unidirectional V2G may be understood as a regulated and grid-responsive charging mode, rather than a conventional charging process determined solely by users' immediate preferences.

To avoid grid congestions at rush hours, unidirectional V2G allows the third party to propel PEV users to charge their vehicles during the entire off-peak period, not just when off-peak moment starts. Therefore, the unidirectional V2G services are able to absorb the surplus power in the valley periods before PEV batteries are fully charged. Nevertheless, it is incapable of sending energy from PEV batteries back into the grid in the absence of the bidirectional power converters. Meanwhile, the extra degradation on PEV batteries would scarcely be incurred in this situation [32]-[33]. This feature is important because it makes unidirectional V2G comparatively attractive to both vehicle users and system operators. On the one hand, it retains much of the flexibility benefit associated with managed charging. On the other hand, it avoids the more direct battery wear and converter complexity that may arise in bidirectional power exchange. As a result, unidirectional V2G is often considered a more acceptable compromise between technical feasibility, operational benefit, and user acceptance.

It has been shown that properly designed unidirectional chargers have the ability to charge PEVs with the surplus power of grids so as to mitigate the impact of large-scale PEVs charging [6]. Consequently, they can effectively support both distribution and transmission systems through active load management of PEVs. What's more, unidirectional V2G services can help consume abundant renewable energy sources such as solar and wind energy through a coordinated charging strategy. In distribution networks with increasing shares of distributed renewable generation, this capability is especially valuable. Instead of allowing EV charging demand to coincide randomly with existing load peaks, unidirectional control makes it possible to shift charging toward periods in which renewable output is relatively high or system demand is relatively low. In this way, EVs can function as flexible loads that improve the temporal matching between electricity demand and cleaner electricity supply. Although unidirectional V2G services still face some obstacles, appropriate technical and market solutions are gradually being developed. These obstacles include the need for communication infrastructure, suitable incentive mechanisms, reliable user participation, and effective scheduling strategies under uncertain charging behavior. Additionally, unidirectional V2G would build a solid foundation to implement bidirectional V2G in the future.

2.4.2 Bidirectional vehicle-to-grid

Bidirectional V2G refers to the technology that allows for the reversal of electric energy stored in PEV batteries to facilitate the balance of electricity while satisfying the charging demands of PEVs. In contrast to unidirectional V2G, this mode enables not only controlled charging from the grid to the vehicle, but also controlled discharging from the vehicle back to the grid. As well as requiring a two-way communication system, this mode also necessitates that the chargers are bidirectional so as to feed the stored energy from PEV batteries to the electric grid during periods of higher demand. On account of its feedback function, bidirectional V2G can provide spinning reserve, frequency regulation, and other ancillary services for the power grid. This makes bidirectional V2G a more active and flexible form of EV-grid interaction, in which PEVs are no longer treated merely as controllable loads, but also as dispatchable distributed energy resources.

In bidirectional V2G, a large number of PEVs serve as a controllable fleet under a PEV aggregator. This fleet has a function similar to that of distributed energy storage systems capable of providing ancillary services to the grid. Hence, the PEV discharging processes are indispensable for energy feedback. From the system perspective, the value of such a fleet lies in its aggregated capacity. While the discharge capability of a single vehicle may be limited, a coordinated fleet of PEVs can form a considerable flexible resource when properly scheduled. In this way, bidirectional V2G can contribute not only to peak shaving and load balancing, but also to the wider coordination of renewable generation, distributed resources, and system reserves. Beyond all doubt, it will lead to the extra adverse impact on the health of PEV batteries. However, these incurred costs might be offset by the benefits generated by the reasonable bidirectional V2G services [34]. In the near future, it is viable and worthwhile to implement bidirectional V2G from the economic perspective [35]. The problem is that, PEV aggregators should build an interest compensation mechanism to promote PEV users to participate in bidirectional V2G.

Under such a mechanism, PEV users may benefit from rewards based on the provision of grid services and energy trading revenue. Therefore, the practical success of bidirectional V2G depends not only on technical feasibility, but also on whether the economic interests of EV users, aggregators, and grid operators can be reasonably aligned.

Generally, bidirectional V2G has a greater advantage than unidirectional V2G on the integration of all kinds of resources including renewable energy and DGs. Because it supports reverse power flow, bidirectional V2G offers a broader range of operational possibilities than controlled charging alone. For example, it can help relieve supply shortages during peak-demand periods, support local balancing under fluctuating renewable output, and provide more direct ancillary services to the grid. In this sense, bidirectional V2G represents the more advanced form of EV participation in power-system operation. In the meantime, it is confronted with more barriers for its practical application. In the past decade, bidirectional V2G has been tested in several projects. But this technology is still in the experimental stage. A number of technical and regulatory issues need to be resolved before it can be widely and effectively used.

2.4.3 Comparison

Table 2.4 lists several main characteristics among these power interaction modes in technical requirements, impacts on grids, and influences on PEV batteries and so on. The uncoordinated charging mode may cause many issues on grids when large-scale PEVs involved. The V2G technology can alleviate and even avoid these problems. In particularly, bidirectional V2G can shave load peaks; however, bidirectional chargers are necessary for energy reversal. Meanwhile, it will cause the serious battery degradation due to the frequent charging or discharging cycles. Additionally, the V2G technology can help the power company decrease the electricity costs in generation, transmission and distributions, thereby reducing the PEV charging costs. As far as the feasibility of V2G implementation is concerned, bidirectional V2G would require significantly higher investments than unidirectional V2G on the hardware devices.

2.5 Scheduling schemes

2.5.1 Centralized scheduling

In the centralized scheduling method, PEV aggregators optimize and control PEV charging based on the grid status and PEV charging demands for a special target. Hence, the controllers should be equipped on the PEV chargers to adjust the charging rates of PEVs in this scenario. Fig. 2.2 shows the general process to implement centralized scheduling method for PEV coordinated charging. It can be seen that PEV aggregators are the crucial bonds to integrate

TABLE 2.4 Characteristic analysis of different power interaction modes.

Mode Feature	Uncoordinated charging mode	V2G mode	
		Unidirectional V2G	Bidirectional V2G
Technical requirements	• No information interaction • No PEV aggregators • No PEV coordination	• Two-way information system • PEV aggregators • Coordinated charging management	• Two-way information system • Bidirectional charger • PEV aggregators • Coordinated charging/ discharging management
Impacts on power grid	• Increases power loss • Drops network voltage • Elevates load peaks • Affects grid frequency • Overloading of lines • Overloading of transformers	• Load-valley filling • Voltage regulation • Frequency regulation • No lines overloading • No transformers overloading	• Load-valley filling • Load-peak shaving • Voltage regulation • Frequency regulation • No lines overloading • No transformers overloading
Influences on PEV batteries	• No extra degradation	• No extra degradation	• Cause extra degradation
Economic cost/benefits	• Increase electricity costs for power company • Increase charging costs for PEV users	• Decrease electricity costs for power company • Decrease charging costs for PEV users	• Decrease electricity costs for power company • Decrease charging costs for PEV users • Benefit from rewards on ancillary services to grids for PEV users
Feasibility	• Have widely existed	• A little difficulty	• Great difficulty

PEVs in the power system and make PEVs participate in electricity markets. At first, each PEV user ought to inform the respective PEV aggregator of all the charging related information including the identity number of the plugged-in charging post, the nominal capacity of PEV batteries, the plugged-out/

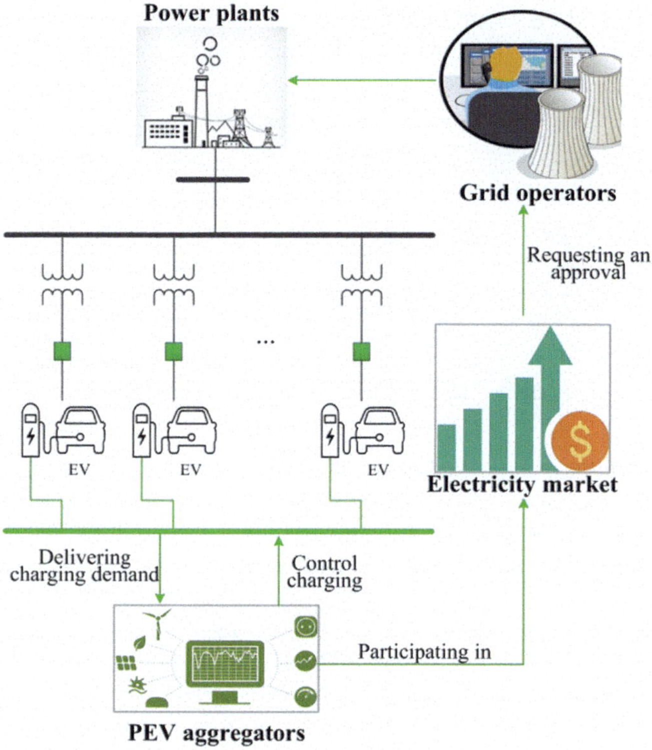

FIGURE 2.2 General framework for centralized scheduling method.

plugged-in time, and the plugged-out/plugged-in *SoC* of PEV batteries. Afterward, PEV aggregators will compute the optimal charging/discharging strategy of each PEV to determine their buy/sell bids in accordance with the received charging demands and the forecasted market behaviors. Then, the bidding instruction should be at once reported to grid operators for approval ahead of schedule. And, grid operators will check whether the biding strategy is a safe and economic operation for power system. If yes, PEV aggregators can proceed to the electricity market; if not, grid operators will ask PEV aggregators to change their bidding strategies until a protocol is reached. Finally, PEV aggregators will charge each PEV in accordance with the optimal load profile of the fleet. On the other hand, PEV aggregators will check out with each PEV user by the online settlement system.

It can be seen that the PEV aggregator is responsible for a group of PEVs to exploit business opportunities in electricity markets. One of the PEV aggregator's targets is to minimize the charging costs of PEVs managed by this aggregator. In the centralized scheduling approach, the overall charging costs resulting from all PEVs are optimized in a programming model for the global optimal solution [36]. This objective function can be formulated as:

$$\min \sum_{n=k}^{N} \sum_{t=l}^{T} \left(P_{t,n}^{RT} \Delta t\right) \lambda_t^{RT} \tag{2.1}$$

subject to

$$P_{min,n}^{EV} \le P_{t,n}^{EV} \le P_{max,n}^{EV}, t = \left[\Gamma_n^s, \Gamma_n^e\right], n \in [1, N] \tag{2.2}$$

$$SoC_{min,n}^{EV} \le SoC_{t,n}^{EV} \le SoC_{max,n}^{EV}, t = \left[\Gamma_n^s, \Gamma_n^e\right], n \in [1, N] \tag{2.3}$$

$$\eta_n \Delta t \sum_{t=\Gamma_n^s}^{\Gamma_n^e} P_{t,n}^{EV} = \left(SoC_n^e - SoC_n^s\right) C_n, n \in [1, N] \tag{2.4}$$

where all variables, parameters and indices are defined in Section 2.6. Eq. (2.2) indicates the limitations of the charging power of the n-th PEV at time t; Eq. (2.3) indicates the limitations of the battery SoC of the n-th PEV at time t; Eq. (2.4) indicates that the charging requirement of the n-th PEV should be satisfied.

Up till now, a general model for minimizing the total charging costs of PEVs in a centralized way is built. The charging strategies of all involved PEVs are optimized from a global perspective. The centralized model can achieve the optimal charging/discharging strategies for PEV charging scheduling. However, there are a huge amount of variables and constraints to handle when a large number of PEVs are involved in this model. Therefore, it may be difficult for the centralized scheduling method to search for the global optimal solution.

As illustrated in Fig. 2.2, there are three major participants in centralized approaches, namely grid operators, PEV aggregators and PEV users. They have extremely different considerations in their respective position. From the view of grid operators, they focus primarily on the grid performance improvement and secondarily the total operational costs reduction. This requires that the charging time of PEVs ought to be scheduled and limited in the off-peak periods as far as possible. In this way will the power system run more safely and efficiently. However, at PEV users level, they require a more convenient and affordable recharging system. It may result in severe conflicts with grid operators on the charging periods. In this context, PEV aggregators are employed to coordinate the both sides' relation. From the PEV aggregator perspective, how to control the PEV chargers to maximize the total operational profits with the presence of satisfying the charging demands of PEV users and grid restrictions is their considered issue. On the other hand, they should design a reasonable compensation strategy to make more PEV users participate in the centralized scheduling method and provide V2G ancillary services.

Many works focus on the optimization methods to make the charging decisions on PEVs in the place of PEV aggregators. The critical issue for PEV aggregators is to develop the so-called load shifting or valley filling strategy so as to alleviate the demand pressure of power grid during the load peak hours. Originally for improving the power network quality, the model with the target of minimizing the overall load variance is presented to smooth out the fluctuation of the power load profiles while satisfying the driving demands of PEVs [37]-[38]. Despite all this, both the PEV charging costs and the aggregator operational profits are not taken into consideration in this sort of methods. In order to remedy these issues, the PEV aggregator in Ref. [39] provides a global optimal solution to minimize the total charging costs of all PEVs by performing charging and discharging during the day. Moreover, the aggregator operational profits are optimized in Ref. [40] to acquire the optimal dispatching strategy for the aggregator participating in the frequency regulation while satisfying the PEV driving demands. Additionally, scheduling PEVs charging properly would help in reducing the carbon emissions [41]. Consequently, the environmental impacts of PEVs participating in the V2G operation are also investigated and optimized in Ref. [42].

Conversely, the feasibility of the centralized approaches has not been paid sufficient attentions. Overall, there are two key issues, viz. the time-space characteristic of PEV charging and the execution time for the optimal solutions in the PEV charging problem. Initially, many scheduling methods for PEVs participating in the day-ahead electricity market are proposed in Ref. [43]-[44]. It can be observed that the execution time is not a big problem at all in the day-ahead time horizon. In spite of this, these methods are generally built on the assumption that the PEV aggregators have a full knowledge of the PEV-involved variables. Evidently, it is almost infeasible to precisely predict or collect the PEV charging demands at least one day in advance due to the stochastic driving pattern. Therefore, the real-time scheduling approaches are employed in Ref. [45]-[46]. In such methods, the solving time of an optimization model is crucially important when a large population of PEVs involved. As a result, some efficient algorithms such as PSO algorithm, the interior point algorithm, and GA are employed to obtain the real-time solutions. It has been proved that the solving time highly depends on the computational complexity of the applied algorithm. In consequence, the key to implement the real-time charging operation is to develop an algorithm with lower computational complexity for PEV coordination.

The centralized approaches require the PEV users to give up their charging authorities to control the PEV charging process. It can be foreseen that PEV aggregators can make a generous profit by scheduling PEVs to charge at valley hours and discharge at peak hours. The critical issue is that, this action will lead to the extra degradation on PEV batteries. Therefore, battery restricted conditions should be added into the scheduling optimization models to relieve the influences on battery lifetime as far as possible [27]-[47]. Nonetheless, it may still not be enough from the benefits of PEV users. On this account, the battery aging costs for PEVs taking part in V2G should be evaluated to compensate PEV users for the extra battery degradation [48].

2.5.2 Decentralized scheduling

The decentralized scheduling method refers to an incentive strategy, wherein, the charging behaviors of PEVs will be indirectly coordinated by the electricity price mechanism. In some simple situations, this only requires grid operators or PEV aggregators broadcast the price information to each PEV user so as to induce the PEV loads from high-demand hours to less congested hours. Hence, it is easy to be implemented in the short term. Fig. 2.3 depicts the general process to implement decentralized scheduling methods for the V2G operation. Firstly, PEV aggregators collect or forecast the PEV charging demands during each period to determine the bidding strategy. Afterward, the bidding information will be transmitted to grid operators for approval in advance just like the centralized scheduling method. When the agreement is reached, PEV aggregators will broadcast the price to each PEV user. Consequently, each PEV user will schedule their vehicles charging as their own willing such as reducing their charging costs.

In the decentralized scheduling approach, PEV aggregators are also employed to manage the charging of PEVs. Each PEV user will respond to the price signals according to their own willingness. And the bidding strategy will be delivered to the PEV aggregators for regulated charging . As for each PEV user, they usually strive for the minimization of their individual charging costs. In general, the objective function to minimize the charging cost of the n-th PEV can be given by Ma et al. [49]:

$$\min \sum_{t=l}^{T} \left(P_{t,n}^{EV} \Delta t \right) \lambda_t^{RT} \tag{2.5}$$

subject to

$$P_{min,n}^{EV} \leq P_{t,n}^{EV} \leq P_{max,n}^{EV}, t = \left[\Gamma_n^s, \Gamma_n^e \right] \tag{2.6}$$

$$SoC_{min,n}^{EV} \leq SoC_{t,n}^{EV} \leq SoC_{max,n}^{EV}, t = \left[\Gamma_n^s, \Gamma_n^e \right] \tag{2.7}$$

$$\eta_n \Delta t \sum_{t=\Gamma_n^s}^{\Gamma_n^e} P_{t,n}^{EV} = \left(SoC_n^e - SoC_n^s \right) C_n \tag{2.8}$$

It can be observed that the decentralized method allows for a distributed way to deal with the charging scheduling optimization for each PEV. The centralized model presented in Section 2.5.1 seems to be divided into many submodels stated above. There are a few variables and constraints in the

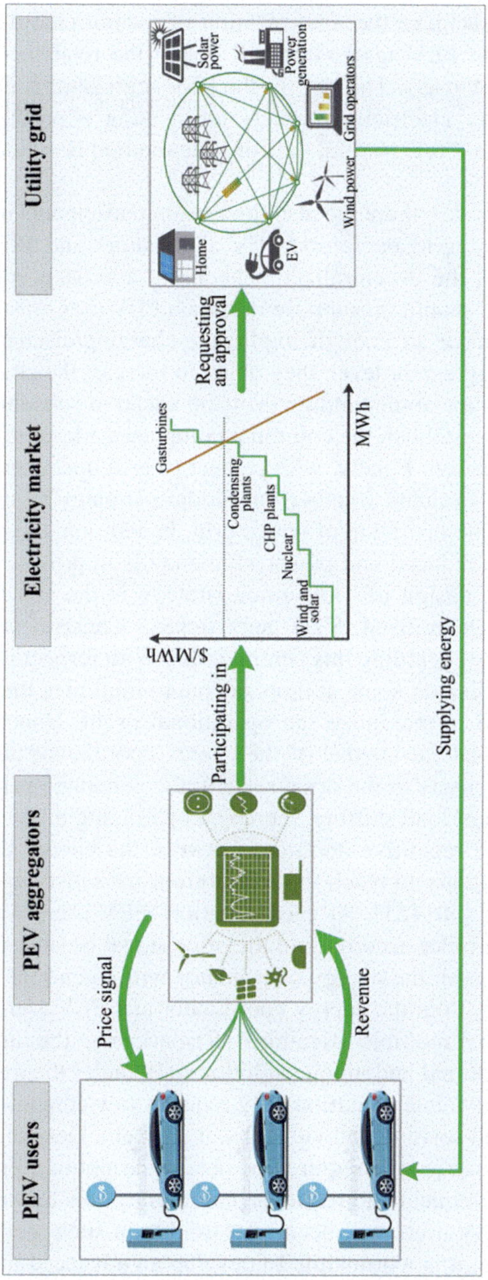

FIGURE 2.3 General framework for decentralized scheduling method.

decentralized model. Therefore, this model is comparatively easy to be solved. Additionally, this model can also attain the optimal solution for a special PEV; however, it cannot achieve the minimization of overall charging costs for all involved PEVs. The PEV aggregator will obtain the total income by energy trading with the PEV users. The key for the PEV aggregator is how to optimize the bidding strategy electricity markets for making a profit. This issue is presented in Section 2.6.1. Herein, the concrete method is not described in this section.

As shown in Fig. 2.3, there are also three main participants in decentralized approaches, namely, grid operators, PEV aggregators and PEV users. From operating principle, the decentralized scheduling approach is a price-signal-based method. This should presuppose that each PEV user must respond to the price signal and make an enough intelligent charging/discharging decision. Then, at the PEV aggregator level, they ought to forecast the charging demands from PEVs so as to make the bidding decision ahead of schedule; on the other hand, they ought to establish the communication network to inform PEV users of the price information. Finally, with respect to grid operators, their primary responsibility is to evaluate whether the bidding strategy of each aggregator will endanger the safe operation of power grid. In addition, they also desire the operating costs are reduced and power efficiency is improved.

Apparently, the design of TOU price strategy is the key to successfully implement the decentralized V2G approaches. Consequently, substantial research efforts exist regarding this central issue. With respect to the objectives of the energy coordinator, some authors devote to minimize the operating cost while others intend to maximize the operational profit. Nonetheless, if only considering the unilateral interest of the energy coordinator and ignoring the responding of PEV users to the price system, the generated price system may result in the failure of load shifting, thereby jeopardizing the security of power grid. Therefore, the repetitive iterations between the energy coordinator and PEV users are necessary to reach the equilibrium for satisfying the interest of both of them [49], [50]-[51]. At each iteration, PEV users will update their optimal charging profiles according to the price signal broadcast by the energy coordinator. Afterward, the energy coordinator will amend the price signal to guide their updates. Both the energy coordinator and PEV users search for the optimal strategy by multiple iterations. Remarkably, the iterative process converges to the optimal and unique solution only under the given conditions. At the PEV side, its implementation only requires low computation capability since each PEV will solve its individual local problem. Despite all this, a large number of iterations are still required to obtain the optimal price scheme. In consequence, the frequent bidirectional communication between the energy coordinator and PEV users will become a trouble in such methods.

As stated above, it is worthwhile to develop an advanced algorithm with a rapid convergence speed for less bidirectional communication and computational effort. Based on this, a decentralized control method is presented to schedule PEV charging loads to achieve valley filling [52]. The iterative process in this method is able to converge within only several iterations, making it very applicable for the real-time implementation of coordinated charging. Apart from only employing the price signal, another pricing scheme is presented in Ref. [51], wherein both price and quantity information is conveyed to PEV aggregators for PEV coordination. The simulation results show that this kind of price/quantity-based approach not only has a superior performance on controlling PEV fleet charging, but also involves 24 times less information exchange than the price-only scheme. On the other hand, the research in Ref. [53] makes an improvement on the required computational effort as compared with Refs. [50] and [49], in which a convex optimization problem must be solved by each PEV at each iteration. Generally speaking, if the proposed methods allow for the use of cheaper communication facilities and computational devices, it will be beneficial for both grid operators and PEV users in the decentralized scheduling method.

Massive works pay great attention to the operational feasibility of the decentralized methods so as to make an improvement on the operation efficiency and power quality. In Ref. [54], a two-level optimal charging algorithm is designed to achieve both load shifting and frequency regulation. However, the load shifting algorithm is built on the basis that PEV aggregators have a full knowledge of four pieces of information: forecasted base load profile, estimated number of plugged vehicles, estimated plug-off time, and the battery *SoC* of the vehicle being charged. Obviously, it is considered impractical in real world. In addition, a more practical way for the decentralized valley-filling strategy is developed in Ref. [55]. Although it does not rely on a bidirectional communication system to collect the PEV charging information, the one-way communication is still required to broadcast the pricing scheme to PEV users. Moreover, a full decentralized controlled PEV charger is designed to mitigate the PEV charging impact on power grids [56]. There are no requirements on a centralized control center and information exchanges between the energy management system and PEV chargers. As a result, the charging scheme for PEV chargers can be effectively achieved.

The PEV users in the decentralized method preserve individual authority on the PEV charging processes according to the price scheme broadcast by the energy coordinator. However, it is possible that a part of PEV users may not be sensitive to the electricity price mechanism, or they do not receive this kind of price information at all. As a consequence, in this situation the unexpectedly plugged-in PEVs will also pose an additional stress on power system. In order to tackle this terrible scenario, in Ref. [57] a practical coordinated PEV charging approach is presented based on the worst node voltage profile. The results show that deferred plugging of vehicles will lead to the detrimental impacts on the grid, and these impacts could be encouraged by introducing higher electricity prices during the peak load hours. In addition, a non-anonymous version of the algorithm is proposed under the communication failures in Ref. [53]. This method is able to handle the PEV charging problem without successful communication. Additionally, the results indicate that the algorithm can converge to the optimal solutions of the charging problem.

2.5.3 Comparison

Table 2.5 shows the main characteristics between centralized and decentralized methods. In the centralized methods, the energy coordinator is responsible to optimize the charging/discharging profile of all PEVs involved, subject to the grid constraints. With the rapid increase of PEV scale, the PEV variables required to be optimized in a centralized manner will become tremendously large. Herein, the computational complexity is remarkably high due to the curse of dimensionality, especially for the NP problem. Therefore, the computational expense for the optimal charging/discharging profile of PEVs will also sharply increase. In particular, it is a more problematic issue in the real-time scheduling operations of large-scale PEVs. On the other hand, in the decentralized scheduling method the design of electricity price mechanism is mainly done by the local grid condition and the overall charging demands of PEVs, and each PEV user solves its individual local problem according to the price scheme. Hence, there is no requirement to optimize such a large number of the PEV variables such as the centralized method even if a huge mass of PEVs is involved. Nevertheless, it is observed that the repeated iterative process is necessary to search for the optimal price system in some decentralized approaches, and convergence cannot be guaranteed. Generally speaking, the decentralized approaches have a significant superiority in the aspect of computational cost and accessibility to the optimal solutions over the centralized approaches for the charging/discharging scheduling strategies of large-scale PEVs.

Theoretically speaking, the centralized scheduling approach is able to obtain an optimal solution by the unified schedule of PEVs for a special

TABLE 2.5 Comparison between centralized and decentralized methods.

Method Feature	Centralized method	Decentralized method
User charging authority	Not granted	Granted
Grid security risk	Almost risk-free	Potential
User concern for privacy	Moderately concerned	Marginally concerned
Computational complexity	High	Low
Optimal solution	Global	Local

objective function. What's more, the charging/discharging processes of all PEVs are directly under the control of the energy coordinator. Consequently, the centralized method, whether in theory or technology, is capable to achieve the optimal energy management between power grid and PEV users. In contrast, the decentralized scheduling approach is a price-signal-based method. In theory, the decentralized methods can provide the optimal solutions for the V2G operations. But the prerequisite is that each PEV user is enough intelligent to charging/discharge their vehicles based on the electricity price just as the energy coordinator desires. Obviously, it may be unattainable in practical applications since PEV users in the same charging situation may have different responding to the price information. Hence, the decentralized method is extremely difficult to attain the optimal V2G operations at implementation stage. In spite of this, it is an economic, effective and practical way to mitigate the negative impacts of PEV charging.

In general, the centralized solutions give more weight on grid performance and operational costs than the individual charging authority of PEV users. In addition, due to the requirement of collecting the charging information from the distributed PEVs, this involves a bidirectional communication network for information exchange. And this will cause the privacy concerns for PEV users on the PEV charging information such as the plug-in and plug-out time, the charging location. In contrast to centralized methods, the decentralized approaches allow for the individual authority on the PEV charging. As a result, it cannot ensure the overall system's optimal operation. Unfortunately it may produce some security concerns on power system when the broadcasted price system fails in shifting PEV charging loads from load peak to valley periods. However, the benefits are that, it only requires low communication and computational expenses. Moreover, it raises much less privacy concerns for PEV users.

2.6 Mathematical foundations

The strategy decision-making for the V2G implementation is essentially a mathematical optimization problem regardless of the centralized or decentralized scheduling method. As a rule, a typical mathematical programming model is composed of three parts, viz. the objective function, the constraint conditions and the solving algorithm. The following section will illuminate the mathematical foundation on the optimization technique for the optimal V2G operation. On this basis, an overall evaluation is conducted on these models. In general, the framework of this section is illustrated in Fig. 2.4.

2.6.1 Objective function

An old Chinese saying goes, "Make the best use of everything and give full scope to the talents". On the basis of this way of thinking, the ultimate objective for V2G implementation could be considered to integrate all the available resources (including but not limited to PEVs, energy storage devices, DGs and renewable energy sources) and fully exploit their respective advantages to fulfill the maximization of the overall energy efficiency. It is believed that in this way composite economic and environmental benefits will be improved greatly. Under the guidance of this ultimate objective, several main views on the V2G operation are summarized as follows: Technically, the PEV scheduling strategy should balance supply and demand of electrical energy in power generation, transmission and distribution. Economically, the designing strategy ought to abide by the principles of considering all multistakeholders, for example, the operational profits of the whole power system are maximized while the PEV charging costs are minimized. Environmentally, the carbon emissions incurred by PEVs should be reduced sharply in a coordinated charging/discharging scheduling scheme under the premise of satisfying the interest desire of multistakeholders.

Inspired by improving the overall energy efficiency, all kinds of methods are explored to pursue the optimal charging patterns of PEVs in the previous literature. With reference to the objective functions, many authors initially try to minimize the energy losses by employing the optimal power flow calculation [37], [58]-[59] while other authors intend to minimize peak demands [60]-[61]. Another distinctive objective is to mitigate the deviation between the purchased energy in the electricity market and the consumed energy for PEV charging [62]. In other words, to achieve the electricity supply–demand balance with the flexible PEV loads is also regarded one of the valid objectives. Meanwhile, some efforts focus on the multiobjective optimization models. For instance, both the operating costs and emissions are minimized in the day-ahead coordinated charging scheduling of PEVs [63]. Another multiobjective example is to minimize both overall load variance and the charging satisfaction of PEV users regarding to energy management strategy [64]. Also, in Ref. [65] the optimal charging scheduling method is developed to minimize both electricity generation costs and carbon emissions. It can be observed that all the above objectives are designed to improve the following aspects: gird performance, economic index and environmental index.

Although these objectives are often closely intertwined in practical scheduling models, they may still be understood from three main perspectives, namely grid performance, economic performance, and environmental performance. Among them, grid-performance-oriented objectives have received particularly extensive attention in the early development of V2G scheduling studies, because the most immediate concern raised by large-scale PEV integration is its impact on power-system operation. In other words, before the economic and environmental value of V2G can be fully realized, the additional charging demand must first be accommodated in a secure, stable, and grid-friendly manner. For this reason, the discussion starts with the objectives related to grid performance.

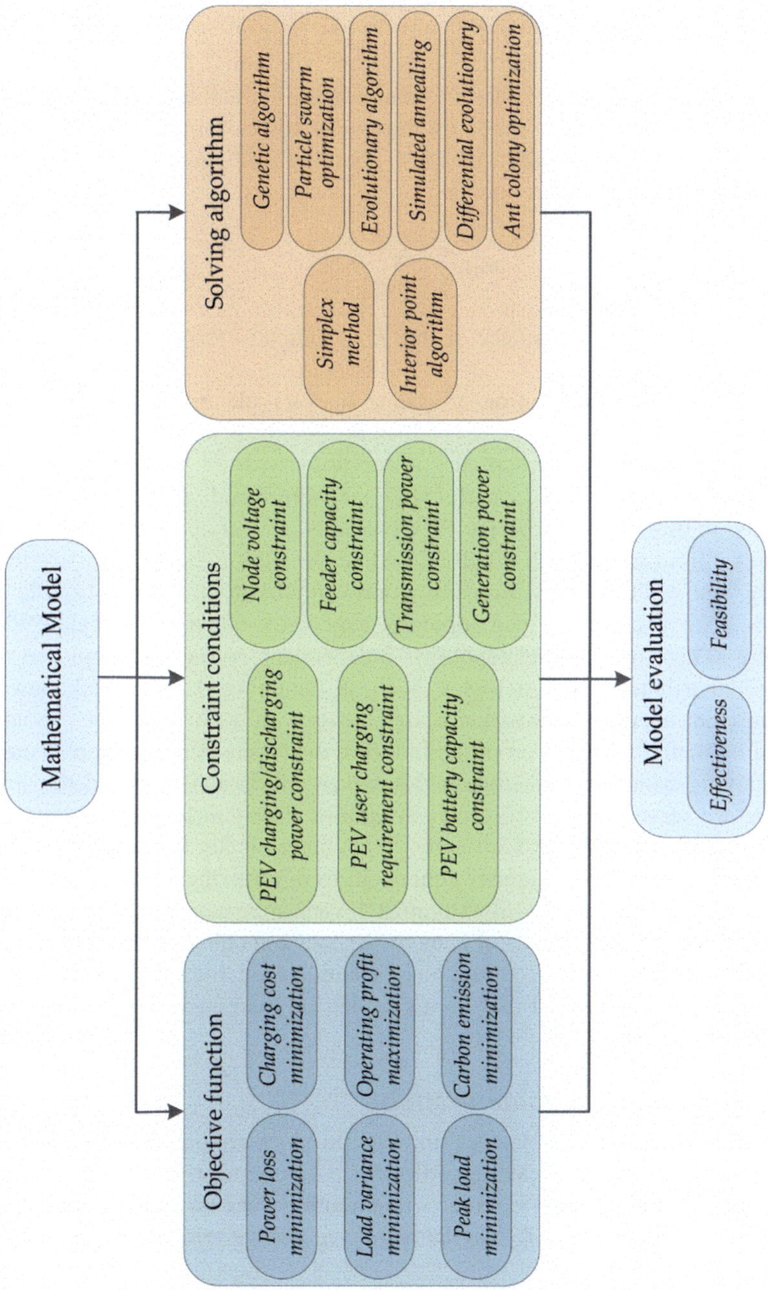

FIGURE 2.4 General framework of mathematical modeling in V2G scheduling.

2.6.1.1 Power loss minimization

The stochastic connections of large-scale PEVs not only increase the power losses, but also affect the power quality. It has been demonstrated that minimizing the distribution system losses with the additional PEVs is capable to improve the voltage profile, thereby maintaining the stability of power grid. Consequently, many works focus on the power losses minimization so as to mitigate the adverse impacts of PEV charging. In general, this kind of objective function can be defined as:

$$\min \sum_{t=1}^{T} \sum_{l=1}^{\text{lines}} R_l I_{l,t}^2 \tag{2.9}$$

where R_l indicates the resistance of line l; $I_{l,t}$ indicates the current of line l at time t.

Obviously, the minimization of system losses is determined by the operating schedule of PEVs. In Ref. [58] a load flow analysis is performed to assess the power losses in the residential grid topology with 34 nodes for PEV uncoordinated charging. Later, in Ref. [37] a coordinated charging method is developed to directly optimize the PEV charging current at the different nodes for losses minimization with the IEEE-node case. Although power losses, load variance and voltage regulation are improved all, it is a great trouble to consider such a complex grid topology especially when large-scale PEVs involved. Therefore, some topology independent ways are explored to decrease distribution system losses, such as flattening out the peak power, reducing the overall load variance and so on. In Ref. [66], the optimal scheduling methodology is proposed to tackle the losses minimization through peak shaving using energy storage. Moreover, the load factor and the load variance are served as the objectives to minimize system losses and improve voltage regulation [38]. Additionally, it is theoretically demonstrated that maximizing system load factor is equivalent to minimizing load variance and minimizing load variance will minimize losses approximately for practical systems. Furthermore, based on Cauchy–Schwarz inequality, the action and importance of flattening the overall load profile in reducing the network loss is analyzed and it was proved that the more flat the load curve is, the more the network loss reduces [59].

2.6.1.2 Load variance minimization

The load variance-based objective function provides an alternative option for coordinated charging of PEVs regardless of the network topology. Evidently, minimizing the overall load variance will smooth out the fluctuation in the load profiles. As a result, it will inevitably reduce the system losses to a great extent. A typical objective function served as load variance minimization is given by:

$$\min \frac{1}{T}\sum_{t=1}^{T}\left(P_t^{Con}+\sum_{n=1}^{N}P_{t,n}^{EV}-\mu_T^H\right)^2 \cdot \text{where } \mu_T^H=\frac{1}{T}\sum_{t=1}^{T}\left(P_t^{Con}+\sum_{n=1}^{N}P_{t,n}^{EV}\right) \tag{2.10}$$

where P_t^{Con} represents the conventional loads without PEVs at time t; $P_{t,n}^{EV}$ represents the charging power of the n-th PEV at time t; T represents the divided number of time period; N represents the number of PEVs.

In theory, the objective function using load variance concept is able to achieve load-valley filling and load-peak shaving with the PEV loads so as to improve the overall energy efficiency. There are many vital issues to be considered such as grid limits, the charging satisfaction of PEV users and the operating costs of system operators while reducing the overall load variance. In Ref. [67] the proposed charging scheduling scheme aims at minimizing the overall load variance of power grid by V2G operation while satisfying the charging requirement of all PEVs. However, the voltage issues and PEV charging costs are not mentioned. By contrast, literature [68] minimizes the load variance and charging costs by using the weighted sum method, as well as reducing the voltage drop. Similarly, Li et al. seeks to minimize the load variance and the satisfaction of all the PEV users with a weighing parameter subject to the limits on the feeder overload of power network [69]. Also in Ref. [70], the load variance and PEV charging costs are balanced by the proposing criteria. Apart from the load variance, the operating cost and pollutant treatment cost are also considered into the objective function in Ref. [71].

2.6.1.3 Peak load minimization

The peak load minimization is also one of the potential ways to effectively integrate PEVs into power system. The cost-benefit analysis has been carried out to implement the V2G operation with the peak load minimization, and it has been demonstrated that the social annual benefit is very likely larger than the total annual costs for the PEV coordination [72]. Therefore, the peak-minimizing objective is developed to reduce the peak load consumption. A simplest form of the peak load minimization is described as:

$$\min \max_{t\in\{1,2,..,T\}}\left\{P_t^{Con}+\sum_{n=1}^{N}P_{t,n}^{EV}-P_t^{PV}\right\} \tag{2.11}$$

where P_t^{PV} indicates the power supply of the photovoltaic array at time t; other variables, parameters and indices are defined as the same in Eq. (2.10).

The peak shaving strategy provides windows of opportunity to shift a part of the electricity demand to valley periods without increasing the grid capacity to the level of high peak loads. Therefore, an extensive research has devoted to minimizing the peak load for the optimal scheduling scheme of PEVs. In Ref. [73], the day-ahead scheduling method for PEV loads is proposed to minimize the system peak while satisfying the technical restrictions of the utility. Although the peak load consumption is decreased dramatically for various PEV penetrations by controlled charging, the economic benefits of PEV users are not involved. Consequently, the financial factor namely PEV charging costs are considered in Refs. [74], [61], [75] while reducing the peak demand. Based on minimization of peak-to-average and minimization of charging cost, a hybrid planning method is proposed to minimize the financial impacts by combining the strengths of two objectives more effectively [74]. In addition, in Ref. [61] a strategy that minimizes network peak load is compared with a strategy to minimize PEV charging costs. The results show that when PEV users make their charging decision only on electricity price, it may lead to high peaks in network load thereby high network investment costs; on the other hand, when PEV charging is controlled with the objective of minimizing peaks in network load, far less reinforcements are required. Moreover, a virtual time-of-use rate based on the load demand is proposed to encourage the PEV users to participate in peak load shaving by providing economic incentives in Ref. [75].

Beyond the objectives related to grid performance, economic considerations also play an important role in V2G scheduling. In practice, the charging behavior of PEVs is closely linked to electricity expenditure, and the minimization of charging cost has therefore become another widely adopted objective in the literature. With this in mind, the following discussion turns to the objective functions associated with charging cost.

2.6.1.4 Charging cost minimization

The economic strategy is regarded as one of the promising solutions for PEV coordinate charging. As a consequence, the TOU scheme is usually assigned with regard to load shifting. The low prices provide an effective incentive to make PEVs be charged during less congested hours. From the PEV user perspective, the minimization of charging costs is often adopted to acquire the cost saving strategy. A most general form of the total charging cost minimization is given by:

$$\min \sum_{n=1}^{N} \sum_{t=1}^{T} \left(P_{t,n}^{EV} \Delta t \right) \lambda_t^{RT} \tag{2.12}$$

where $P_{t,n}^{EV}$, T and N are defined as the same in Eq. (2.2); Δt indicates the length of each time interval; λ_t^{RT} indicates the real-time price of electricity at time t.

Eq. (2.12) aims to minimize the real-time charging costs of PEVs. The PEV aggregator will adjust the charging strategies of PEVs for cost saving. This objective function is established on the basis of a known real-time electricity price. However, this issue may be complex if considering the charging cost minimization of PEVs in a day-ahead horizon due to the uncertainty in electricity markets and charging behaviors. In Ref. [76], an optimal bidding strategy for PEV aggregators participating in electricity markets is developed. The objective is the minimization of the cost of purchased energy so as to reduce the charging costs of PEVs. The objective function can be given by:

$$\min\sum_{t}\lambda_t^{DA}E_t^{DA}+\sum_{w}\pi_w\sum_{t}\left[-\lambda_{t,w}^{RT}\Delta E_{t,w}+\left(\mathrm{Pnlty}_{t,w}^{U,Up}-\mathrm{Pnlty}_{t,w}^{U,Dn}\right)\right] \quad (2.13)$$

where λ_t^{DA} indicates the day-ahead price of electricity at time t; E_t^{DA} indicates the day-ahead energy bidding at time t; π_w indicates the scenario w probability of occurrence; $\lambda_{t,w}^{RT}$ indicates the real-time price at time t for the scenario w; $\Delta E_{t,w}$ indicates the deviation between real-time energy consumption and day-ahead energy bid; $\mathrm{Pnlty}_{t,w}^{U,Up}$ indicates the penalty for up uninstructed energy deviation higher than threshold; $\mathrm{Pnlty}_{t,w}^{U,Dn}$ indicates the penalty for down uninstructed energy deviation higher than threshold.

A significant body of research has demonstrated that the charging-cost-minimizing objectives have the potential of filling the valleys in electric load profiles under a reasonable pricing scheme. However, it also takes the risks of causing another peak demand at low-price hours if PEV users only consider the reduction of their charging costs based on the pricing scheme. Literature [77] shows the PEV charging strategy, based on the price signal only, will lead to significant overloading of power system as well as severe voltage issues in power grid. Therefore, they present an optimal PEV charging strategy to minimize the charging costs, at the same time, avoid any overloading problem. However, this method may be impractical because it requires that the arrivals of PEVs during the target day are able to be predicted in advance [78]. As a result, in Ref. [79] a local scheduling optimization problem is formulated to minimize the total cost of all PEVs in the current ongoing PEV set within the local group. Also in Ref. [80], the optimal energy management for PEVs is developed to reduce their charging costs under time-varying electricity price signal while considering the stochastic arrivals and departures of PEVs. Moreover, the system load factor and PEV users' satisfaction are improved by minimizing the electricity costs required for charging [81]. It can be observed that the charging cost-based objective function provides a theoretical foundation for the design of the electricity pricing mechanism from the utility perspective.

2.6.1.5 Operating profit maximization

Ordinarily, the PEV aggregator managing a fleet of PEVs participates in energy and ancillary services markets. As for the PEV aggregators, they are responsible for modulating the PEV charging level so as to maximize their profits. In general, they make an optimal bidding strategy in the day-ahead market on the basis of the PEV charging demand forecast and sell the energy for PEV charging in real-time market. On the one hand, the PEV aggregator should pay for the energy delivery at time t as

$$\lambda_t^{DA} E_t^{DA} - \lambda_t^{RT} \left(E_t^{DA} - E_t^{RT} \right) \tag{2.14}$$

where λ_t^{DA} indicates the day-ahead price of electricity at time t; E_t^{DA} indicates the day-ahead energy bidding at time t; E_t^{RT} indicates the real-time energy consumption at time t.

On the other hand, they obtain the total income by energy trading with the PEV users at time t given by

$$\lambda_t^{RT} \sum_{n=1}^{N} \left(P_{t,n}^{EV} \Delta t \right) \tag{2.15}$$

As a result, the objective function of the total operating profit minimization is simplified to (7) when no deviations penalty and no participation in regulation market are taken into account [76].

$$\max \sum_{t=1}^{T} \left(\lambda_t^{RT} \sum_{n=1}^{N} \left(P_{t,n}^{EV} \Delta t \right) - \left(\lambda_t^{DA} E_t^{DA} - \lambda_t^{RT} \left(E_t^{DA} - E_t^{RT} \right) \right) \right) \tag{2.16}$$

where all variables, parameters and indices are defined above.

There are extensive literature regarding the maximization of operating profits from the PEV aggregator perspective. Remarkably, as far as the PEV aggregator's benefit is concerned, the optimal operation is hardly achieved regardless of the PEV charging behaviors on cost-saving. Consequently, the PEV users with driving patterns and willingness to participate should be considered. When bidding in the electricity market, the PEV aggregators should allow for the uncertainty on PEV availability, the utility limits, their own operational risk and other issues. In existing cases, the potential of PEV aggregators in providing the regulation services has been analyzed, calculated and evaluated. In Ref. [82], a PEV-based battery storage models are proposed for power regulation by a V2G system, whereas how to maximize the PEV aggregator profit is not involved. In the later work, the coordinated dispatching method for PEVs is presented to maximize the PEV aggregator profit while regulation is performed by adjusting the PEV charging rate around a set point [6]. But the set point only varies between zero and the maximum charging power. This manifests that energy only flows from grid to vehicles. In fact, the PEV aggregators can provide the bidirectional V2G regulation to pursue more profits, wherein a PEV charges its battery for regulation-down and discharges its battery for regulation-up. However, owing to the additional battery degradation and utilizing the PEV batteries for regulation services, the PEV aggregators should compensate the PEV users by sharing a part of their profit. Furthermore, in consideration of the uncertain nature of electricity market and

PEV charging behavior, stochastic programming model is employed to tackle these uncertain parameters [83]. Additionally, conditional value-at-risk measure is also adopted to maximize the PEV aggregator profit for optimal coordination of a PEV fleet [84].

Besides grid performance and economic considerations, environmental objectives have also attracted growing attention in V2G scheduling. As EV deployment is closely associated with low-carbon energy transition, it is necessary to consider how charging and discharging strategies can contribute to emission reduction and broader environmental benefits.

2.6.1.6 Carbon emission minimization

The environmental objective function is also employed to coordinate PEV charging so as to reduce the carbon emissions of power plants. Actually, electricity production emissions levels are highly related to the national legislation, regulatory structure, and electricity market [41]. Meanwhile, emissions deriving from the PEV usage can vary greatly depending on the region and charging time periods [42]. As a consequence, to manage PEV charging during the low-emission periods as far as possible will help the grid minimize the total carbon emissions. One representative form of the carbon emission minimization can be defined as:

$$\min \sum_{t=1}^{T} \sum_{n=1}^{N} \left(P_{t,n}^{EV} \Delta t \right) MER_t \tag{2.17}$$

where MER_t indicated the marginal carbon emissions rate per kWh at time t; the other symbols are defined as the same in Eq. (2.10).

The environmental impacts of PEV charging have been investigated when controlled and uncontrolled charging is taken into account respectively. It has been demonstrated that uncontrolled charging of massive PEVs may dramatically elevate the emission levels in highly carbonized power systems. On the other hand, smart charging almost doesn't result in additional carbon emissions in some specific cases; moreover, the application of bidirectional V2G technology could reduce the generator carbon emissions substantially [85]. Nevertheless, a little research focuses on the carbon emission minimization by scheduling PEV charging or discharging in an appropriate way. In Ref. [86], the total CO_2 production from both vehicles and generation units are minimized by determining the number of each EV type. However, the optimal dispatching strategy for PEV charging is not considered. In Ref. [87], an optimal scheduling for PHEV charging/discharging is presented to minimize the total fuel cost and emission for the integration of PHEVs in an electrical power system. The results show that the proposed method not only has a significant superiority over other methods in the regard of reducing total operating costs and emission, but it also increases profit, reserve and reliability.

Apart from the above objective functions, appropriate modeling of the battery degradation cost incurred in the V2G operation is necessary since that battery accounts for a remarkable portion of total vehicle cost. The battery degradation is related to the charge/discharge rate, ambient temperature, and depth of discharge. In general, it is hard to accurately weigh the battery degradation resulting from a charging event for PEVs. In Ref. [88], the battery degradation cost of PEVs is specified by a function of a charging power. Besides, the battery degradation cost is calculated into the total charging costs given by:

$$\min \sum_{n=1}^{N} \sum_{t=1}^{T} \left(\left(P_{t,n}^{EV} \Delta t \right) \lambda_t^{RT} + f\left(P_{t,n}^{EV} \right) \right) \tag{2.18}$$

where $f\left(P_{t,n}^{EV}\right)$ indicates the battery degradation cost of the n-th PEV at time t under a charging power $P_{t,n}^{EV}$; and other variables, parameters and indices are defined as the same in Eq. (2.12).

2.6.2 Scheduling constraints

Before a scheduling strategy can be regarded as feasible, it must satisfy a series of operational constraints arising from both the EV side and the power-system side. The former are directly related to the physical characteristics and usage requirements of PEVs, such as battery capacity, state-of-charge limits, charging/discharging power bounds, and users' travel demands. These constraints define the basic operating boundaries within which coordinated charging and discharging can be carried out. Accordingly, the following discussion begins with the constraints originating from the EV side.

2.6.2.1 PEV charging/discharging power constraint

A huge experimental results show that too high charging/discharging power will accelerate the aging processes inside the PEV battery under the existing battery technology [89]. As for each type of PEV, there is usually an allowable charging range for the onboard charger. In order to enable a longer battery life, the charging power of the n-th PEV at time t should be limited as:

$$P_{\min,n}^{EV} \leq P_{t,n}^{EV} \leq P_{\max,n}^{EV} \tag{2.19}$$

where $P_{min,n}^{EV}$, $P_{max,n}^{EV}$ denote the minimum and maximum charging power of the n-th PEV, respectively.

2.6.2.2 PEV battery capacity constraint

Apparently, the battery degradation is related to the depth of discharge/charge. There is evidence that a high depth of discharge will increase the battery degradation to some extent. In addition, when the battery is overcharged, the battery wear may occur deriving from a loss of active cathode materials [90]. As a consequence, the PEV battery capacity should be restricted into a reasonable range to prevent overcharge and over discharge. For convenience, the battery SoC is introduced to indicate the ratio of the remaining capacity to the nominal capacity of the battery. Therefore, the battery SoC of the n-th PEV at time t is limited as:

$$SoC^{EV}_{min,n} \leq SoC^{EV}_{t,n} \leq SoC^{EV}_{max,n} \tag{2.20}$$

where $SoC^{EV}_{min,n}$, $SoC^{EV}_{max,n}$ are the allowable minimum and maximum SoC value of the n-th PEV, respectively.

2.6.2.3 PEV user charging requirement constraint

PEVs are served as a kind of transportation tool. The remaining driving range of a PEV is determined by the SoC of its battery pack. When a PEV user plugs his/her vehicle into the charging post, he/she has a desire that this vehicle is charged to his/her expected SoC so as to ensure his/her next journey. Consequently, as an effort to improve the charging satisfactions, the PEV user charging requirement should be satisfied by:

$$\eta_n \Delta t \sum_{t=\Gamma^s_n}^{\Gamma^e_n} P^{EV}_{t,n} = \left(SoC^e_n - SoC^s_n\right) C_n \tag{2.21}$$

where η_n denotes the charger efficiency of the n-th PEV; Γ^s_n, Γ^e_n denote the plugged-in and plugged-out time of the n-th PEV, respectively; SoC^e_n, SoC^s_n are the plugged-in and plugged-out SoC of the n-th PEV, respectively; C_n denotes the onboard battery capacity of the n-th PEV.

While the constraints from the EV side define the basic operating boundaries of charging and discharging, they are not sufficient on their own to ensure the feasibility of V2G scheduling in practical power systems. Since PEVs are ultimately connected to the grid through charging infrastructures and local networks, their coordinated operation must also comply with the technical limits of the power system. These grid-side constraints are essential for maintaining secure, stable, and efficient system operation, and therefore constitute another important part of the scheduling model. On this basis, the following discussion turns to the constraints arising from the grid side.

2.6.2.4 Node voltage constraint

In order to maintain the stability of power grid, the voltage constraints of the girds should be given by setting the upper and lower bounds. In general, it is acceptable for 95% of the time that the voltage has a 10% deviation according to the mandatory EN50160 standard. Hence, the gird voltage at time t at node k should be limited in an allowable scope as:

$$V_{min} \leq V_{t,k} \leq V_{max} \tag{2.22}$$

where V_{min}, V_{max} denote the minimum and maximum voltage value at each node, respectively.

2.6.2.5 Feeder capacity constraint

In practice, there is a maximum allowable load of each feeder of the distribution network. When PEVs are plugged into power gird, the total load capacity of each feeder should be restricted not exceeding their maximum load capacity. Consequently, the inequality constraints can be given by:

$$P_l^{Con} + \sum_{n=1}^{N} P_{t,n}^{EV} \leq P_{l,max} \tag{2.23}$$

where P_l^{Con} denotes the conventional load without charging load through the feeder l; $P_{l,max}$ denotes the maximum load capacity supported by the feeder l.

2.6.2.6 Transmission power constraint

When considering the secure running of regional or whole grids, the transmission power between the microgrid and the main grid should be considered. Overall, the transmission power at time t cannot exceed the limits by:

$$-P_{L,\max} \leq P_{\text{grid},t} \leq P_{L,\max} \tag{2.24}$$

where $P_{L,max}$ denotes the upper limit of transmission power between the microgrid and the main power grid.

2.6.3 Solving algorithm

The solving process of an optimization model is usually an intractable task since it involves a bulk of variables when a large scale of PEVs joins in the V2G implementation. In addition, the established model may consist of the nonlinear objective function or the nonlinear constrain. From the mathematical point of view, the PEV scheduling models in the present literature include LP, NP, MIP and other models. Generally speaking, the LP model provides a simple and efficient approach to solve the PEV charging problem by using the traditional solving techniques, such as simplex method. Nevertheless, these traditional techniques have difficulties to deal with the NP or MIP models. Therefore, other solving techniques have been developed and used in the NP and MIP model.

In the existing works, several LP models are developed in terms of PEV scheduling [65], [81], [91]. These models are able to determine a global optimal solution for PEV charging in a fairly easy way, but are just limited to linear objectives and constrains. Apparently, it is not enough for a more complicated V2G system. Hence, NP and MIP models are applied to tackle the nonlinear objective functions or the nonlinear integer constrains. The QP model is the typical one of the NP models [37]-[92], [58], [67]. According to the convexity, the QP models can fall into the convex and nonconvex optimization models. In general, the convex optimization model will consume a less solving time than the nonconvex optimization model. The reason is that the convex programming model is able to be solved by the interior point algorithm efficiently. Moreover, there are almost no complete and robust algorithms to solve the MIP models or other nonconvex models. As a result, many metaheuristic algorithms are used to serve as an alternative option in solving the complex models.

The most popular algorithms for the optimal V2G strategy are PSO [93] and GA [94]. In Ref. [93], GA is employed on the MINP models for the PEV penetration maximization. In Ref. [94], PSO is adopted to optimize the nonlinear parameters for the power consumption minimization and the PEV satisfaction maximization. Additionally, the improved GA and PSO are presented in Ref. [95]-[96] to search for the optimal solution in less time. Apart from GA and PSO, the metaheuristic algorithms also include evolutionary algorithm [97], simulated annealing [98], differential evolutionary [99], ant colony optimization [100] and so on. Most of them have the advantage that they require less computational time than the traditional optimization methods. In spite of this, they almost cannot ensure seeking the globally optimal solutions.

2.6.4 Computational performance evaluation

In order to integrate PEVs in power grid, various objective functions and constraints are formulated into distinct programming models from different perspectives. As shown in Table 2.6, seven representative models are selected so as to compare with each other. It can be observed that different objectives are employed in these models. It signifies that the multifarious effectiveness on either the grid performance or operational benefits is able to be achieved. On the other hand, the enormous differences with regard to the feasibility also exist depending on the structure of power system and the properties of programming models. It is evident both the effectiveness and feasibility of a model are the key factors for the V2G successful implementation. Therefore, it is valuable to make an overall evaluation for these models on the basis of effectiveness and feasibility.

In Table 2.6, the achieved effectiveness on power grid consists of reducing power losses, filling load valley, shaving load peak, improving voltage profile, improving three-phase unbalance, providing ancillary services and so on. Similarly, the achieved effectiveness on economic benefits includes PEV charging costs reduction, system operating costs reduction, aggregator operating profits increase and so on. It is deemed to be reasonable that a model considering multiple aspects is superior to the model only considering the single aspect. For example, the method proposed by Maigha et al. not only

TABLE 2.6 Outline and evaluation among seven typical programming models.

Model outline				Model evaluation	
Objective	Main constrains	Programming model	Centralized/ decentralized	Effectiveness	Feasibility
Minimize power loss [37]	**(a)** Charging demand limit **(b)** Charging power limit **(c)** Voltage limit	Quadratic programming	Centralized	**(a)** Reduce power loss **(b)** Improve voltage profile	**(a)** High computation cost **(b)** Depend on grid topology
Minimize load variance [67]	**(a)** Battery SoC limit **(b)** Charging power limit **(c)** Charging demand limit	Quadratic programming	Centralized	**(a)** Fill load valley **(b)** Shave load peak	**(a)** High computation cost **(b)** Topology independent
Minimize system peak [73]	**(a)** Battery SoC limit **(b)** Voltage limit **(c)** Current limit	Mixed integer nonlinear programming	Centralized	**(a)** Reduce load peak **(b)** Improve three-phase unbalance	**(a)** Very high computation cost **(b)** Depend on grid topology
Minimize PEV charging cost [81]	**(a)** Charging power limit **(b)** Battery SoC limit **(c)** Charging demand limit	Linear programming	Centralized	**(a)** Fill load valley **(b)** Reduce charging cost **(c)** Demand response	**(a)** Low computation cost **(b)** Topology independent

Maximize aggregator profits [91]	**(a)** Battery SoC limit **(b)** System load limit **(c)** Charging demand limit	Linear programming	Decentralized	**(a)** Achieve load shifting **(b)** Provide ancillary services **(c)** Reduce charging cost	**(a)** Low computation cost **(b)** Topology independent
Minimize system operating costs [101]	**(a)** Voltage limit **(b)** Current limit **(c)** DGs load limit **(d)** Charging demand limit	Mixed integer linear programming	Decentralized	**(a)** Reduce operating costs **(b)** Improve voltage profile **(c)** Improve load unbalance	**(a)** Very high computation cost **(b)** Depend on grid topology
Minimize generation cost and carbon emissions [65]	**(a)** Charging demand limit **(b)** Charging power limit	Linear programming	Decentralized	**(a)** Mitigate impact of PEV charging **(b)** Reduce generation cost and carbon emission	**(a)** Low computation cost **(b)** Topology independent

considers filling the load valley, but also minimizing PEV charging costs. This type of methods is better than the ones that only allow for filling the load valley such as in Refs. [102], [38]. As for feasibility, these models can be assessed on the computation cost and whether considering the complex grid topology or not. Obviously, the topology independent models with low computation cost have a significant superiority over the models with high computation cost depending on the grid topology.

There are all kinds of mathematical descriptions to deal with the problem of integrating PEVs into power grids friendly. In general, they can be applied to the special application scenario for the V2G implementation. In the majority of existing works, the mathematical programming model is established under the conceptual operating framework with the energy and information exchange. The hypothetical framework is always perfect for their models' establishment regardless of some operating environmental issues such as the TOU price. In the coordinated V2G operation, the electricity price system plays a very important role for the PEV participation. In some regions, for example, in China and North Korea, the design and application of the TOU electricity price system is a rather crucial issue even concerned with upper-level economic reform and government policy making. More importantly, their electricity market is not mature currently and there are no electricity retailers to offer fixed prices or variable for electricity to their customers and manage the risk involved in purchasing electricity. The State Grid Corporation of China is in charge of the power generation, transmission and distribution. It can be foreseen that the grid company is more likely to become the conductor of the V2G implementation if the smart grid technology is employed in the future. Therefore, to directly improve the grid performance may be their primary target. In this sense, the gird performance index or environmental index may be taken as the objective function of the mathematical programming model.

Comparatively speaking, the electricity market environment in America and most European countries is open and flexible with fierce competition. The electricity price level will encourage more efficient use of energy, and market prices will encourage more demand response. In this situation, the PEV aggregators are allowed to exist for managing PEV charging and bargaining with electricity retailers in electricity markets to make a profit. Thus, PEV aggregators may become the energy aggregator for charging PEVs in a coordinated way. In this context, the operating cost minimization may be their major aims. As a consequence, the economic index is likely to be taken as the objective function of the mathematical description for charging scheduling optimization.

2.7 Concluding remarks

As an effort to present the relevant works for integrating PEVs into the electric grid friendly, the power interaction mode between PEVs and grids, and the mainstream methodology for the V2G implementation are overviewed comprehensively. Firstly, the power interaction mode between PEVs and grids can fall into two categories: uncoordinated charging and coordinated charging/discharging mode. The uncoordinated charging mode operates regardless of the status of the gird. In the coordinated charging/discharging mode (viz. the V2G mode), the grid operator will establish an intelligent communication link with PEV users so as to make them better response to grid requirements. In this regard, the discharging mode is allowed for PEVs to feedback the stored energy to the grid. Therefore, the incurred battery degradation cost should be taken the total charging costs into account in the charging optimization.

The operating framework for the V2G implementation is crucially important. Generally speaking, the scheduling optimization for PEV charging can be operated in a centralized or decentralized way. Due to the unified charging management of PEVs involved in this system, the centralized scheduling approach can achieve the overall optimal operation; however, it is difficult to preserve the individual charging authority of PEV users. In contrast, the decentralized scheduling approach is a price-signal-based method. Each PEV user will response to price signals for determining the charging profile for their individual vehicles. In this sense, it is likely to be widely appreciated due to a higher acceptance of PEV users on charging authority; however, it is hard to attain the optimal energy management between PEVs and grids.

In essence, the acquisition of the optimal V2G strategy is ultimately to build and solve a mathematical optimization model. All kinds of objective functions are employed for coordinated charging/discharging, and LP, NP, and MIP models are developed in terms of PEV scheduling. The LP model can be solved more efficiently than the NP and MIP model. Additionally, the grid topology independent model with lower computation cost has a significant superiority over the models with high computation cost depending on the grid topology. As a consequence, it is meaningful to develop a grid topology independent model with lower computation cost for the practical implementation of V2G technology in the future.

References

[1] Chan CC, Wong YS. The state of the art of electric vehicles technology. In: 2004 IEEE Power Electronics and Motion Control Conference; 2004 [Xi"an].

[2] Sierzchula W, Bakker S, Maat K, van Wee B. The influence of financial incentives and other socio-economic factors on electric vehicle adoption. Energy Policy 2014;68:183–94.

[3] Flores R, Shaffer B, Brouwer J. Electricity costs for an electric vehicle fueling station with Level 3 charging. Appl Energy 2016;169:813–30.

[4] Global EV Outlook 2018-Towards cross-modal electrification. International Energy Agency (IEA); May 2018.
[5] Chen N, Tan CW, Quek TQS. Electric vehicle charging in smart grid: Optimality and valley-filling algorithms. IEEE J Sel Top Signal Process 2014;8(6):1073–83.
[6] Sortomme E, El-Sharkawi MA. Optimal charging strategies for unidirectional vehicle-to-grid. IEEE Trans Smart Grid 2011;2(1):131–8.
[7] Mahmud K, Town GE, Morsalin S, et al. Integration of electric vehicles and management in the internet of energy. Renew Sustain Energy Rev 2018;82:4179–203.
[8] Dubarry M, Devie A, Mckenzie K. Durability and reliability of electric vehicle batteries under electric utility grid operations: Bidirectional charging impact analysis. J Power Sources 2017;358:39–49.
[9] Kempton W, Tomić J. Vehicle-to-grid power fundamentals: Calculating capacity and net revenue. J Power Sources 2005;144(1):268–79.
[10] Hassan F. Utilizing solar and wind energy in plug-in hybrid electric vehicles. Energy Convers Manag 2018;156:317–28.
[11] Clement-Nyns K, Haesen E, Driesen J. The impact of charging plug-in hybrid electric vehicles on a residential distribution grid. IEEE Trans Power Syst 2010;25(1):371–80.
[12] Zhan K, Hu Z, Song Y, et al. A probability transition matrix based decentralized electric vehicle charging method for load valley filling. Elec Power Syst Res 2015;125:1–7.
[13] Global top 20 - June 2018, ://ev-sales.blogspot.com/2018/
[14] Fernandez LP, Roman TGS, Cossent R, et al. Assessment of the impact of plug-in electric vehicles on distribution networks. IEEE Trans Power Syst 2011;26(1):206–13.
[15] Akhavan-Rezai E, Shaaban MF, El-Saadany EF, et al. Uncoordinated charging impacts of electric vehicles on electric distribution grids: Normal and fast charging comparison. In: 2012 IEEE power and energy society general meeting, San Diego; 2012.
[16] Deilami S, Masoum AS, Moses PS, et al. Voltage profile and THD distortion of residential network with high penetration of plug-in electrical vehicles. In: 2010 IEEE Innovative Smart Grid Technologies Conference Europe, Gothenberg; 2010.
[17] Moller F, Muller S, Meyer J, et al. Impact of electric vehicle charging on unbalance and harmonic distortion-field study in an urban residential area. In: 23rd International Conference on Electricity Distribution, Lyon; 2015.
[18] En50160. Voltage characteristics of electricity supplied by public distribution systems. 1999.
[19] Muñoz ER, Razeghi G, Zhang L, et al. Electric vehicle charging algorithms for coordination of the grid and distribution transformer levels. Energy 2016;113:930–42.
[20] Shao S, Pipattanasomporn M, Rahman S. Challenges of PHEV penetration to the residential distribution network. In: 2009 IEEE power & energy society general meeting, calgary; 2009.
[21] Taylor J, Maitra A, Alexander M, et al. Evaluation of the impact of plug-in electric vehicle loading on distribution system operations. In: 2009 IEEE power & energy society general meeting, Calgary; 2009.
[22] Razeghi G, Zhang L, Brown T, et al. Impacts of plug-in hybrid electric vehicles on a residential transformer using stochastic and empirical analysis. J Power Sources 2014;252:277–85.
[23] Trovao L, Jorge HM. Power demand impacts of the charging of electric vehicles on the power distribution network in a residential area. In: 2011 IEEE International Youth Conference on Energetics, Leiria; 2011.

[24] Darabi Z, Ferdowsi M. Impact of plug-in hybrid electric vehicles on electricity demand profile. IEEE Trans Sustain Energy 2011;2(4):501–8.

[25] Tuttle D, Fares R, Baldick R, Webber M. Plug-in vehicle to home (V2H) duration and power output capability. In: Proc. IEEE Transp. Electr. Conf. Expo (ITEC); June 2013.

[26] Monteiro V, Exposto B, Ferreira JC, Afonso JL. Improved vehicle-to-home (iV2H) operation mode: Experimental analysis of the electric vehicle as off-line UPS. IEEE Trans Smart Grid 2017;8(6):2702–11.

[27] Mozafar MR, Amini MH, Moradi MH. Innovative appraisement of smart grid operation considering large-scale integration of electric vehicles enabling V2G and G2V systems. Elec Power Syst Res 2018;154:245–56.

[28] Liu C, Chau KT, Wu D, Gao S. Opportunities and challenges of vehicle-to-home, vehicle-to-vehicle, and vehicle-to-grid technologies. Proc IEEE 2013;101(11):2409–27.

[29] Jian L, Zheng Y, Shao Z. High efficient valley-filling strategy for centralized coordinated charging of large-scale electric vehicles. Appl Energy 2017;186:46–55.

[30] Green RC, Wang L, Alam M. The impact of plug-in hybrid electric vehicles on distribution networks: A review and outlook. Renew Sustain Energy Rev 2011;15(1):544–53.

[31] Zheng Y, Jian L. Smart charging algorithm of electric vehicles considering dynamic charging priority. In: 2016 IEEE International Conference on Information and Automation; 2016. Ningbo.

[32] Wang D, Coignard J, Zeng T, et al. Quantifying electric vehicle battery degradation from driving vs. vehicle-to-grid services. J Power Sources 2016;332:193–203.

[33] Yoshio M, Brodd RJ, Kozawa A. Lithium-ion batteries: Science and technologies. Springer; 2009.

[34] Uddin K, Jackson T, Widanage WD, et al. On the possibility of extending the lifetime of lithium-ion batteries through optimal V2G facilitated by an integrated vehicle and smart-grid system. Energy 2017;133:710–22.

[35] Pelletier S, Jabali O, Laporte G, et al. Battery degradation and behavior for electric vehicles: Review and numerical analyses of several models. Transp Res Part B 2017;103:158–87.

[36] Nguyen VL, Tranquoc T, Bacha S, et al. Charging strategies to minimize the peak load for an electric vehicle fleet. In: 2015 IEEE Industrial Electronics Society Conference. Istanbul; 2015.

[37] Zhan K, Song Y, Hu Z, et al. Coordination of electric vehicle charging to minimize active power losses. Proceedings of the Csee 2012;32(31):11–8.

[38] Sortomme E, Hindi MM, Macpherson SDJ, et al. Coordinated charging of plug-in hybrid electric vehicles to minimize distribution system losses. IEEE Trans Smart Grid 2011;2 (1):198–205.

[39] He Y, Venkatesh B, Guan L. Optimal scheduling for charging and discharging of electric vehicles. IEEE Trans Smart Grid 2012;3(3):1095–105.

[40] Peng C, Zou J, Lian L, et al. An optimal dispatching strategy for V2G aggregator participating in supplementary frequency regulation considering EV driving demand and aggregator's benefits. Appl Energy 2017;190:591–9.

[41] Tikka V, Lassila J, Haakana J, et al. Electric vehicle smart charging aims for CO2 emission reduction?. In: IEEE Pes Innovative Smart Grid Technologies Conference Europe, Ljubljana; 2017.

[42] Hoehne CG, Chester MV. Optimizing plug-in electric vehicle and vehicle-to-grid charge scheduling to minimize carbon emissions. Energy 2016;115:646–57.

[43] Soares J, Ghazvini MAF, Vale Z, et al. A multi-objective model for the day-ahead energy resource scheduling of a smart grid with high penetration of sensitive loads. Appl Energy 2015;162:1074–88.
[44] Sundström O, Binding C. Planning electric-drive vehicle charging under constrained grid conditions. In: 2010 IEEE International Conference on Power System Technology, Hangzhou; 2010.
[45] Soares FJ, Almeida PMR, Lopes JAP. Quasi-real-time management of electric vehicles charging. Elec Power Syst Res 2014;108(108):293–303.
[46] Kang J, Duncan SJ, Mavris DN. Real-time scheduling techniques for electric vehicle charging in support of frequency regulation. Procedia Comput Sci 2013;16(1):767–75.
[47] Fernández IJ, Calvillo CF, Sánchez-Miralles A, et al. Capacity fade and aging models for electric batteries and optimal charging strategy for electric vehicles. Energy 2013;60 (1):35–43.
[48] Lunz B, Walz H, Sauer DU. Optimizing vehicle-to-grid charging strategies using genetic algorithms under the consideration of battery aging. In: 2011 IEEE Vehicle Power and Propulsion Conference, Chicago; 2011.
[49] Ma Z, Callaway DS, Hiskens IA. Decentralized charging control of large populations of plug-in electric vehicles. IEEE Trans Control Syst Technol 2010;21(1):67–78.
[50] Gan L, Topcu U, Low SH. Optimal decentralized protocol for electric vehicle charging. IEEE Trans Power Syst 2013;28(2):940–51.
[51] Xi X, Sioshansi R. Using price-based signals to control plug-in electric vehicle fleet charging. IEEE Trans Smart Grid 2014;5(3):1451–64.
[52] Zhan K, Hu Z, Song Y, et al. A probability transition matrix based decentralized electric vehicle charging method for load valley filling. Elec Power Syst Res 2015;125:1–7.
[53] Andres S, Martinze S. A hierarchical algorithm for optimal plug-in electric vehicle charging with usage constraints. Automatica 2016;68:119–31.
[54] Ahn C, Li CT, Peng H. Optimal decentralized charging control algorithm for electrified vehicles connected to smart grid. J Power Sources 2011;196(23):10369–79.
[55] Zhang K, Xu L, Ouyang M, et al. Optimal decentralized valley-filling charging strategy for electric vehicles. Energy Convers Manag 2014;78(2):537–50.
[56] Teng JH, Liao SH, Wen CK. Design of a fully decentralized controlled electric vehicle charger for mitigating charging impact on power grids. IEEE Trans Ind Appl 2016;2 (53):1497–505.
[57] Moghbel M, Masoum MAS, Shahnia F. Coordinated charging of PEVs in unbalanced residential network based on worst node voltage profile. In: 2013 IEEE power and energy society general meeting, Vancouver; 2013.
[58] Clement K, Haesen E, Driesen J. Coordinated charging of multiple plug-in hybrid electric vehicles in residential distribution grids. In: 2009 IEEE Power Systems Conference and Exposition, Seattle; 2009.
[59] Chen JS, Zhang JH, Lin JY, et al. Strategies for electric vehicle charging with aiming at reducing network loss. Proc Chin Soc Univ Electr Power Syst Autom 2012;24(3):139–44.
[60] Mehboob N, Cañizares C, Rosenberg C. Day-ahead dispatch of PEV loads in a residential distribution system. In: IEEE Pes General Meeting Conference & Exposition, National Harbor; 2014.
[61] Veldman E, Verzijlbergh RA. Distribution grid impacts of smart electric vehicle charging from different perspectives. IEEE Trans Smart Grid 2014;6(1):333–42.

[62] Bessa RJ, Matos MA. Global against divided optimization for the participation of an EV aggregator in the day-ahead electricity market. Part I: Theory. Elec Power Syst Res 2013;95(95):309–18.

[63] Zakariazadeh A, Jadid S, Siano P. Multi-objective scheduling of electric vehicles in smart distribution system. Energy Convers Manag 2014;79(3):43–53.

[64] Kang J, Duncan SJ, Mavris DN. Real-time scheduling techniques for electric vehicle charging in support of frequency regulation. Procedia Comput Sci 2013;16(1):767–75.

[65] Ahn C, Li CT, Peng H. Decentralized charging algorithm for electrified vehicles connected to smart grid. In: 2011 IEEE American Control Conference; 2011. San Francisco.

[66] Kalkhambkar V, Kumar R, Bhakar R. Energy loss minimization through peak shaving using energy storage. Perspect Sci 2016;8:162–5.

[67] Jian L, Zheng Y, Xiao X, et al. Optimal scheduling for vehicle-to-grid operation with stochastic connection of plug-in electric vehicles to smart grid. Appl Energy 2015;146:150–61.

[68] García-Villalobos J, Zamora I, Knezović K, et al. Multi-objective optimization control of plug-in electric vehicles in low voltage distribution networks. Appl Energy 2016;180:155–68.

[69] Li J, Li C, Wu Z, et al. Sparsity-promoting distributed charging control for plug-in electric vehicles over distribution networks. Appl Math Model 2017;58.

[70] Moon SK, Kim JO. Balanced charging strategies for electric vehicles on power systems. Appl Energy 2017;189:44–54.

[71] Lu X, Zhou K, Yang S, et al. Multi-objective optimal load dispatch of microgrid with stochastic access of electric vehicles. J Clean Prod 2018;195:187–99.

[72] Luo Z, Hu Z, Song Y, et al. Optimal coordination of plug-in electric vehicles in power grids with cost-benefit analysis—part I: Enabling techniques. IEEE Trans Power Syst 2013;28 (4):3546–55.

[73] Mehboob N, Cañizares C, Rosenberg C. Day-ahead dispatch of PEV loads in a residential distribution system. In: IEEE Pes General Meeting Conference & Exposition; 2014. National Harbor.

[74] Mehta R, Srinivasan D, Trivedi A, et al. Hybrid planning method based on cost-benefit analysis for smart charging of plug-in electric vehicles in distribution systems. IEEE Trans Smart Grid 2017;(99):1.

[75] Tan J, Wang L. Integration of plug-in hybrid electric vehicles into residential distribution grid based on two-layer intelligent optimization. IEEE Trans Smart Grid 2014;5 (4):1774–84.

[76] Vagropoulos SI, Bakirtzis AG. Optimal bidding strategy for electric vehicle aggregators in electricity markets. IEEE Trans Power Syst 2013;28(4):4031–41.

[77] Mahat P, Handl M, Kanstrup KR, et al. Price based electric vehicle charging. In: 2012 IEEE power and energy society general meeting, San Diego; 2012.

[78] Mahmoodi M, Mcdonough M, Shamsi P, et al. Peak shaving and minimum cost operation of an electric vehicle charging station based on multi-port power electronic interface. In: 2015 IEEE Transportation Electrification Conference and Expo, Dearborn; 2015.

[79] He Y, Venkatesh B, Guan L. Optimal scheduling for charging and discharging of electric vehicles. IEEE Trans Smart Grid 2012;3(3):1095–105.

[80] Nguyen VL, Tran-Quoc T, Bacha S, et al. Charging strategies to minimize the energy cost for an electric vehicle fleet. In: 2015 IEEE Innovative Smart Grid Technologies Conference Europe. Istanbul; 2015.

[81] Crow M, Maigha M. Cost-constrained dynamic optimal electric vehicle charging. IEEE Trans Sustain Energy 2016;8(2):716–24.

[82] Pillai JR, Bak-Jensen B. Vehicle-to-grid systems for frequency regulation in an Islanded Danish distribution network. In: 2010 IEEE Vehicle Power and Propulsion Conference, Lille; 2010.

[83] Shafie-Khah M, Moghaddam MP, Sheikh-El-Eslami MK, et al. Modeling of interactions between market regulations and behavior of plug-in electric vehicle aggregators in a virtual power market environment. Energy 2012;40(1):139–50.

[84] Al-Awami AT, Sortomme E. Coordinating vehicle-to-grid services with energy trading. IEEE Trans Smart Grid 2012;3(1):453–62.

[85] Sioshansi R, Denholm P. Emissions impacts and benefits of plug-in hybrid electric vehicles and vehicle-to-grid services. Environ Sci Technol 2009;43(4):1199–204.

[86] Barazesh M, Saebi J, Javidi DBMH. Optimal distribution of electric vehicle types for minimizing total CO2 emissions. In: IEEE electrical engineering; 2015. Tehran.

[87] Kumar A, Bhalla V, Kumar P. Unit commitment in a smart grid with plug-in hybrid electric vehicles—a cost-emission optimization. In: 2017 IEEE Transportation Electrification Conference. Pune; 2017.

[88] Ma Z, Zou S, Liu X. A distributed charging coordination for large-scale plug-in electric vehicles considering battery degradation cost. IEEE Trans Control Syst Technol 2015;23 (5):2044–52.

[89] Kavousi-Fard A, Abunasri A, Zare A, et al. Impact of plug-in hybrid electric vehicles charging demand on the optimal energy management of renewable micro-grids. Energy 2014;78:904–15.

[90] Lam L. A practical circuit-based model for state of health estimation of Li-on battery cells in electric vehicles. Lifepo4. 2011.

[91] Escudero-Garzas JJ, Garcia-Armada A, Seco-Granados G. Fair design of plug-in electric vehicles aggregator for V2G regulation. IEEE Trans Veh Technol 2012;61(8):3406–19.

[92] Jian L, Zhu X, Shao Z, et al. A scenario of vehicle-to-grid implementation and its double-layer optimal charging strategy for minimizing load variance within regional smart grids. Energy Convers Manag 2014;78(2):508–17.

[93] Alonso M, Amaris H, Germain JG, et al. Optimal Charging scheduling of electric vehicles in smart grids by heuristic algorithms. Energies 2014;7(4):2449–75.

[94] Wang Z, Wang L, Dounis A, et al. Integration of plug-in hybrid electric vehicles into energy and comfort management for smart building. Energy Build 2012;47:260–6.

[95] Cheng YH, Lai CM, Teh J. Optimization of control strategy for hybrid electric vehicles based on improved genetic algorithm. In: 2017 IEEE Vehicle Power and Propulsion Conference, Belfort; 2017.

[96] Yang J, He L, Fu S. An improved PSO-based charging strategy of electric vehicles in electrical distribution grid. Appl Energy 2014;128(3):82–92.

[97] Ahmad MR, Musirin I, Othman MM. Optimal charging strategy for plug-in hybrid electric vehicle using evolutionary algorithm. In: 2014 IEEE Power Engineering and Optimization Conference; 2014. Langkawi.

[98] Valentine K, Temple WG, Zhang KM. Intelligent electric vehicle charging: Rethinking the valley-fill. J Power Sources 2011;196(24):10717–26.

[99] Zhang L, Lin C, Niu X. Optimization of control strategy for plug-in hybrid electric vehicle based on differential evolution algorithm. In: 2009 IEEE Asia-Pacific Power and Energy Engineering Conference, Wuhan; 2009.

[100] Ghanbarzadeh T, Goleijani S, Moghaddam MP. Reliability constrained unit commitment with electric vehicle to grid using hybrid particle swarm optimization and ant colony optimization. In: 2011 IEEE power and energy society general meeting, detroit; 2011.

[101] Franco JF, Rider MJ, Romero R. A mixed-integer linear programming model for the electric vehicle charging coordination problem in unbalanced electrical distribution systems. IEEE Trans Smart Grid 2017;6(5):2200–10.

[102] Jian L, Xue H, Xu G, et al. Regulated charging of plug-in hybrid electric vehicles for minimizing load variance in household smart microgrid. IEEE Trans Ind Electron 2013;60 (8):3218–26.

Chapter 3

Industrial standards and infrastructure technologies

3.1 Introduction

The transportation sector is one of the primary producers of greenhouse gas emissions [1]. According to statistics, the CO_2 emission of the global transportation sector is more than seven billion tons, accounting for about 25% of the total in the world [2]. The feasible solution to carbon emissions, fossil fuels crisis, and climate changes would involve the widespread application of electric vehicles (EVs) [3]. Compared to traditional internal combustion engine vehicles (ICEVs), EVs feature zero carbon emissions, high operation efficiency, lower maintenance requirements, and noise [4]. In addition to these direct advantages, EVs also provide an important opportunity to reshape the relationship between transportation systems and electricity systems. Unlike conventional fuel vehicles, EVs are inherently connected to the power grid through charging infrastructures, which means that their large-scale deployment may influence not only mobility patterns but also load behavior and power-system operation. With the policy incentives and technological advancements, the EV industry is expected to experience continual growth in the following decades [5]. In 2021, the sales of EVs had reached 6.5 million worldwide, with 3.2 million sold in China. Along with the fast surge of EVs, various charging infrastructures have also been deployed widely, thus providing multiple charging services for EV users, such as slow, fast, or ultrafast charging [6]. The diversity of charging modes further indicates that EV integration is no longer a simple matter of adding new electrical loads. Different charging powers, different user expectations, and different installation contexts lead to heterogeneous impacts on the distribution grid, thereby making the architecture of EV-grid integration increasingly important.

The substantial increment in EVs application also seriously affects power grids, especially the distribution grid [7]. Generally, the distribution grid is designed with a limited safety margin and overloading capacity, while the uncoordinated charging of large-scale EVs raised from random behavior of EV users would dramatically elevate load peaks of distribution grids during rush

Smart Penetration of Electric Vehicles in Distribution Grids. https://doi.org/10.1016/B978-0-443-27706-1.00010-6

periods [8]. Additional loads would increase the risk of power lines and transformers overload, accompanied by extra energy losses and power quality degradation [9]. These impacts are particularly severe in areas where EV charging demand is concentrated at specific times and locations, such as residential districts in the evening or commercial charging hubs during peak hours. In such cases, the local distribution grid may face simultaneous stresses in thermal loading, voltage deviation, and power-flow imbalance. To address the increasing peak power and grid security risk, one possible solution is to reinforce the power grid. However, this scheme needs numerous investments in upgrading the existing power structure and deploying new power capacities [10]. As a cost-effective alternative, the vehicle-to-grid (V2G) technology could mitigate adverse impacts of the random EV charging behavior via the coordinated power flow dispatch between the power grid and parked EVs [11]. Moreover, the large-scale V2G dispatching conducted by EV aggregators could reduce the grid load variance, and provide a series of auxiliary services for the power grid to enhance the grid's operation economy, stability, and security [12]. These auxiliary services contain frequency regulation, voltage support, spinning reserve, congestion management, and so forth [13]. The concept of V2G can be extended to more specific objects, for example, the bidirectional power flow between multiple EVs (vehicle-to-vehicle, V2V), between EV and home (vehicle-to-home, V2H) or building (vehicle-to-building, V2B) [14]. This evolution from simple charging to multi-form energy interaction illustrates that EVs are increasingly being treated as active participants in distributed energy systems rather than passive electricity consumers.

In recent years, the carbon reduction concern and energy crisis also enhance the proliferation of distributed generations of electricity from various renewable energy sources (RESs), including photovoltaic (PV), wind generation, tidal power, etc [15]. In 2021, China was ramping up its push for distributed rooftop PV installations with a minimum of 20% of all residential rooftops and 30% of commercial and industrial structures in rural areas. Like EVs, the power generation of RESs-based distributed generators (DGs) is intermittent, uncertain, and random in nature, easily resulting in a sharp fluctuation of feeder power and voltage violation if no compensation measures are to be executed [16]. As a result, the contemporary distribution grid is facing a dual transformation. On the demand side, EVs introduce flexible but potentially disruptive charging loads, while on the supply side, RES-based DGs introduce environmentally beneficial but fluctuating power injections. The coexistence of these two trends greatly increases the operational complexity of distribution systems, while also opening up new possibilities for coordinated energy management.

This confluence of volatile renewable generation and flexible EV loads creates both a challenge and an opportunity, demanding grid architectures that are inherently more resilient, flexible, and active. To mitigate the adverse effects of high penetrated RESs integration and realize effective management of EV charging and V2G operation, the notion of the active distribution grid was proposed more than a decade ago [17]. According to the shared global definition by the CIGRE C6.19 working group, the distribution grid is a power network with active control capabilities for contained RESs, energy storage, and other distributed energy units to achieve an economical and safe system operation [18]. The active distribution grid is usually classified into several autonomous subsystems which have self-management capability and could collaborate to make the active distribution grid operate in an economic and coordinated manner [19]. Compared with the traditional passive distribution network, the active distribution grid emphasizes controllability, observability, and coordinated dispatch at multiple levels. Its significance lies not only in accommodating more distributed resources, but also in enabling these resources to interact constructively with one another under a common operational framework. As an extensive participator in the active distribution grid, the microgrid merges a variety of local DGs and loads, and optimizes its operation to satisfy the power demands of targeted scenarios, such as residential buildings, hospitals, fast-charging stations, etc [20]. Toward a more specific and discriminative division of active distribution grids, the microgrid concept could be scaled down and extended further to form two new concepts, viz., nanogrid and cluster, respectively. The nanogrid is a narrower concept than a microgrid, referring to small-scale systems like residential houses [21]. The cluster is usually pointed to the interconnected multiple microgrids. The cluster could mitigate operation issues concerned by the single microgrid, such as limited penetration of RESs and relatively low inertia through mutual power coordination between microgrids [22]. This hierarchical structuring—from nanogrid to microgrid to cluster—represents a paradigm shift from centralized, passive distribution to a decentralized, active, and multi-layered grid topology. A series of novel power architectures in active distribution grids have been explored widely, for example, the DC-coupled architecture and hybrid AC/DC-coupled architecture, aiming to achieve desired system performance in efficiency, functionalities, reliability, and compactness.

More recently, much of the current literature on V2G pays particular attention to scheduling and control schemes, aiming to realize desired operation objectives, for example, peak shaving of charging loads, frequency and voltage regulations, etc [23]. In Ref. [8], a high efficient V2G strategy for centralized charging coordination is introduced to suppress the elevated peak loads of power grids. The introduced scheduling strategy is of low computational complexity. In Ref. [24], a multi-objective dispatching strategy is proposed to minimize load variance and maximize the revenue of EV users under considering EV traffic behavior characteristics. The genetic algorithm is adopted to solve the formulated optimization problem. In Ref. [25], a bargaining game that belongs to the decentralized scheme is proposed to accommodate the benefits of EV users and EV aggregators. These studies clearly demonstrate that the success of V2G depends heavily on how charging and discharging decisions are coordinated in time, across users, and under different system objectives. They also show that economic incentives, control logic, and computational efficiency are all important dimensions of V2G implementation. Although the above-mentioned schemes provide valuable inspiration for the V2G implementation, the characteristics of specific power architectures are not considered, which is much concerned with the V2G operation performance, such as possible V2G restriction owning to node voltage violation.

There has not been much literature considering the characteristics of power architectures in the study of V2G operation. In Ref. [12], the voltage profile of the AC distribution campus grid is considered in the design of the V2G dispatching algorithm. By coordinating EVs' charging/discharging behavior, the voltage override at specific distribution nodes could be mitigated. In Ref. [1], it has been demonstrated that the low voltage concern is usually encountered at the end of radial AC feeders owing to the high power charging. The studies [1,12] focused on the typical AC distribution system. Under the hybrid AC/DC-coupled architecture, Rahman et al. proposed a vehicle-to-microgrid framework comprising the centralized economic dispatch and distributed EV coordination strategy. The modified control of the interlinking converter (ILC) enables proper power transfer between the AC and DC subgrid with simultaneous AC and DC bus voltage regulations. Similarly, in Ref. [26], Rahman et al. introduced a centralized power management scheme to coordinate the AC and DC type EV charging points in an islanded hybrid microgrid. The voltage and frequency regulation, as well as active/reactive power management between two subgrids, are realized under the proposed scheme. The novel power electronic device, namely, the solid state transformer (SST), is deemed a key player in the smart grid thanks to its high efficiency and controllable power control capability. In Ref. [27], an SST-based EV charging station architecture is established. With the help of the introduced V2G dispatching algorithm, the architecture could provide a real-time bidirectional frequency regulation service for the utility grid. Qin et al. proposed a novel multiport DC–DC SST topology for EV charging stations. Assisted by the introduced DC-bus signaling algorithm, the autonomous V2G operation could be implemented without communication links [28]. The previous literature [26–28] has shown that the emergence of various novel power architectures brings new challenges and opportunities for EV-grid integration and V2G operation. More importantly, these studies imply that the technical performance of V2G cannot be fully understood without considering the underlying hardware and interfacing architecture. Different architectures lead to different constraints on converter control, voltage regulation, power-transfer paths, and communication requirements, all of which can materially affect the feasibility and quality of EV-grid coordination.

While these studies are pioneering, they remain isolated cases. A systematic framework for comparing, evaluating, and selecting appropriate grid architectures for specific V2G applications is conspicuously absent from the literature. A comprehensive survey on EV reviews is summarized in Table 3.1. These literature makes a good contribution to the research of V2G technology, covering the content of EV charger topologies, grid-connection standards, V2G auxiliary services, economic costs and revenues, V2G application challenges, and so forth. The main research gap is that these literature reviews about V2G are based on the framework of traditional AC distribution grids without considering novel power architectures. Additionally, there is no comprehensive review involving all the hardware perspectives of EV-integration systems, which involve connection standards, charging infrastructures, converter topologies, and interfaced power architectures. Moreover, previous studies have failed to illustrate the related infrastructures, common standards, and power converter topologies when EVs play different roles in interaction with power grids, like EVs as loads or energy storages.

This chapter aims to make a comprehensive overview of distribution grid architectures, grid connection infrastructures and standards, and typical applications from the perspective of EV-grid integration and V2G operation. The key contributions of this work are summarized below.

1. The grid connection infrastructures, common standards, and power converter topologies are elaborated when EVs as the role of loads, energy resources, and storages in the interaction with power grids.
2. The conventional AC-powered and emerging DC-powered converter topologies of bidirectional/unidirectional charging facilities are identified and discussed.
3. A thorough review of typical nanogrid, microgrid, and cluster architecture with typical applications under the active distribution grid is presented.
4. A comprehensive assessment of EV-grid integration and V2G operation performance under different power architectures is performed considering charging demand compatibility, power quality severity, V2G availability, architecture scalability, local control complexity, and technology maturity. To the best of our knowledge, this is the first time to review distribution grid architectures from the viewpoint of EV-grid interaction.

The remaining part of this chapter is organized as follows. Section 3.2 briefly introduces the typical nanogrid, microgrid, and cluster with EVs integration. Considering the power exchanges between EVs and various power architectures, the corresponding interconnection standards are reviewed in Section 3.3. In Section 3.4, the typical electric vehicle charging circuits are breifly reviewed. Section 3.5 describes typical "electric vehicle-distribution grid" integration architectures. In Section 3.6, a comprehensive assessment of various power architectures for nanogrid, microgrid, and cluster is

TABLE 3.1 Comparison between the existing reviews and this chapter.

Ref.	Study Goals
[29]	This paper makes a comprehensive review of managed EV charging in terms of the acquired revenues, existing implementations, and costs, as well as methodologies in related analysis/modeling studies.
[23]	This paper presents an overview of V2G technology referring to provided ancillary services, possible faced challenges, and market potential.
[30]	This paper reviews the main technical, economic, regulatory, and user-related frameworks and faced barriers of the active integration of EVs into distribution grids.
[31]	This paper reveals that more EVs could be integrated into distribution grids without grid reinforcements if coordinated charging could be implemented.
[32]	A comprehensive review of current EV states, EV charging and grid interconnection standards, and grid integration impacts is conducted in this paper.
[33]	In this paper, the grid impact of EV fast-charging stations is clarified in terms of the current state-of-the-art technology, standards for fast charging, power quality issues, and mitigation measures.
[34]	This paper thoroughly discusses the current status and development trend of industrial electronics related to electric transportation, such as charger classification and standards, converter topologies for AC chargers, and characteristics of Lithium-ion batteries.
[35]	An overview of extreme fast charging technology is elaborated concerned with DC fast-charging standards, system configuration, and converter topologies for DC fast chargers.
[36]	An intensive review of the EV-grid integration infrastructure, DERs integration with EVs, and the feasibility of a smart V2G system is presented in this paper.
[37]	This paper makes a comprehensive overview of energy management strategies considering the V2G operation and DERs integration.
[38]	The contemporary trend of power electronics converters for EVs charging is surveyed.
[39]	This paper reviews the multilevel EV aggregator based on substation level, feeder level, and MG level in terms of system modeling and operation strategy.
This chapter	This chapter conducts a comprehensive review of distribution grid architectures, grid connection infrastructures and standards, and typical applications from the perspective of EV-grid integration and V2G operation for the first time.

performed in terms of multiple evaluation indexes. Finally, Section 3.7 concludes this paper.

3.2 Active distribution grids

Figure 3.1 depicts a typical active distribution grid. Unlike the traditional passive distribution grid, the active distribution grid is featured by bidirectional power flow and information exchange between the utility grid and local grid, enabling active management and control capability for multiple units [40]. The specific definition and discussion are elaborated below.

3.2.1 Nanogrid

According to Ref. [41], the nanogrid is defined as a DGs-integrated power distribution system for a single house/small building with clearly defined

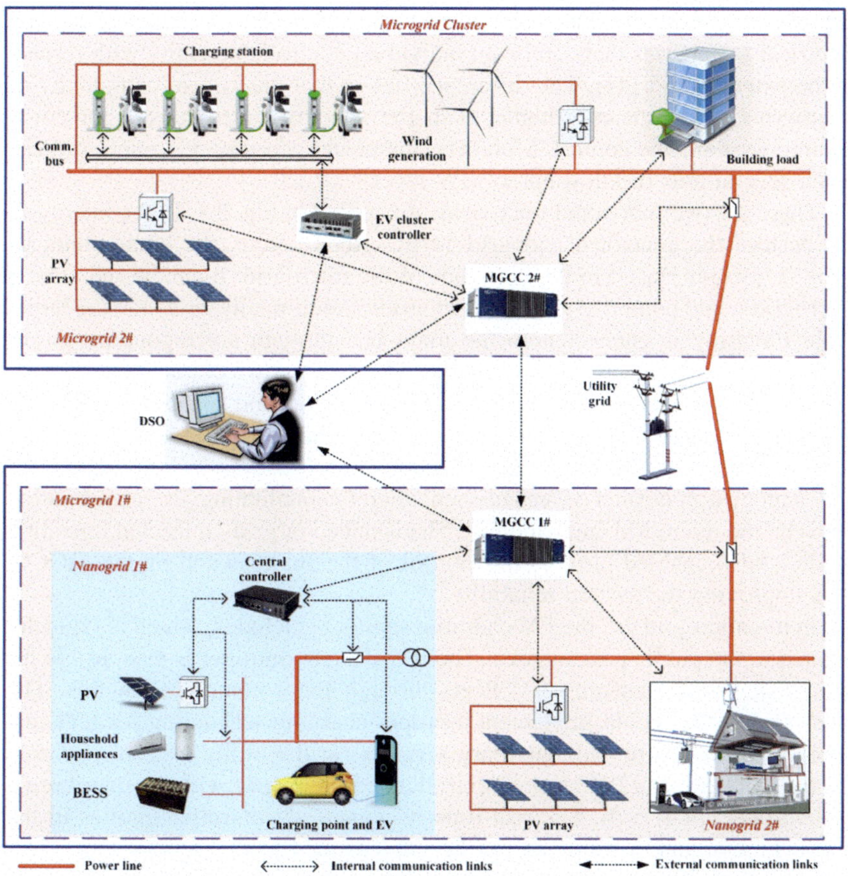

FIGURE 3.1 Overall structure of a typical active distribution grid.

electrical boundaries with other power entities. The nanogrid features low power and complexity and is technologically simpler than a microgrid. There is no strict power level requirement to define the nanogrid. In general, the total power capacity of a nanogrid varies from several kilowatts to tens of kilowatts [42]. The nanogrid 1# shown in is a typical home nanogrid consisting of local PV, battery energy storage system (BESS), and residential appliances like air conditioners. Especially, the EV, as a kind of new-type power electronic load, can also be plugged into the home nanogrid through dedicated residential charging points. The supervision and power management of the whole nanogrid is commonly implemented by a nanogrid central controller, which coordinates multiple resources and loads for an economic operation and realizes the smooth switch between the islanded and grid-connected mode [43].

3.2.2 Microgrid

The definition between the nanogrid and microgrid is very similar. According to the definition given by the United States Bureau of Energy, a microgrid is defined as a group of interconnected DGs and loads within clearly defined electrical boundaries that can be an individual controllable entity with respect to the external grid [44]. The difference lies in that the system complexity of microgrids is significantly higher than the nanogrid in terms of protection, communication, and control. Moreover, the power capacity of a microgrid can reach several tens of kilowatts to MW level.

There are two individual microgrids illustrated in Fig. 3.1. In the microgrid 1#, besides the contained nanogrid 1# described before, the home nanogrid two# is also interfaced to the main bus of the microgrid. It can be seen that a microgrid could comprise multiple nanogrids. Additionally, as a DG, the large-scale PV array is connected to the main bus through corresponding power converters for the local power supply. The microgrid central controller 1# (MGCC 1#) is utilized for the overall supervision and management of the microgrid through high-performance computing units and dedicated communication links [45]. Notably, the communication link between the MGCC 1# and nanogrid controller is established, thereby facilitating the coordination between the microgrid and nanogrid. The islanded or grid-connected operation mode is also enabled through the control of the static switch by the MGCC, thus improving the system reliability.

In the microgrid 2#, the EV charging station is included, which is typically located in the public parking area. The EV cluster controller is responsible for supervising and managing EV fleets through the communication bus. The cluster controller could implement a proper charging scheduling for EVs and coordinate EVs to provide auxiliary services for the utility grid. In addition, the two MGCCs and EV cluster controllers communicate with the distribution system operator (DSO). The real-time responses of controlled entities to the dispatching signal of the DSO can be implemented [46].

3.2.3 Cluster

A cluster is defined as a group of interconnected microgrids in close vicinity, which can increase the efficiency and reliability of each other by the proper power distribution [47]. Depending on the cluster architecture, the mutual power exchanges between microgrids can be realized in DC or AC power types. In, microgrid 1# and microgrid 2# are interconnected via the point-to-point way. Under the optimal power dispatching executed by the EV cluster controller, the EVs could provide power support for external microgrids [48]. It can be seen that the concept of the cluster aims to strengthen the interconnection at the microgrid level, thereby establishing the Energy Internet in a larger area [49]. Table 3.2 summarizes the definition, power ratings, and typical applications of nanogrid, microgrid, and cluster.

3.3 Industrial standards

It is a fundamental and pressing issue to fulfill the ever-increasing EV charging demands. According to the difference in coupling method, the EV charging method can be classified as conductive and inductive charging [50]. The scope of this article focuses on the mainstream conductive charging technology.

As shown in Fig. 3.2, the conductive charging uses direct contact between the EV connector and the charge inlet. The electric cable can be fed from a standard electrical outlet or a charging pile. Depending on the location of the power conversion unit, the conductive charging is further categorized as on-

TABLE 3.2 Definition, power rating, and typical applications of nanogrid, microgrid, and cluster.

	Definition	Power Rating	System Complexity	Typical application
Nanogrid	A DGs-integrated power distribution system for a single house/small building	Several kW to a few tens of kW	L	Home; small building
Microgrid	A group of interconnected DGs and loads within clearly defined electrical boundaries to the external grid	Few tens of kW to several MW	M	Commercial building; Industrial building; Charging station
Cluster	A group of interconnected microgrids in a close vicinity	Hundreds of kW to tens of MW	H	Community power grid

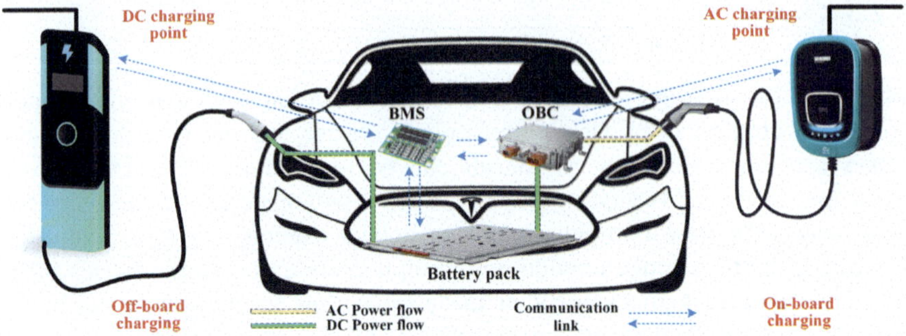

FIGURE 3.2 Typical EV conductive charging configuration.

board charging and off-board charging [51]. As for the on-board charging, due to EVs' space and weight constraints, the on-board charger (OBC) inside the EV is of small power ratings and is mainly used for slow charging. The rated power is generally less than the maximum household power (e.g., 19-kW upper bound in the United States) [52]. As for the fast high-power charging (generally over 20 kW), deploying a high-power charger on EVs would not be feasible due to the increased cost, volume, and weight. In light of this, the off-board charging utilizes the external charger to execute the high-power DC fast charging for EV battery packs directly [53]. The desired fast-charging power could climb up to hundreds of kilowatts, thus shortening the charging time significantly. The conductive charging has been widely applied globally and standardized in terms of the power level, connection provision, and communication protocol.

The EV could undertake different roles in the interaction with the utility grid, as shown in Fig. 3.3. In a conventional sense, an EV is simply deemed a load without considering reverse power flow. When the V2G is conducted, an EV can be regarded as an energy resource or energy storage device. Its specific role is determined by the unidirectional or bidirectional power flow in practical use. As there have been no large-scale commercial applications of the V2G technology, dedicated grid integration standards for the V2G application have not been developed. However, the essence of power importing to the grid via power electronic devices reveals that the current grid integration standards for distributed energy resources (DERs) have a high referential value to guide the V2G operation.

3.3.1 Charging standards

There have been four mainstream standards related to EV charging infrastructures in the world. These standards mainly deal with the charging port/connector design, output voltage/current restriction, communication protocol, etc. Detailed information about the charging infrastructure standards is shown

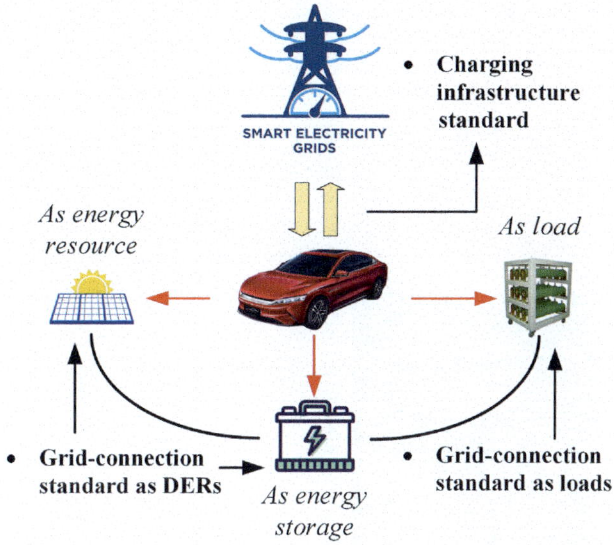

FIGURE 3.3 EV's triple roles when interacting with utility grid.

in Table 3.3. SAE J1772 standard is commonly used in the United Sates, which is released by a professional association named Society of Automotive Engineers (SAE). It defines two charging levels for the AC and DC charging demand, respectively, as well as corresponding ports/connectors criteria. Notably, the communication of AC charging under mentioned standards including SAE is realized using a control pilot (CP) and proximity pilot (PP) [54]. The pulse width modulation (PWM) signal released by the control pilot is used to indicate the maximum output current of EV chargers. According to the PWM signal on CP, the charging current is determined by EVs considering the current battery status.

As for the DC charging under SAE standards, the adopted communication protocol is based on the power line communication (PLC). To accommodate the AC slow charging and DC fast charging simultaneously by the same port/connector, the Combined Charging System (CCS) standard has been proposed and authorized by the European Automobile Manufacturers Association. The CCS standard is compatible with SAE J1772 and IEC 62,196-2, and its Combo 1 type is specific for the North American market. For DC charging, Japan has its charging standard named CHAdeMO, which is proposed by an association of the same name formed by five major Japanese automakers. IEC 62,196–2 standard is vastly used in Europe, which defines three types of AC charging. The CCS Combo 2 standard, which is compatible with AC and DC charging requirements, has been widely adopted in the European and Australian markets. China owns the biggest EV market worldwide and also defines the corresponding standard named GB/T 20,234 (issued by the Standardization

TABLE 3.3 Summary of mainstream standards of EV charging infrastructures.

Item	 No Permission Required	 No Permission Required	 No Permission Required	 No Permission Required
Countries or regions	USA	JAPAN	EU	CHINA
AC charging standard	**SAE J1772 AC** **Level 1:**120 V 1-16 A **Level 2:**240 V 1-80 A	**SAE J1772 AC** **Level 1:**120 V 1-16 A **Level 2:**240 V 1-80 A	**IEC 62196–2** **Type 1:** 250 V 1–32 A **Type 2:** 250 V 1–70 A 480 V 3–63 A **Type 3:** 250 V 1–32 A 480 V 3–63 A	**GB/T 20,234 (AC)** 250 V 1–32 A

Typical AC EV charging connector/ port	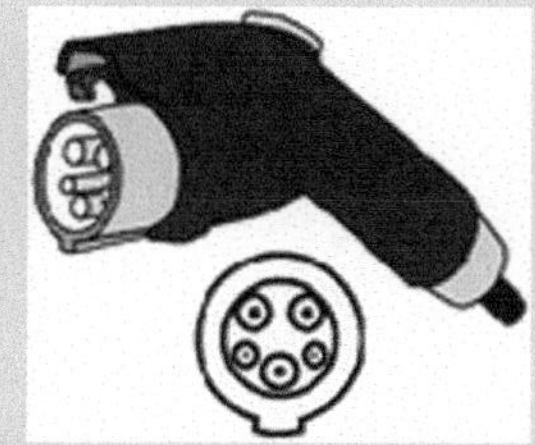 No Permission Required	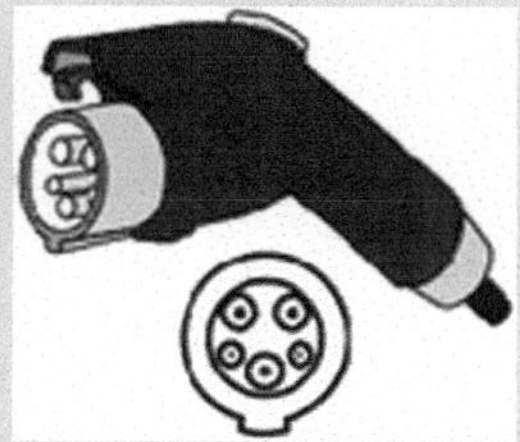 No Permission Required	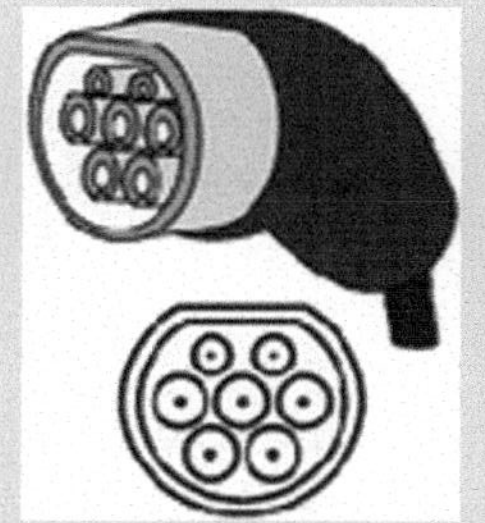 No Permission Required	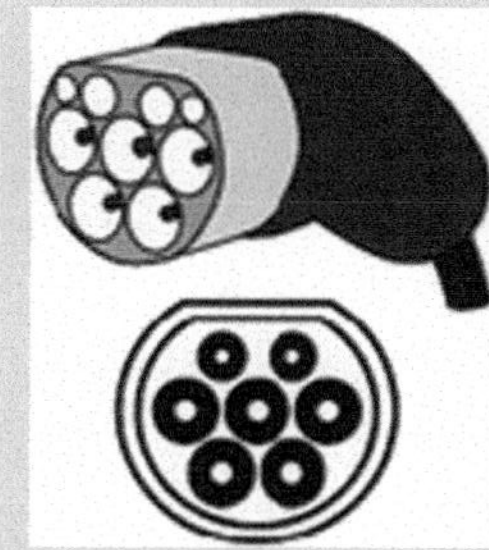 No Permission Required
Comm. Protocol for AC charging	PWM over control piolt	PWM over control piolt	PWM over control piolt	PWM over control piolt
DC charging standard	• **SAE J1772 DC** **Level 1:**450 V DC 80 A **Level 2:**450 V DC 200 A • **CCS Combo 1** 1000V DC 200 A	**CHAdeMo** 1000 V DC 400 A	**CCS Combo 2** 1000 V DC 200 A	**GB/T 20,234 (DC)** 1000 V DC 250 A

Continued

TABLE 3.3 Summary of mainstream standards of EV charging infrastructures.—cont'd

Item	 No Permission Required	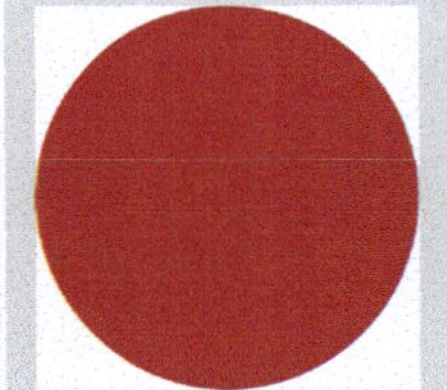 No Permission Required	 No Permission Required	 No Permission Required
Typical DC EV charging connector/port	 No Permission required	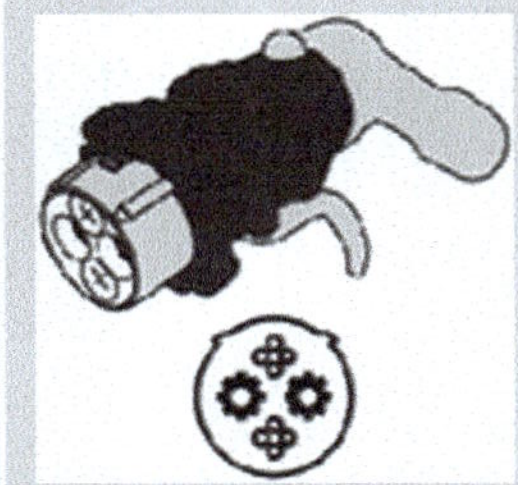 No Permission required	 No Permission required	 No Permission required
Comm. Protocol for DC charging	PLC	CAN	PLC	CAN

Administration of China). PLC and CAN communication methods are adopted for DC charging in Europe and China, respectively.

Regarding the V2G operation, the CP communication-based AC charging has no provision for initiating V2G since the power regulation command from EV chargers is not defined in the current protocol. The CCS and GB/T 20,234 (DC) standards have the potential to achieve V2G operation since the bidirectional communication is established between EV chargers and EVs. However, specific details in related provisions like EV response time should be clearly defined to facilitate the V2G application [54]. Notably, CHAdeMO is the only charging standard worldwide so far with the V2G functionality enabled. Under this standard, EV only sets the maximum current for the charging and discharging, and between the upper and lower bounds, the EV power could be regulated flexibly by EV chargers considering current energy prices and renewable generation, etc. Thus, EVs could provide varied auxiliary services for the utility grid. To promote homogeneity in charging standards and infrastructures, the CHAdeMO Association and China Electricity Council have co-developed a new DC fast charging standard named Chaoji with supporting 900 kW charging power at most. Chaoji standard's compatibility with CHAdeMO and GB/T standards has been realized and aids for the promotion of V2G further.

3.3.2 Integration standards: EV as a load

The EV as charging loads is the most common situation. A series of power quality standards should be followed in the EV charging process to guarantee the power quality of distribution grids and the normal operation of charging facilities.

In Table 3.4, the widely used power quality standards for EVs integration are listed, which are adopted in the United Sates, European, and China. Related Harmonic current limits in standards are illustrated in Tables 3.5 and 3.6, respectively. The most common power quality issues of EV charging encountered are also summarized. Specifically, it has been demonstrated that the single-phase charging would cause asymmetries in the distribution network that potentially causes a violation of voltage balance limitation [42]. Although an evenly distributed deployment of charging points in each phase feeder could be implemented, the random charging demand in each phase would lead to a phase unbalance [55]. The voltage unbalance factor is used to indicate the unbalance degree, defined as the ratio of the modulus of the negative-sequence to the positive-sequence components of the voltage at the fundamental frequency. The IEEE series and GB/T series set the voltage unbalance factor as 2%, yet the limitation is 3% in EN 50,160. In addition, the uncoordinated charging behavior tends to increase the load at peak hours and causes voltage deviations across the distribution line [56]. Related standards have specified

TABLE 3.4 Summary of mainstream standards for integrating EVs as loads.

Item	 No Permission Required	 No Permission Required	 No Permission Required
Countries or regions	USA	EU	CHINA
Connecting standards	IEEE 1159/141/519	EN 50160	GB/T 36,278/ GB/T 29,316
Voltage unbalance factor	2%	3%	2%
Voltage range for normal operation	LV grid: $V_{rated} \leq 1$ kV MV grid: 1 kV $\leq V_{rated} \leq 69$ kV 90% $V_{rated} < V < 110\% V_{rated}$	LV grid: $V_{rated} \leq 1$ kV MV grid: 1 kV $\leq V_{rated} \leq 35$ kV 90% $V_{rated} < V < 110\% V_{rated}$	For 220 V level 90% $V_{rated} < V < 107\%$ V_{rated} For 380 V and 10(20) V level 93% $V_{rated} < V < 107\%$ V_{rated}

PF restriction	\	\	For 10 kV power supply, PF ≥ 0.95 in peak load period; For 380 V or below power supply, PF ≥ 0.9 in peak load period;
Voltage harmonics	*THD*<5%	*THD*<8%	*THD*<5%
Current harmonics	Table 3.5	\	Table 3.6

TABLE 3.5 Harmonic current limits in GB/T 29,316.

Rated Voltage (kV)	Standard Short circuit capacity (MVA)	Harmonic order and allowable harmonic current I_{hp} (A)																						
		2	3	4	5	6	7	8	9	10	11	12	13	14	15	16	17	18	19	20	21	22	23	24	25
0.38	10	78	62	39	62	26	44	19	21	16	28	13	24	11	12	9.7	18	8.6	16	7.8	8.9	7.1	14	6.5	12
6	100	43	34	21	34	14	24	11	11	8.5	16	7.1	13	6.1	6.8	5.3	10	4.7	9.0	4.3	4.9	3.9	7.4	3.6	6.8
10	100	26	20	13	20	8.5	15	6.4	6.8	5.1	9.3	4.3	7.9	3.7	4.1	3.2	6.0	2.8	5.4	2.6	2.9	2.3	4.5	2.1	4.1
35	250	15	12	7.7	12	5.1	8.8	3.8	4.1	3.1	5.6	2.6	4.7	2.2	2.5	1.9	3.6	1.7	3.2	1.5	1.8	1.4	2.7	1.3	2.5
66	500	16	13	8.1	13	5.4	9.3	4.1	4.3	3.3	5.9	2.7	5.0	2.3	2.6	2.0	3.8	1.8	3.4	1.6	1.9	1.5	2.8	1.4	2.6
110	750	12	9.6	6.0	9.6	4.0	6.8	3.0	3.2	2.4	4.3	2.0	3.7	1.7	1.9	1.5	2.8	1.3	2.5	1.2	1.4	1.1	2.1	1.0	1.9

TABLE 3.6 Harmonic current limits in IEEE 519.

	$V_{pcc} \leq 69$ kV					
$I_{sc}/I_L(h)$	$3 \leq h < 11$	$11 \leq h < 17$	$17 \leq h < 23$	$23 \leq h < 35$	$35 \leq h < 50$	*TDD*
<20	4.0	2.0	1.5	0.6	0.3	5.0
20–50	7.0	3.5	2.5	1.0	0.5	8.0
50–100	10.0	4.5	4.0	1.5	0.7	12.0
100–1000	12.0	5.5	5.0	2.0	1.0	15.0
>1000	15.0	7.0	6.0	2.5	1.4	20.0

the bounds of acceptable distribution voltage under different voltage levels. The power factor (PF) signifies how effectively the electricity is used. In GB/T series, the PF bound for EVs integration is specified. The harmonic wave should be mitigated as it could result in increased equipment and conductor heating and abnormal working of equipment like motors and generators [57]. The discussed mainstream standards limit the harmonic distortion of different orders.

Note: The limits for even harmonics are 25% of the odd harmonic limits; I_L denotes the maximum demand load current; I_{SC} denotes the maximum short circuit current at PCC.

3.3.3 Integration standards: EV as an energy resource/storage

Concerning the V2G operation, an EV acts as an energy resource/storage when unidirectional/bidirectional power flow exchange is executed with the utility grid. In Table 3.7, three mainstream standards for DERs, viz., North American IEEE 1547, European IEC 61,727, and China GB/T 33,593, are elaborated. It can be seen that the strict grid-connected voltage and frequency restrictions at the point of common coupling (PCC) are specified to guarantee safe and stabilized system operation. During the process of power imported to the utility grid, the DC injection is a severe issue, which could cause the saturation of transformers, leading to line heating and trip [58]. Reactive power compensation provided by EVs is more efficient and economical owning to reducing the installation and maintenance costs of compensation devices [59]. It has been demonstrated that the reactive power exchange does not affect the EV battery charging, as well as the battery state of charge (SoC) or battery lifetime [59]. The operation range of EVs for reactive power compensation can be expressed as Equ. 3.1 and illustrated in Fig. 3.4.

TABLE 3.7 Summary of mainstream standards for integrating EVs as DERs.

Item	 No Permission Required	 No Permission Required	 No Permission Required
Countries or regions	USA	EU	CHINA
Grid connection standards	IEEE 1547	IEC 61727	GB/T 33,593 GB/T 37,408
Limitation of DC injection	<0.5% I_{rated}	<1% I_{rated}	<0.5% I_{rated}
Voltage unbalance factor	2%	3%	2%
Frequency range for normal operation	59.3 Hz $< f <$ 60.5 Hz	f_n-1 $< f < f_n$+1	49.5 Hz $< f <$ 50.2 Hz
Voltage range for normal operation	89% $V_{rated} < V <$110% V_{rated}	85% $V_{rated} < V <$110% V_{rated}	85% $V_{rated} < V <$110% V_{rated}

Power factor (PF) and reactive power injection	No power factor requirement Encourage reactive power support		Requirement of $PF > 0.9$ Ind.uctive for the load over 50%		Requirement of PF adjustable from 0.98 Ind.uctive to 0.98 capacitive	
Current Harmonics	THD	*Threshold*	THD	*Threshold*	THD	*Threshold*
		$<5\% I_{rated}$		$<5\% I_{rated}$		$<5\% I_{rated}$
	Odd Harmonics	*Threshold*	*Odd Harmonics*	*Threshold*	*Odd Harmonics*	*Threshold*
	<9	$<4\% I_{rated}$	<9	$<4\% I_{rated}$	<9	$<4\% I_{rated}$
	11 to 15	$<2\% I_{rated}$	11 to 15	$<2\% I_{rated}$	11 to 15	$<2\% I_{rated}$
	17 to 21	$<1.5\% I_{rated}$	17 to 21	$<1.5\% I_{rated}$	17 to 21	$<1.5\% I_{rated}$
	23 to 33	$<0.6\% I_{rated}$	23 to 33	$<0.6\% I_{rated}$	23 to 33	$<0.6\% I_{rated}$
	>35	$<0.3\% I_{rated}$	>35	/	>35	$<0.3\% I_{rated}$
	Even Harmonics	*Threshold*	*Even Harmonics*	*Threshold*	*Even Harmonics*	*Threshold*
	2 to 8	$<1\% I_{rated}$	2 to 8	$<1\% I_{rated}$	2 to 8	$<1\% I_{rated}$
	10 to 14	$<0.5\% I_{rated}$	10 to 14	$<0.5\% I_{rated}$	10 to 14	$<0.5\% I_{rated}$
	16 to 20	$<0.375\% I_{rated}$	16 to 20	$<0.5\% I_{rated}$	16 to 20	$<0.375\% I_{rated}$
	22 to 34	$<0.15\% I_{rated}$	22 to 34	$<0.5\% I_{rated}$	22 to 34	$<0.15\% I_{rated}$
	>36	$<0.075\% I_{rated}$	>36	/	>36	$<0.075\% I_{rated}$

$$P^2 + Q^2 \leq S_{max}^2 \tag{3.1}$$

where *P*/*Q* denotes the output active power/reactive power of EV chargers, respectively; *S*max denotes the rated capacity of EV chargers.

As shown in Fig. 3.4, four quadrants operation capability of EVs could be achieved at most, which is associated with the power converter topologies of EV chargers. Related details will be discussed in Section 3.3. Referring to the power factor requirements for DERs, GB/T 33,596 & 37,408 prefer a power factor adjustment capability from 0.98 Ind.uctive to 0.98 capacitive. Thus, the reactive power support from DREs is to be achieved under the dispatching of the distribution operator. Since DERs commonly work as current sources, the possible generated harmonic current should be addressed. The three standards have the same THD restrictions, and a slight difference for the high-order harmonic limitation is encountered.

3.4 Electric vehicle charging circuits

The power converter serves as a significant interface between the power grid and EV battery, aiming to achieve voltage scaling/conversions and reliable charging/discharging regulation. Guided by the varied requests of weight, volume, power density, and control functions, there have been a series of proposed novel power converter topologies for EV chargers [38]. Classified by the number of conversion stages, these topologies can be divided into one-stage and two-stage types [39]. For the one-stage type, the direct AC/DC power conversion enables the EV charging with more control simplicity, and

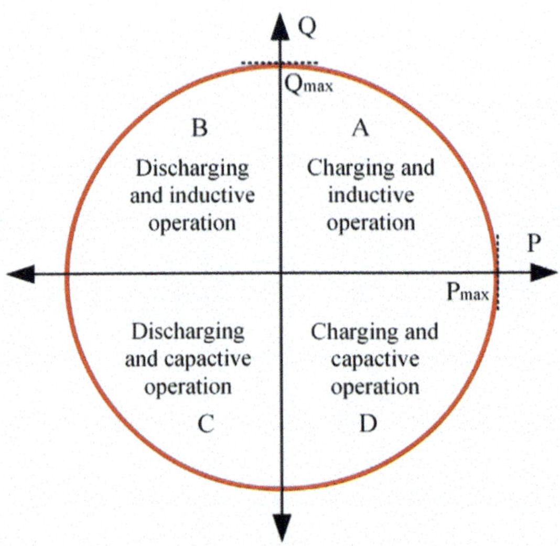

FIGURE 3.4 *P-Q* plane illustrating operation capability of EVs.

higher efficiency. However, this design lacks the capability of wide voltage range output, easily leading to a mismatch with various EV battery voltages. Moreover, the generated large low-frequency ripple in the output current would negatively affect the safe charging and life of EV batteries [40]. Considering the isolation and safety requirements of related standards (e.g., ISO, EN, etc.), galvanic isolation is essential in designing power circuits [60]. Therefore, the two-stage type is the mainstream power topology of EV chargers.

With the widespread deployment of DERs and the rapid development of power conversion technology, the DC-based distribution system would embrace the promising application potential due to a series of merits over its AC counterpart, such as the natural integration with DC sources, no synchronization, reactive power issue, etc [61]. Although there have been no established standards for EV chargers with DC power input, it is worthy and meaningful to investigate corresponding power topologies.

3.4.1 Power converter for on-board charger

As shown in Fig. 3.5, the schematic block diagram of the on-board charging facility is illustrated. The power source is connected to the EV battery pack via the corresponding protection unit and on-board charger. The protection unit prevents damage to related devices caused by short circuits or overcurrent [36]. And the on-board charger is utilized for the charging/discharging control of batteries. Due to the feature of slow charging, the on-board charger generally exploits the solution of single-phase AC input. Moreover, the typical two-stage topology that comprises the first stage AC/DC PFC converter and second DC/DC converter is adopted for on-board charger [62]. The AC/DC PFC converter is used to convert the AC voltage to the intermediate dc-link voltage, as well as to shape the input AC current for the power factor correction and harmonic reduction, thereby meeting the grid standard of power

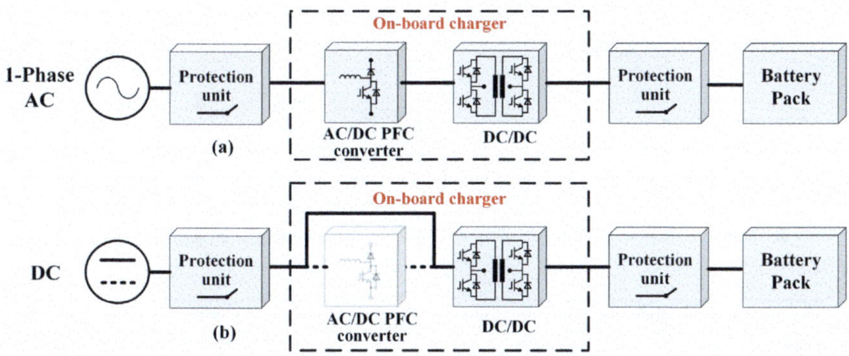

FIGURE 3.5 On-board charging facility. (a) 1-Phase AC power supply; (b) DC power supply.

quality [63]. And the DC/DC converter after the AC/DC front end is to produce the regulated voltage for battery charging and maintain reliable isolation between the grid and EV battery from the safety concern [35].

An isolated DC/DC converter stage is needed for the DC-powered on-board EV charger. Thus, the DC-powered on-board charger eliminates the AC/DC PFC converter unit, thus increasing the overall conversion efficiency and reducing the control complexity and device cost. The basic power topologies for the 1stfront-end PFC rectifier and the second back-end DC/DC converter are elaborated below. According to the requirement of unidirectional or bidirectional power flow control, the corresponding unidirectional or bidirectional topologies need to be adopted.

A first stage Front-end PFC rectifier.

A.1 Unidirectional topology

1) Conventional power factor correction (PFC) boost converter: This converter is the most popular and widespread topology for PFC applications. In Fig. 3.6 (a), the PFC boost converter contains a single-phase full-bridge diode rectifier to rectify the input voltage and a boost converter to match the high voltage of EV batteries. Additionally, the power factor correction is implemented by the boost converter, which makes the input AC current to be sinusoidal and in phase with the corresponding AC voltage by controlling the duty ratio of the controllable switch S1 [64]. This topology

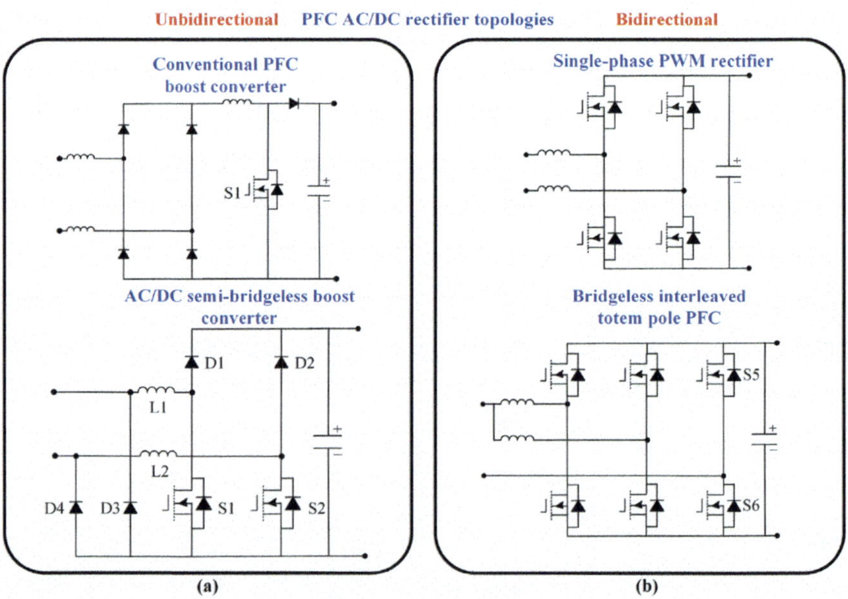

FIGURE 3.6 Typical first stage PFC topologies for on-board chargers. (a) Unidirectional topology; (b) bidirectional topology.

features relatively simple in PFC control and adopts common control strategies like the peak current control [57]. The main undesirable characteristic of this converter is the low efficiency due to the usage of three semiconductor devices during the operation [54]. The conduction losses and reverse-recovery losses of output diodes would climb significantly, accompanied by the increase of switching frequency. Accordingly, this topology is not proper to be applied in high-power devices [63].

2) AC/DC semi-bridgeless boost converter: By minimizing the number of semiconductor components in the line current path, the produced semi-bridgeless boost converter (as shown in Fig. 3.6 (a)) is proven to be more efficient than a conventional PFC converter [65]. To reduce the common-mode noise of the conventional bridgeless PFC boost converter, two additional slow diodes, v.z., D3 and D4, connect the input to the PFC ground [66]. In addition to the inductor L3 and L4, two dc/dc boost circuits are constructed, of which the first one, the L1-S1-D1 circuit loop is active through diode D3 and the second one, the L2-S2-D2 is active through diode D4. It should be noted that the switches S1 and S2 can be driven by the same PWM signal. Thus, this feature simplifies the design of PWM driver circuits as both switches are connected to the common ground [64]. On the other hand, the possibility of short-circuit through one leg is eliminated. A disadvantage of the semi-bridgeless boost converter lies in requiring two inductors.

A.2 bidirectional topology

1) Single-phase PWM rectifier: To realize the bidirectional power flow between the external power grid and intermediate DC bus of EV chargers, the full-bridge PWM rectifier is the most common topology adopted due to the simplicity in the circuit layout and control [67]. This topology consists of four switches and two boost inductors on the AC side, enabling a low-harmonic input current and a sinusoidal current waveform. Due to the capability of the four-quadrant operation, the PWM rectifier could operate as an active rectifier in the grid-to-vehicle (G2V) mode, whereas operating as an inverter to power the grid in V2G mode [68]. Aiming for a smaller AC filter size, the high-frequency PWM schemes are typically utilized, yet the generated high-frequency dc-side leakage current would impact the normal operation of DC capacitors in the intermediate bus.

2) Totem pole PFC: To reduce the high conduction loss of the conventional boost PFC and realize the bidirectional V2G operation, the totem pole PFC is widely applied [69]. In Fig. 3.5 (b), the illustrated totem-pole PFC contains two boost interleaved phases driven with a phase deviation of 180°. The interleaved design is used to smooth the current ripple and decrease the inductance size [70]. The synchronous rectification is realized by the phase leg (S5 and S6). Notably, if only one boost phase remains, the

topology of the totem pole PFC keeps the same with the single-phase full-bridge.

B second back-end DC/DC converter.

B.1 Unidirectional topology

1) Phase-shift full-bridge (PSFB) converter: for the unidirectional power flow, a possible solution is to adopt a PSFB converter as shown in Fig. 3.7 (a). This topology comprises two bridges, which are isolated by a high-frequency transformer. In the primary side, the full-bridge contains four active switches [71]. And in the secondary side, four diode forms the rectifier circuit. This converter works on the phase-shift PWM mode, and its active switches could operate in the zero-voltage-switching (ZVS) turn-on state to decrease switching losses [72]. A prominent drawback of this topology is the switching losses of the output diodes.
2) LLC converter: The LLC converter is widely used in EV charging applications due to its simpler structure and high efficiency [73]. Its topology is similar to the PSFB converter but has extra inductance and capacitor to form a resonant circuit. The output voltage is regulated by changing the impedance ratio of the resonant tank to equivalent load by the regulation of the switching frequency. Compared to the PSFB converter, the LLC converter could realize ZVS and zero-current-switching (ZCS) performance under all load conditions, and has no voltage spike on the rectifier in the secondary side [74]. The disadvantages of the LLC converter lay in the

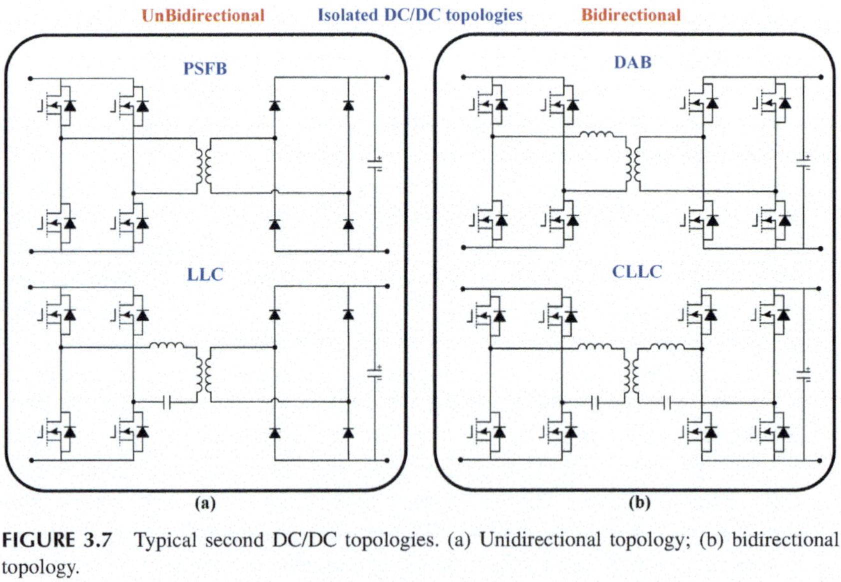

FIGURE 3.7 Typical second DC/DC topologies. (a) Unidirectional topology; (b) bidirectional topology.

limited power regulation capability in light-load conditions and narrow ZVS operation range [35].

B.2 Bidirectional topology

1) Dual active bridge (DAB) converter: The DAB converter is also a common scheme for EV charging due to its high power efficiency and density, low device stress, small filter sizes, and so forth [75]. The DAB converter contains two full bridges and a transformer in series with an inductance. For the DAB converter, the bidirectional power flow is realized by adjusting the phase shift between the AC square-wave voltage of the primary and secondary bridge [76]. However, the phase shift control may cause a narrow soft-switching range and high backflow current in light-load conditions, thus leading to decreased conversion efficiency.
2) CLLC converter: To realize the bidirectional power control capacity of the LLC converter, the resonant capacitor and inductance are added to each side of the transformer. This is the so-called CLLC converter [77]. The CLLC converter is controlled by a variable switching frequency and could be operated with a much wider soft-switching region. However, the control of the CLLC converter is more complicated as the output voltage gain curve against frequency is to be steady in specific frequency ranges [78].

3.4.2 Power converter for off-board chargers

Fig. 3.8 depicts the schematic block diagram of the off-board charging facility, which comprises the power sources, protection units, and the off-board charger. The power source can be a 3-phase AC source or DC source. The functionalities of the first stage PFC unit and second stage isolated DC/DC

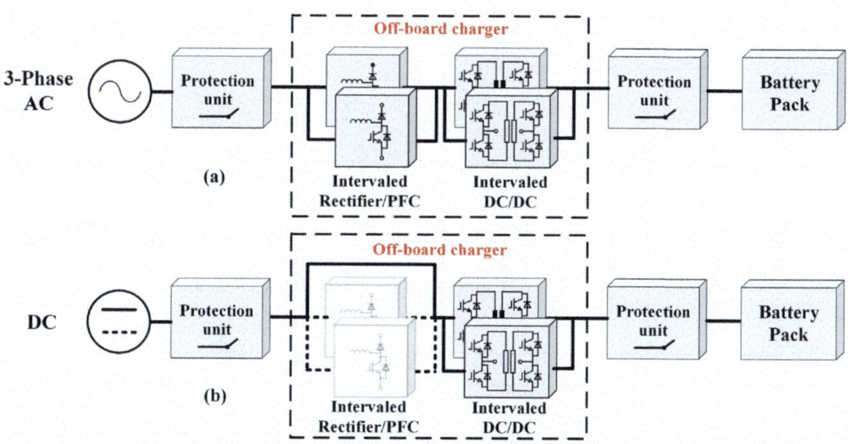

FIGURE 3.8 Off-board charging facility. (a) 3-Phase power supply; (b) DC power supply.

unit keep the same with the on-board charger. Notably, when a single PFC or DC/DC module could not meet the power requirement of the fast-charging, multiple identical modules can be interconnected in parallel to satisfy the power demand. The second stage DC/DC module in the off-board charger keeps the same as the one in the on-board charger. The detailed common topology of the first PFC stage is discussed below.

A first stage Front-end PFC rectifier.

A.1 Unidirectional topology

1. Conventional 3-phase PFC converter: In Fig. 3.9 (a), the conventional 3-phase PFC converter is the fundamental topology that comprises a 3-phase diode rectifying bridge and a boost converter. Although a relatively low current distortion could be achieved, the high voltage and current stress on the main switches would increase the switching losses, accompanied with the significant conduction losses of output diodes. In addition, a unity power factor cannot be remained in a wider range [79].
2. Vienna rectifier: The Vienna rectifier is widely adopted for the fast-charging of EVs in the industry due to its relatively lower cost, higher power factor, and lower harmonics [80]. The Vienna rectifier is a three-level solution with reducing the number of active switches and decreasing the dV/dt of active switches, thus the switching losses could be

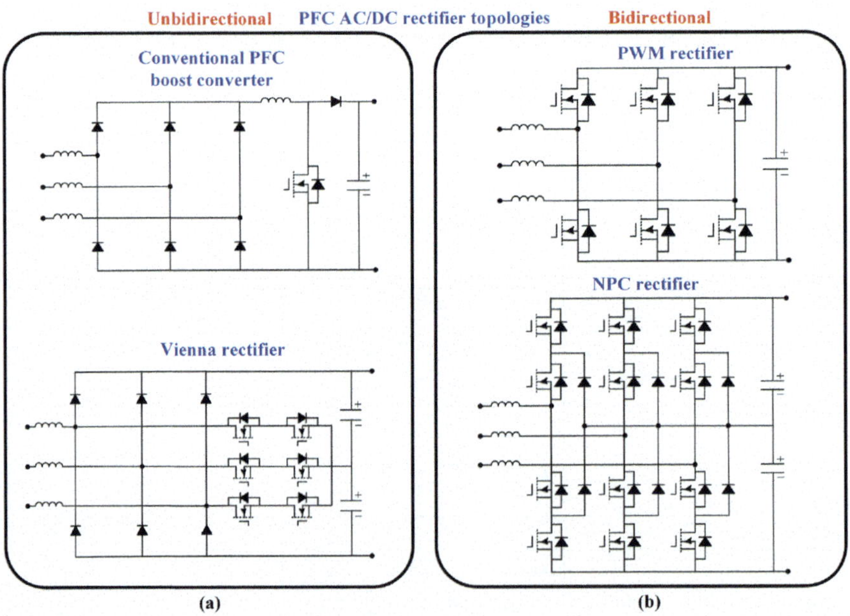

FIGURE 3.9 Typical PFC topology for off-board chargers. (a) Unidirectional; (b) bidirectional topology.

decreased. Moreover, due to the three-level structure, the low-voltage switching devices could be adopted in Vienna rectifier, thereby decreasing the costs dramatically. The major drawback of the Vienna rectifier is the unidirectional power flow and the limited reactive power control capability [35].

3. PWM rectifier: The widely used 3-phase PWM rectifier features the simplest topology in mentioned PFC converter for off-board chargers [81]. It offers high reliability, low complexity, power factor, and voltage regulation capability. The PWM converter could generate low-harmonic input currents and provide bidirectional power flow control capability. As the PWM rectifier belongs to the two-level topology, relatively higher switching losses would generate compared to the three-level topology [82]. Moreover, the poor tolerance to shoot-through of the rectifier bridge leg is inherent due to the direct parallel of the DC bus and rectifier bridge, thus raising the risk of short circuit of the DC bus.
4. NPC rectifier: Owing to high efficiency, compactness, and reliability, the neutral-point-clamped (NPC) converter has gained wide attention and application in both industrial and civil areas [83]. The NPC converter belongs to the three-level topology. Thus, the utilization of power devices with lower switching losses and costs could be achieved. In addition, lower current harmonics and dV/dt could be achieved derived from the three-level topology [84]. What is more, due to the existence of clamping diodes, this converter is immune to the short circuit of the output voltage in case of control errors or switch faults. Thus, the fault tolerance and reliability of systems are improved. The undesirable aspects of the NPC rectifier lay in more switching elements and high control complexity [85].

The reactive power compensation capability is associated with the power converter topologies of EV chargers. For unidirectional power topologies, the adopted EV can only operate at the positive x-axis of the P-Q frame (shown in Fig. 3.4). Thus, the limited reactive power support could be executed in the charging mode of EVs. As for bidirectional power topologies, the four-quadrant operation is enabled by providing full reactive power compensation capability. Thus, although the unidirectional topologies are commonly applied, they are not desirable for reactive power compensation compared to bidirectional ones [86].

3.5 Typical "electric vehicle-distribution grid" integration architectures

Recently, novel power architectures for nanogrid and microgrid have received growing attention and extensive research. In this section, comprehensive power architectures are reviewed and evaluated in view of the EV-grid integration and V2G operation.

3.5.1 AC-coupled architecture

The typical AC-coupled architecture, shown in Fig. 3.10, is the dominant structure in active distribution grids due to its natural compatibility with the widespread AC systems and high technology maturity [87]. In Fig. 3.10, several sources and loads (e.g., PV and EV load) are connected to the common AC bus (bus-2) through corresponding power conversion units. Additionally, there also exist other buses to interface dedicated loads. The multi-bus design favors the connection and management of load clusters, as well as the isolation of electrical faults, thus improving the system reliability [88].

3.5.2 DC-coupled architecture

The typical DC-coupled architecture is depicted in Fig. 3.11. This architecture essentially belongs to the single polarity-type structure and contains multiple 2-wire DC buses with different voltage levels to interface corresponding DGs and loads. In the case of one bus in power deficit or surplus, the power exchanges between bus-1 and bus-2 could be realized by the control of the isolated DC/DC converter (Conv.). Additionally, when a bus is faced with electrical faults like overvoltage or short circuit, the corrupted bus could easily be isolated from the system without affecting other buses' normal operation, thus increasing the system redundancy and reliability [89]. The DC/AC voltage source converter (VSC) provides bidirectional power flow between the AC utility grid and DC system. Determined by the desired power exchange capability between the utility grid and the DC bus, the multiple VSCs could be paralleled to increase the transfer power rating and reliability [43]. The BESS can be directly connected to the main bus or interfaced by the dedicated converter to the main bus. For the former way, although it achieves inherent dynamic stability of the system, the uncontrollable bus voltage would occur since the battery terminal voltage varies with the current SoC and output

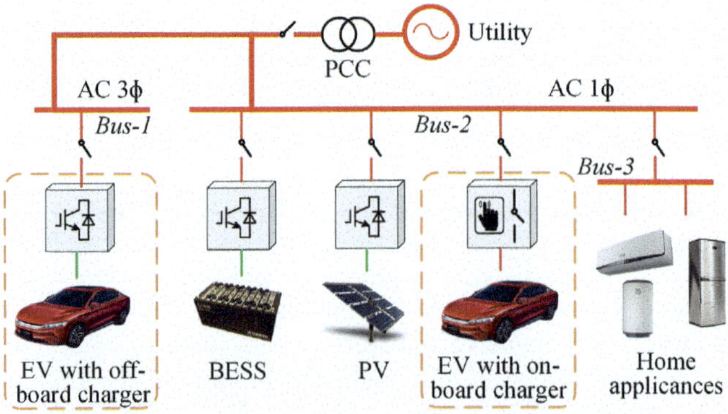

FIGURE 3.10 Typical AC-coupled architecture.

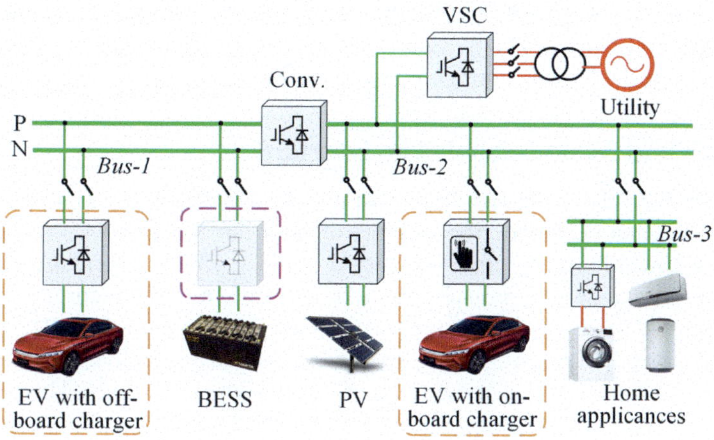

FIGURE 3.11 Multi-bus DC-coupled architecture.

current. The converter interfaced scheme could achieve regulated battery charging and allow more control flexibility and possibility of connecting more buses, thus enhancing the system reliability or powering loads in a broad region [90].

A slightly complex topology called bipolar DC-coupled architecture is presented in Fig. 3.12. The 3-wire configuration comprising positive, negative, and neutral lines could provide multiple voltage levels. For example, as described in Ref. [91], 370 V, +170 V, or −170 V can be chosen by load side DC/DC converters. Therefore, the reliability of the system is enhanced, since in the condition of faults of one line, the power supply could also be provided

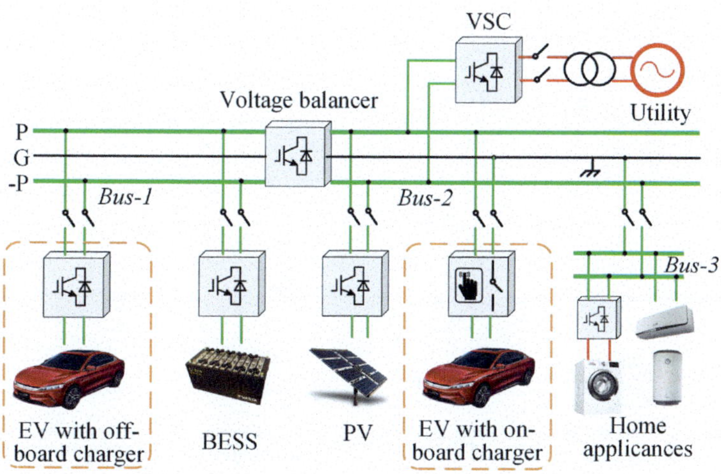

FIGURE 3.12 Bipolar DC-coupled architecture.

via the other two lines [89]. Unlike the single polarity-type structure, the phase unbalance problem should be concerned in the bipolar structure. A voltage balancer near the VSC is configured to balance positive and negative voltages.

3.5.3 Hybrid AC/DC-coupled architecture

The hybrid AC/DC-coupled architecture is an emerging power distribution architecture that combines the benefits of both the AC and DC-coupled architectures mentioned before. As depicted in Fig. 3.13, the hybrid architecture always encompasses an AC subgrid, a DC subgrid, and bidirectional AC/DC ILCs. The AC and DC components, such as the wind turbine, BESS, and residential appliances, are segregated and linked to respective subgrids primarily according to corresponding interface characteristics [92]. This kind of integration feature brings an improved overall efficiency due to reducing the number of power conversion stages [93]. The ILC generally adopts the VSC structure, and it is deemed a critical component in interconnecting the AC and DC subgrids. Through the proper power dispatching, the ILC could realize mutual power support between subgrids, improve the utilization of RESs, and

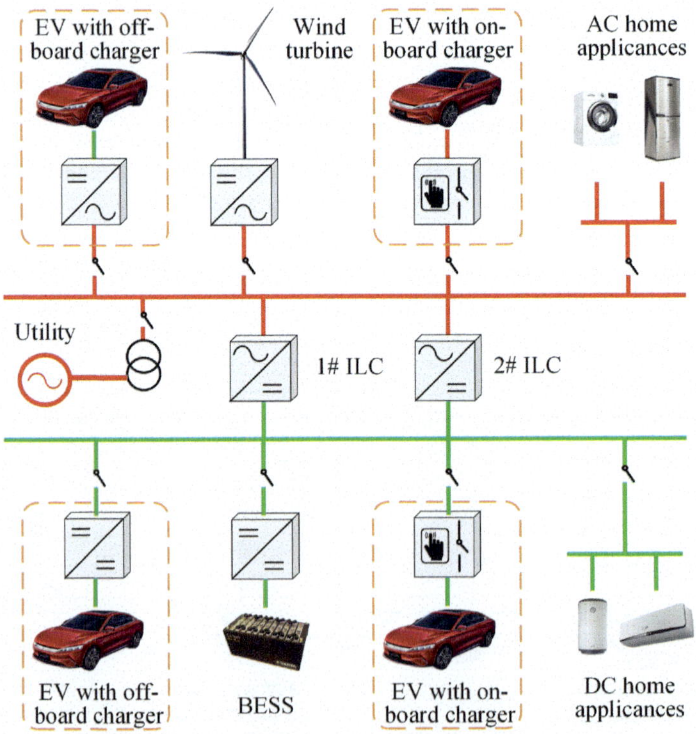

FIGURE 3.13 Typical hybrid AC/DC-coupled architecture.

increase the system reliability of power supply [94]. Concerned about the limited power capacity of the single ILC, multiple ILCs can be in parallel operation to enhance the total power capacity and increase the system redundancy.

3.5.4 Advanced ILC-based hybrid AC/DC-coupled architecture

To facilitate the construction of Energy Internet, the advanced ILC-based architecture with the virtue of high integration, plug-and-play interfaces, and enhanced control flexibility is preferred in future smart grid [95]. This kind of architecture facilitates the incorporation of DGs and loads, improves power transfer efficiency, and realizes efficient energy management via flexible power flow control [96].

The emerging SST has received widespread attention in recent years to achieve higher reliability, modularity, and controllability of distribution systems. The most prominent merit of SST is the elimination of the bulky 50/60 Hz transformer for interfacing medium-voltage AC grid via high-density power conversion, thereby decreasing the needed installation spaces and weight of systems dramatically [97]. Moreover, extensive control flexibility, such as flexible power flow, harmonics and transient compensation, as well as direct integration of DGs, could also be achieved by SST [98]. This kind of device is believed to play an essential role in the future Energy Internet. The typical SST-based hybrid AC/DC-coupled architecture is drawn in Fig. 3.14. It can be seen that the SST comprises three stages, viz., the MV input stage, isolation stage, and low voltage output stage [98]. And the two DC buses are provided to access different loads with varied voltage levels. In the MV input stage, the cascaded multilevel rectifier is generally adopted to realize the AC/DC conversion. The modular design facilitates the extension to higher voltages and power ratings, as well as the isolation of fault modules [99]. Several DABs can form the isolation stage to scale the voltage and current, and obtain the electric isolation and bidirectional power flow control [100]. The DABs can be interconnected in series at input terminals and parallel at output terminals [101]. The low voltage output stage consists of VSCs to achieve DC/AC conversion. The rise of advanced power electronic architectures brings more opportunities for V2G services in a broad sense.

3.6 Assessment and comparison

Under different power architectures, the large-scale EV-grid integration and V2G operation pose varied challenges and opportunities for system operation. In addition, the satisfaction level of the charging demand and V2G application is related to the architecture topology. Thus, it's necessary to make a

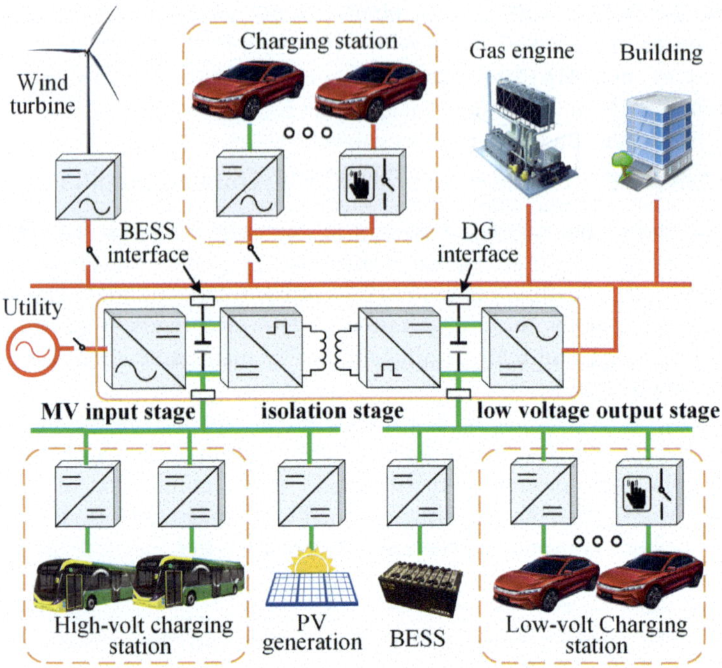

FIGURE 3.14 SST-based hybrid AC/DC coupled architecture.

comprehensive assessment of power architectures within the active distribution grids in view of the EV-grid integration and V2G operation. In light of this, we define the following six main indexes, viz., charging demand compatibility, power quality severity, V2G availability, architecture scalability, local control complexity, and technology maturity. Three levels, viz., high, medium, and low, are defined to assess the above indexes. The radar map of evaluation for above mentioned four architectures is illustrated in Fig. 3.15.

3.6.1 Charging demand compatibility

This index defines the capability of different power architectures to accommodate various charging demands by corresponding AC-powered and DC-powered charging facilities. Since the coexistence of AC and DC buses in typical hybrid architecture and advanced ILC-based hybrid architecture, the conventional AC-powered EV charging facility and the emerging DC-powered EV charging facility are supported. Thus, in, the charging demand compatibility is high for these two architectures. Regarding the DC-coupled architecture, the conventional AC-powered charging facility could not be supported, leading to low compatibility.

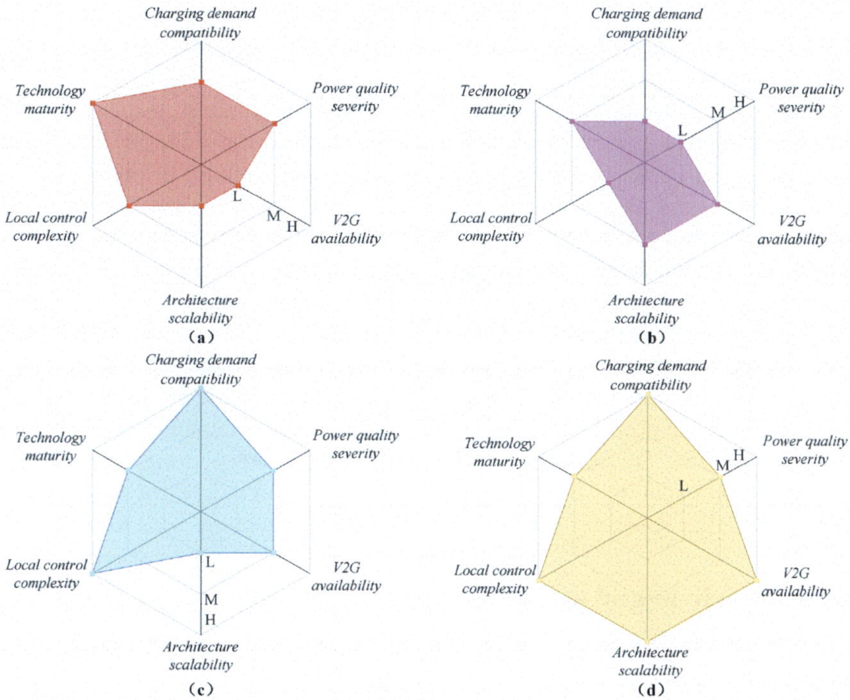

FIGURE 3.15 Radar map of power architectures evaluation. (a) Typical AC-coupled architecture; (b) typical DC-coupled architecture; (c) typical hybrid AC/DC-coupled architecture; (d) advanced ILC-based hybrid architecture.

3.6.2 Power quality severity

This index is set to discuss the faced long-duration power quality issues (defined over 1 min) inherited from large-scale EV charging and V2G operation in varied power architectures.

For the AC-coupled architecture, the EV charging loads gathered in a single phase would create phase unbalance [55], thus impacting the normal operation of power grids. Moreover, the phase balance condition would be more complex considering the varied V2G power output. The AC-coupled architecture usually faces the voltage drop concern when widespread EV charging is conducted, especially at the end of distribution feeders [102]. The power quality severity is set to medium.

Compared to the AC-coupled architecture, the DC-coupled architecture could alleviate the power quality issues raised by the high-power EV charging [103]. This is inherited from the DC nature. According to the definition of harmonics, viz., sinusoidal waveforms having frequencies that are integer multiples of the power frequency, the DC system has no harmonic issues as the fundamental frequency is zero [104]. Additionally, there is no phase unbalance issue for the single polarity-type architecture. Owing to the isolation of the VSC, the voltage or frequency fluctuation in the AC side has no impact on the voltage or power in the DC side [105]. The DC-bus voltage deviation may be encountered due to the high charging current flowing through resistive transmission lines. The power losses would increase as a lower voltage level leads to an increased current transfer to distribute the same amount of power [106]. In conclusion, the assessment of power quality severity for the DC-coupled architecture is set to low.

For the typical hybrid architecture and advanced ILC-based architecture, the above power quality issues in AC and DC systems are contained. Notably, the power quality in these two heterogeneous subgrids would be affected by each other, thereby causing more complicated power quality issues. For example, the AC unbalanced voltage would introduce double-frequency power oscillations at the power output of ILC accompanied by a voltage ripple in the DC bus of the DC subgrid [107]. Meanwhile, considering the power compensation capability of ILCs (elaborated in section 4.2.3), the assessment of power quality severity for the hybrid coupled architecture is set to medium.

3.6.3 Vehicle-to-grid availability

The V2G availability refers to the potential of EVs to provide auxiliary services for the power grid. Inherited from different features of power architectures, such as power quality issues and power flow paths, the V2G availability would be extended or restricted. Three aspects, including the active power quality compensation capability, power flow control capability at PCC, and power coordination capability between subgrids, are concerned with assessing the V2G availability.

Table 3.8 illustrates the comparison of V2G availability for different architectures. The corresponding assessment results are depicted in Fig. 3.15. The enabled typical auxiliary services contain the energy arbitrage, spinning reserve, power backup, and active power support [39]. The realizations of the above auxiliary services rely on the active and coordinated power supply/consumption of EVs. However, these actions would be hampered by the current distorted grid states. In Table 3.8, the typical AC-coupled architecture lacks the active power quality compensation capability. Thus, other compensation devices like voltage regulators and DSTATCOMs are generally adopted [108–110]. However, these devices' installation is affected by the current location and space, inevitably increasing the investment and maintenance cost. In addition, the related compensation devices need to be updated as the changes in system configurations, reflecting relatively low adaptability. The hybrid AC/DC-coupled architecture (contains the typical and advanced ILC-based architecture) is of a specific capability for active power quality compensation [111]. This is because the ILC could detect the power quality conditions of both AC and DC subgrids in real time, and it is not operating at

TABLE 3.8 Comparison of V2G availability for different architectures.

Architecture	Active power quality compensation capability	Power flow control capability at PCC	Power coordination capability between subgrids
Typical AC-coupled architecture	×	×	×
Typical DC-coupled architecture	×	√	×
Typical hybrid architecture	√	×	√
Advanced ILC-based hybrid architecture	√	√	√

full rating all the time. Accordingly, the available ratings of ILC can be intelligently controlled to mitigate the power quality issues. A series of extra functions like reactive power support, AC/DC voltage recovery, and harmonic mitigation can be implemented by the SST, thereby guaranteeing the normal operation of V2G [112]. In Ref. [113], a novel control scheme for an ILC is proposed to maintain a sinusoidal PCC voltage in nonlinear load conditions in islanded hybrid microgrids. The multi-resonant and repetitive controller is used to deal with high-order harmonics. Concerning the active compensation of the PCC unbalanced voltage, the paper [114] proposed novel strategies by using the virtual negative sequence impedance to direct unbalanced negative current flowing into the ILC side, thus improving the PCC voltage. Although the ILC is a promising candidate for power quality control, its adjustment capability is related to the remaining power ratings.

Regarding V2G-enabled auxiliary services provided for the utility grid, the steady and regulated power flow at PCC is favorable, as the expected power exchange is predefined between the power architecture and utility grid [115]. However, since there is no capability for power regulation at PCC in the typical AC-coupled architecture and typical hybrid architecture, the desired power flow would deviate from the expected value when loads or DGs power output change. Power coordination with other devices like BESS is needed to address this issue. The VSC/ILC in DC/advanced ILC-based architecture could provide a steady power output at PCC, thus facilitating the implementation of auxiliary services. As for the hybrid architecture, the power coordination between AC and DC subgrids could be realized via the V2G operation, thus enabling the V2V service and increasing the reliability of the power supply. The V2G availability could also be improved.

3.6.4 Architecture scalability

This index defines the capability of systems to accept new elements. Two aspects are to be concerned in evaluating the system scalability: 1) the ability to interface growth of components inside systems; 2) the ability to interconnect with external individual systems.

In Fig. 3.15, the advanced ILC-based hybrid architecture is of high scalability due to the feature of modularity, multiple plug-and-play interfaces, and diverse functionalities. In the internal scope of the architecture, the plug-and-play interfaces with open and standard-based protocol could recognize the connected devices automatically. The necessary information like DG's capacity can be instantly acquired, thus facilitating the management of various devices. In the outer scope of this architecture, the uniform architecture is helpful in constructing the Energy Internet by undertaking the dynamic energy flow adjustment and real-time communication among grid elements [116]. In this way, the SST becomes an intelligent node in the grid that could execute parameter sensing, operation optimization, communication with other SSTs, and initiate intelligent coordination [117]. The flexible information and power flow enable peer-to-peer trading in the future electricity market [22]. The potential of the V2G operation could be fully exploited in the envisioned Energy Internet.

Regarding the typical AC-coupled and hybrid AC/DC-coupled architecture, the corresponding scalability is relatively low due to a lack of dedicated interfaces and uniform converters to tie with the outer grid. For the DC-coupled architecture, the existence of the VSC enables the controllable power exchange with the external AC grid. Yet, owing to lacking plug-and-play interfaces, the scalability of the DC-coupled architecture is set to medium.

3.6.5 Local control complexity

The local control is generally realized by local units (e.g., battery storage) in distribution systems, aiming to guarantee stable system operation. The local control also achieves the fundamental power coordination and forms the basis of the long-timescale optimal dispatching [118].

The local control complexity of the advanced ILC-based hybrid architecture is relatively high since the core unit of these architectures is the multi-stage ILC. For the SST, the control for the multi-stages and multi-modules topology requires real-time communication and synchronization [90]. As depicted in Fig. 3.13, the DC bus at each stage provides decoupling of power transients and should be in a steady state for the operation stability concern [119]. The high-volt DC bus can be regulated by the medium voltage (MV) converter in the MV input stage. In the isolation stage, the isolated DC/DC converters could maintain the voltage of the low-volt DC bus through the closed-loop control. A power-balancing strategy is usually required in the inner control loop to realize the even sharing of the total power among each DC/DC module [120]. The DGs and BESS connected to the low-volt DC bus can also control the corresponding DC voltage accompanied with the DC/DC converters in an isolated stage working on the power control mode. In the low voltage output stage, the VSC could support the voltage and frequency of the AC subgrid. An alternative way to generate the AC voltage and frequency is using the DGs and BESS in the AC subgrid, the droop control mode or voltage and frequency control mode is to be utilized [121]. In this context, the VSC could remain in the power control mode for the power dispatch control. The overview of local control strategies of SST-based hybrid architecture is shown in Fig. 3.16.

In hybrid AC/DC-coupled architecture, since more DGs and loads are connected to AC and DC buses, the coordination between the AC and DC subgrids should be concerned in the design of local control strategies. Thus, the local control is more complex than the control in individual AC and DC-coupled architecture. In the grid-connected scenario, the frequency and voltage in the AC side are supported by the utility grid. Thus, as illustrated in Fig. 3.17, there are two schemes to stabilize the voltage in the DC subgrid. For the first one, the ILC operates in the voltage control mode to keep the DC subgrid voltage at its desired value [122]. Meanwhile, to produce a regulated V2G output for the external grid, the DGs&BESS in both subgrids should be in power coordination to maintain the power balance in the architecture. For the

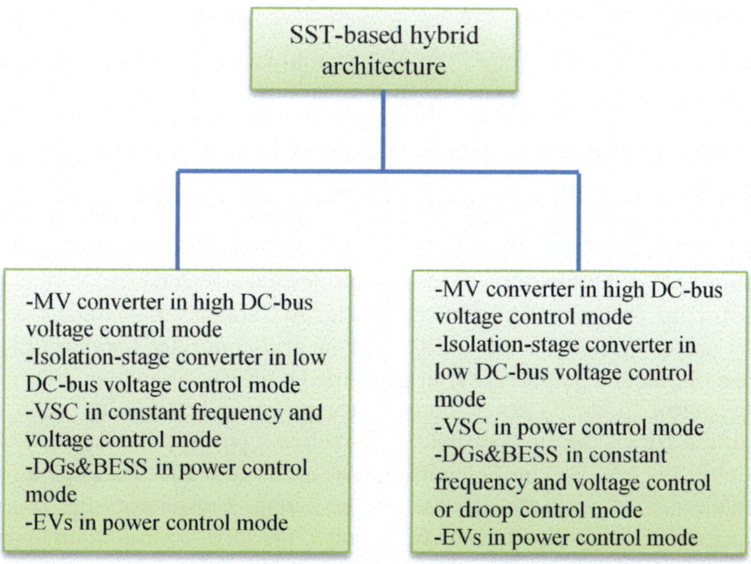

FIGURE 3.16 Local control strategies for SST-based hybrid architecture in grid-connected mode.

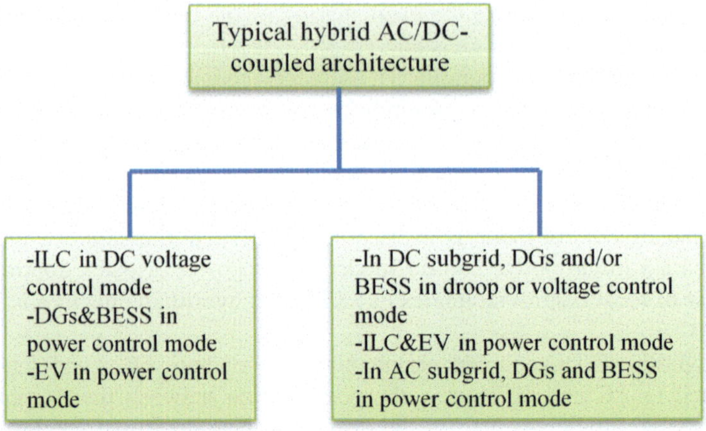

FIGURE 3.17 Local control strategies for typical hybrid architecture in grid-connected mode.

second one, the DGs and/or BESS in DC subgrid regulate the DC bus voltage (droop control mode or voltage control mode). And the ILC is in the power control mode [123]. In this context, the power dispatch of ILC is activated when the V2G service is enabled in the DC subgrid. Meanwhile, the DGs and BESS in AC subgrid should also be coordinated to realize a dispatched output power at PCC.

For the typical DC-coupled architecture, the local control is straightforward thanks to the simple topology. To guarantee the effective V2G operation, the DC bus voltage could be stabilized by two schemes as illustrated in Fig. 3.18. In the first scheme, the VSC controls the DC bus voltage, thus balancing the power surplus/deficit in the DC bus. In the second scheme, the BESS could be operated in the constant voltage control mode for voltage stabilization by covering the DC bus's power unbalance. Notably, the SoC and the power limit of the BESS should be concerned when the BESS remains in the bus voltage control mode [124]. The transition from the voltage control to the power control mode of the BESS is to be executed when the corresponding SoC or power output reaches the designed boundaries.

As for the AC-coupled architecture, the AC voltage magnitude and frequency are supported by the utility grid in grid-connected mode [125]. To realize a dispatched output power at PCC, the DGs&BESS could operate in the power control mode. When it comes to a power deficit/surplus within the system, the BESS can be coordinated to realize a power balance considering the current energy capacity, response time, etc [126]. The V2G operation is also facilitated to reach the desired power output. Considering the AC synchronization and coordination in AC systems, the local control complexity of the typical AC-coupled architecture is set to medium.

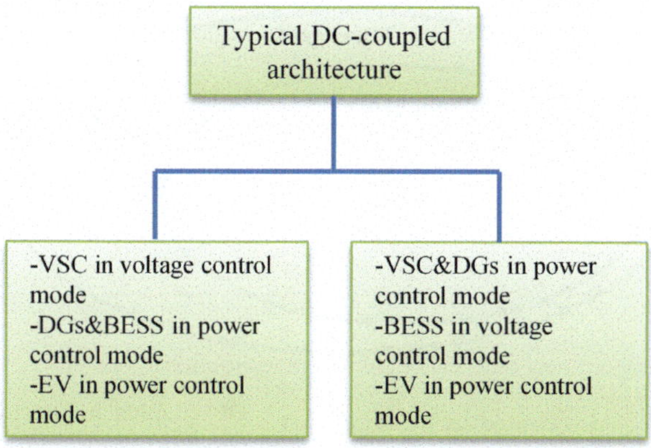

FIGURE 3.18 Local control strategies for typical DC-coupled architecture in grid-connected mode.

3.6.6 Technology maturity

This index aims to illustrate the corresponding setup standards and cost factors of different architectures. The typical AC-coupled architecture is of the most maturity among the studied architectures. It has a broad market perspective due to its natural compatibility with current widespread AC systems and relatively low cost [92]. Compared to the AC counterpart, there have been no unified DC standards for the distribution grid worldwide so far deriving from immature market and technology [127]. Thus, the technology maturity of the DC-coupled architecture and hybrid architecture is relatively lower. Especially, for the SST-based hybrid architecture, it has been proven to have 10% efficiency below the conventional transformer. Moreover, the results in Ref. [128] reveal that the SST could be up to 4.5 times the cost of equivalent conventional power transformers. Thus, there still remains an improvement in cost, efficiency, and standardization for advanced ILC-based architectures.

In Fig. 3.15, it can be seen that the advanced ILC-based hybrid architecture has a high comprehensive performance in EV-grid integration and V2G operation. This architecture exhibits merits in terms of charging demand compatibility, V2G availability, and architecture scalability. Yet, there still remains an improvement in terms of technology maturity concerning the cost, efficiency, and standardization.

According to specific requirements of nanogrid and microgrid, the classification of discussed power architectures is illustrated in Fig. 3.19. Since the nanogrid is served for small-scale application scenarios, the low-cost and straightforward architecture is preferred. For the microgrid application, besides the architecture item 1, 2, 3, concerning additional requirements of multiple control functions, varied components integration, and system networking, the relatively complex topologies can be adopted, that is, the item 4, 5.

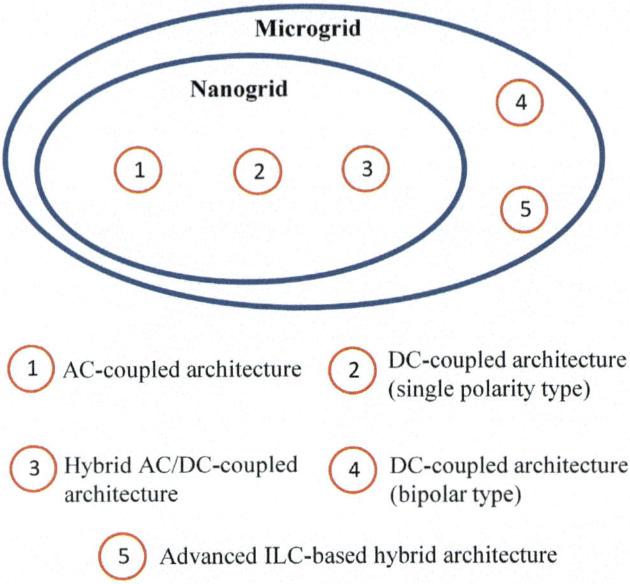

FIGURE 3.19 Power architectures suitable for nanogrid and microgrid applications.

3.7 Concluding remarks

With increasingly stringent restrictions for pollutants and carbon emissions, researchers have shown an increased interest in EV and V2G technology. The previous reviews about V2G are usually conducted in the framework of AC distribution grids while ignoring a series of novel power architectures. However, these novel architectures are much concerned with the performance of EV integration and V2G operation. Additionally, there is no comprehensive review involving all the hardware perspectives of EV-integration systems, which involves connection standards, charging infrastructure, converter topologies, and interfaced power architectures. In light of this, this chapter presented a comprehensive review of distribution grid architectures, grid connection standards, and interface topologies, as well as typical grid applications from the perspective of EV-grid integration and V2G operation for the first time. Firstly, the concept of nanogrid, microgrid, and cluster is identified, and their characteristics are described and compared. Secondly, with a particular focus on the EV integration in these three key participators under active distribution grids, the triple roles of EVs, namely, the loads, energy resources, and energy storage, are unveiled and explained. The currently valid standards of EV grid connection associated with different roles are discussed elaborately. Considering the bidirectional/unidirectional power flow control demand of EVs and power supply type, the conventional AC-powered charger converter topologies and emerging DC-powered converter topologies are reviewed, which provides high referential value for related researchers to select suitable interface converter topologies. Thirdly, for nanogrid and microgrid, a comprehensive assessment of the EV-grid interaction performance for different power architectures is performed in terms of charging demand compatibility, power quality severity, V2G availability, architecture scalability, local control complexity, and technology maturity. The advanced ILC-based hybrid architecture has been proven to satisfy diverse charging demands and facilitate the V2G operation owing to controllable power control capability at PCC and power coordination capability between subgrids, thus having a high comprehensive performance in EV-grid integration and V2G operation. As for clusters, the hybrid parallel-series cluster architecture accommodates the requirements of low complexity, low cost, high redundancy, and reliability. Finally, a brief discussion of practical applications of typical nanogrid, microgrid, and clusters is given. This chapter provides a clear picture of the EV-grid interaction to academia and industry from the connection standards, infrastructures, power architecture, and real applications.

References

[1] Rahman S, Khan IA, Khan AA, Mallik A, Nadeem MF. Comprehensive review & impact analysis of integrating projected electric vehicle charging load to the existing low voltage distribution system. Renew Sustain Energy Rev 2022;153:111756.

[2] Mustapa SI, Bekhet HA. Analysis of CO2 emissions reduction in the Malaysian transportation sector: an optimisation approach. Energy Policy 2016;89:171–83.

[3] Lopes JAP, Soares FJ, Almeida PMR. Integration of electric vehicles in the electric power system. Proc IEEE 2010;99:168–83.

[4] Wang L, Qin Z, Larumbe LB, Bauer P. Python supervised co-simulation for a day-long harmonic evaluation of EV charging. Chinese Journal of Electrical Engineering 2021;7:15–24.

[5] Yu H, Shang Y, Niu S, Cheng C, Shao Z, Jian L. Towards energy-efficient and cost-effective DC nanaogrid: a novel pseudo hierarchical architecture incorporating V2G technology for both autonomous coordination and regulated power dispatching. Appl Energy 2022;313:118838.

[6] Foley A, Tyther B, Calnan P, Gallachoir BO. Impacts of electric vehicle charging under electricity market operations. Appl Energy 2013;101:93–102.

[7] Huda M, Koji T, Aziz MJE. Techno economic analysis of vehicle-to-grid (V2G) integration as distributed energy resources in Indonesia power system. Energies 2020;13:1162.

[8] Jian L, Zheng Y, Shao Z. High efficient valley-filling strategy for centralized coordinated charging of large-scale electric vehicles. Appl Energy 2017;186:46–55.

[9] Clement-Nyns K, Haesen E, Driesen J. The impact of charging plug-in hybrid electric vehicles on a residential distribution grid. IEEE Trans Power Syst 2009;25:371–80.

[10] Li Y, Crossley PA. Impact of electric vehicles on LV feeder voltages. In: 2014 IEEE PES general meeting Conference & Exposition: IEEE; 2014. p. 1–5.

[11] Zheng Y, Niu S, Shang Y, Shao Z, Jian L. Integrating plug-in electric vehicles into power grids: a comprehensive review on power interaction mode, scheduling methodology and mathematical foundation. Renew Sustain Energy Rev 2019;112:424–39.

[12] Shang Y, Liu M, Shao Z, Jian L. Internet of smart charging points with photovoltaic integration: a high-efficiency scheme enabling optimal dispatching between electric vehicles and power grids. Appl Energy 2020;278:115640.

[13] Peng C, Zou J, Lian L, Li L. An optimal dispatching strategy for V2G aggregator participating in supplementary frequency regulation considering EV driving demand and aggregator's benefits. Appl Energy 2017;190:591–9.

[14] Tostado-Véliz M, León-Japa RS, Jurado F. Optimal electrification of off-grid smart homes considering flexible demand and vehicle-to-home capabilities. Appl Energy 2021;298: 117184.

[15] Hoarau Q, Perez Y. Interactions between electric mobility and photovoltaic generation: a review. Renew Sustain Energy Rev 2018;94:510–22.

[16] Yu H, Niu S, Zhang Y, Jian L. An integrated and reconfigurable hybrid AC/DC microgrid architecture with autonomous power flow control for nearly/net zero energy buildings. Appl Energy 2020;263:114610.

[17] Baharvandi A, Aghaei J, Nikoobakht A, Niknam T, Vahidinasab V, Giaouris D, et al. Linearized hybrid stochastic/robust scheduling of active distribution networks encompassing PVs. IEEE Trans Smart Grid 2019;11:357–67.

[18] Yang T. ICT technologies standards and protocols for active distribution network. In: Smart power distribution systems: Elsevier; 2019. p. 205–30.

[19] Marvasti AK, Fu Y, DorMohammadi S, Rais-Rohani M. Optimal operation of active distribution grids: a system of systems framework. IEEE Trans Smart Grid 2014;5: 1228–37.

[20] Katiraei F, Iravani R, Hatziargyriou N, Dimeas A. Microgrids management. IEEE Power Energy Mag 2008;6:54–65.

[21] Javaid S, Kato T, Matsuyama T. Power flow coloring system over a nanogrid with fluctuating power sources and loads. IEEE Trans Ind Inf 2017;13:3174–84.

[22] Yu H, Niu S, Shao Z, Jian L. A scalable and reconfigurable hybrid AC/DC microgrid clustering architecture with decentralized control for coordinated operation. Int J Electr Power Energy Syst 2022;135:107476.

[23] Ravi SS, Aziz M. Utilization of electric vehicles for vehicle-to-grid services: progress and perspectives. Energies 2022;15:589.

[24] Li T, et al. V2G multi-objective dispatching optimization strategy based on user behavior model. Front Energy Res 2021:494.

[25] Wang Y, Wang X, Shao C, Gong N. Distributed energy trading for an integrated energy system and electric vehicle charging stations: a Nash bargaining game approach. Renew Energy 2020;155:513–30.

[26] Rahman MS, Hossain M, Lu J. Coordinated control of three-phase AC and DC type EV–ESSs for efficient hybrid microgrid operations. Energy Convers Manag 2016;122: 488–503.

[27] Chen Q, Liu N, Hu C, Wang L, Zhang J. Autonomous energy management strategy for solid-state transformer to integrate PV-assisted EV charging station participating in ancillary service. IEEE Trans Ind Inf 2016;13:258–69.

[28] Qin D, Sun Q, Wang R, Ma D, Liu M. Adaptive bidirectional droop control for electric vehicles parking with vehicle-to-grid service in microgrid. CSEE J Power Energy Syst 2020;6:793–805.

[29] Anwar MB, Muratori M, Jadun P, Hale E, Bush B, Denholm P, et al. Assessing the value of electric vehicle managed charging: a review of methodologies and results. Energy Environ Sci 2022;15:466–98.

[30] Venegas FG, Petit M, Perez Y. Active integration of electric vehicles into distribution grids: Barriers and frameworks for flexibility services. Renew Sustain Energy Rev 2021;145: 111060.

[31] Clement-Nyns K, Haesen E, Driesen J. The impact of vehicle-to-grid on the distribution grid. Elec Power Syst Res 2011;81:185–92.

[32] Das HS, Rahman MM, Li S, Tan CW. Electric vehicles standards, charging infrastructure, and impact on grid integration: a technological review. Renew Sustain Energy Rev 2020;120:109618.

[33] Wang L, Qin Z, Slangen T, et al. Grid impact of electric vehicle fast charging stations: trends, standards, issues and mitigation measures-an overview. IEEE Open Journal of Power Electronics 2021;2:56–74.

[34] Williamson SS, Rathore AK, Musavi F. Industrial electronics for electric transportation: current state-of-the-art and future challenges. IEEE Trans Ind Electron 2015;62:3021–32.

[35] Tu H, Feng H, Srdic S, Lukic S. Extreme fast charging of electric vehicles: a technology overview. IEEE Trans Transp Electrif 2019;5:861–78.

[36] Mwasilu F, Justo JJ, Kim E-K, Do TD, Jung J-W. Electric vehicles and smart grid interaction: a review on vehicle to grid and renewable energy sources integration. Renew Sustain Energy Rev 2014;34:501–16.

[37] Alsharif A, Tan CW, Ayop R, Dobi A, Lau KY. A comprehensive review of energy management strategy in Vehicle-to-Grid technology integrated with renewable energy sources. Sustain Energy Technol Assessments 2021;47:101439.

[38] Habib S, Khan MM, Abbas F, Ali A, Faiz MT, Ehsan F, et al. Contemporary trends in power electronics converters for charging solutions of electric vehicles. CSEE J Power Energy Syst 2020;6:911–29.

[39] Solanke TU, Khatua PK, Ramachandaramurthy VK, Yong JY, Tan KM. Control and management of a multilevel electric vehicles infrastructure integrated with distributed resources: a comprehensive review. Renew Sustain Energy Rev 2021;144:111020.

[40] Lasseter RH. Smart distribution: coupled microgrids. Proc IEEE 2011;99:1074–82.

[41] Burmester D, Rayudu R, Seah W, Akinyele D. A review of nanogrid topologies and technologies. Renew Sustain Energy Rev 2017;67:760–75.

[42] Boroyevich D, Cvetković I, Dong D, Burgos R, Wang F, Lee F. Future electronic power distribution systems a contemplative view. In: 2010 12th International Conference on optimization of electrical and electronic equipment: IEEE; 2010. p. 1369–80.

[43] Nejabatkhah F, Li Y. Overview of power management strategies of hybrid AC/DC microgrid. IEEE Trans Power Electron 2014;30:7072–89.

[44] Feng W, Jin M, Liu X, et al. A review of microgrid development in the United States–A decade of progress on policies, demonstrations, controls, and software tools. Appl Energy 2018;228:1656–68.

[45] Vasquez JC, Guerrero JM, Miret J, Castilla M, De Vicuna LG. Hierarchical control of intelligent microgrids. IEEE Ind Electron Mag 2010;4:23–9.

[46] Liu C, Chau K, Wu D, Gao S. Opportunities and challenges of vehicle-to-home, vehicle-to-vehicle, and vehicle-to-grid technologies. Proc IEEE 2013;101:2409–27.

[47] Han Y, Pu Y, Li Q, Fu W, Chen W, You Z, et al. Coordinated power control with virtual inertia for fuel cell-based DC microgrids cluster. Int J Hydrogen Energy 2019;44: 25207–20.

[48] Yaqub R, Ahmad S, Ali H, Asar AJI. AI and blockchain integrated billing architecture for charging the roaming electric vehicles. IOT 2020;1:382–97.

[49] Yu H, et al. A Reconfigurable Hybrid AC/DC microgrid clustering architecture with decentralized control for coordinated operation. In: 2021 IEEE 5th Conference on energy Internet and energy system integration (EI2). IEEE; 2021. p. 3685–91.

[50] Habib S, Khan MM, Abbas F, Sang L, Shahid MU, Tang H. A comprehensive study of implemented international standards, technical challenges, impacts and prospects for electric vehicles. IEEE Access 2018;6:13866–90.

[51] Niu S, Xu H, Sun Z, Shao Z, Jian L. The state-of-the-arts of wireless electric vehicle charging via magnetic resonance: principles, standards and core technologies. Renew Sustain Energy Rev 2019;114:109302.

[52] Nicholas M, Hall D. Lessons learned on early electric vehicle fastcharging deployments. In: White paper, Int. Council CleanTransp., Washington D.C., USA; July 2018.

[53] Kesler M, Kisacikoglu MC, Tolbert LM. Vehicle-to-grid reactive power operation using plug-in electric vehicle bidirectional offboard charger. IEEE Trans Ind Electron 2014;61: 6778–84.

[54] Mouli GRC, Kaptein J, Bauer P, Zeman M. Implementation of dynamic charging and V2G using Chademo and CCS/Combo DC charging standard. In: 2016 IEEE transportation Electrification Conference and Expo (ITEC): IEEE; 2016. p. 1–6.

[55] Ul-Haq A, Cecati C, Strunz K, Abbasi E. Impact of electric vehicle charging on voltage unbalance in an urban distribution network. Intell Ind Syst 2015;1:51–60.

[56] Yilmaz M, Krein PT. Review of the impact of vehicle-to-grid technologies on distribution systems and utility interfaces. IEEE Trans Power Electron 2012;28:5673–89.

[57] Garcia O, Cobos JA, Prieto R, Alou P, Uceda J. Single phase power factor correction: a survey. IEEE Trans Power Electron 2003;18:749–55.

[58] Baocheng W, Xiaoqiang G, Herong G, Qiang M, Weiyang W. Real-time DC injection measurement technique for transformerless PV systems. In: The 2nd International Symposium on power electronics for distributed generation systems: IEEE; 2010. p. 980–3.

[59] Kisacikoglu MC, Kesler M, Tolbert LM. Single-phase on-board bidirectional PEV charger for V2G reactive power operation. IEEE Trans Smart Grid 2014;6:767–75.

[60] Nasir M, Jin Z, Khan HA, Zaffar NA, Vasquez JC, Guerrero JM. A decentralized control architecture applied to DC nanogrid clusters for rural electrification in developing regions. IEEE Trans Power Electron 2018;34:1773–85.

[61] Chen D, Xu L. Autonomous DC voltage control of a DC microgrid with multiple slack terminals. IEEE Trans Power Syst 2012;27:1897–905.

[62] Yilmaz U, Turksoy O, Teke A. Intelligent control of high energy efficient two-stage battery charger topology for electric vehicles. Energy 2019;186:115825.

[63] Khaligh A, Dusmez S. Comprehensive topological analysis of conductive and inductive charging solutions for plug-in electric vehicles. IEEE Trans Veh Technol 2012;61: 3475–89.

[64] Figueiredo J, Tofoli F, Silva B. A review of single-phase PFC topologies based on the boost converter. In: 2010 9th IEEE/IAS International Conference on industry applications-Induscon 2010: IEEE; 2010. p. 1–6.

[65] Huber L, Jang Y, Jovanovic MM. Performance evaluation of bridgeless PFC boost rectifiers. IEEE Trans Power Electron 2008;23:1381–90.

[66] Musavi F, Eberle W, Dunford WG. A phase-shifted gating technique with simplified current sensing for the semi-bridgeless AC–DC converter. IEEE Trans Veh Technol 2012;62:1568–76.

[67] Gou B, Ge X, Wang S, et al. An open-switch fault diagnosis method for single-phase PWM rectifier using a model-based approach in high-speed railway electrical traction drive system. IEEE Trans Power Electron 2015;31:3816–26.

[68] Vaideeswaran V, Sankar N. Control techniques of three phase PWM rectifier. Int J Eng Adv Technol 2018;8:148–52.

[69] Liu Z, Lee FC, Li Q, Yang Y. Design of GaN-based MHz totem-pole PFC rectifier. IEEE J Emerg Sel Top Power Electron 2016;4:799–807.

[70] Ribeiro E, Cardoso A, Boccaletti C. Open-circuit fault diagnosis in interleaved DC–DC converters. IEEE Trans Power Electron 2013;29:3091–102.

[71] Wang H, Shang M, Khaligh A. A PSFB-based integrated PEV onboard charger with extended ZVS range and zero duty cycle loss. IEEE Trans Ind Appl 2016;53:585–95.

[72] Zeljkovic S, Reiter T, Gerling D. Analysis of rectifier topologies for automotive HV to LV phase shift ZVT DC/DC converter. *2012 15th International power electronics and motion control Conference (EPE/PEMC): IEEE; 2012. P. DS1b. 4-1-DSb.* 4-7.

[73] Mu M, Lee FC. Design and optimization of a 380–12 V high-frequency, high-current LLC converter with GaN devices and planar matrix transformers. IEEE J Emerg Sel Top Power Electron 2016;4:854–62.

[74] Lee WS, Kim JH, Lee JY, Lee IO. Design of an isolated DC/DC topology with high efficiency of over 97% for EV fast chargers. IEEE Trans Veh Technol 2019;68:11725–37.

[75] Nakakohara Y, Otake H, Evans TM, Yoshida T, Tsuruya M, Nakahara K. Three-phase LLC series resonant DC/DC converter using SiC MOSFETs to realize high-voltage and high-frequency operation. IEEE Trans Ind Electron 2015;63:2103–10.

[76] Shi H, Wen H, Hu Y, Yang Y, Wang Y. Efficiency optimization of DC solid-state transformer for photovoltaic power systems. IEEE Trans Ind Electron 2019;67:3583–95.

[77] Jung JH, Kim HS, Ryu MH, Baek JW. Design methodology of bidirectional CLLC resonant converter for high-frequency isolation of DC distribution systems. IEEE Trans Power Electron 2012;28:1741–55.

[78] Chen W, Rong P, Lu Z. Snubberless bidirectional DC–DC converter with new CLLC resonant tank featuring minimized switching loss. IEEE Trans Ind Electron 2009;57:3075–86.

[79] Qiao C, Smedley K. A general three-phase PFC controller for rectifiers with a parallel-connected dual boost topology. IEEE Trans Power Electron 2002;17:925–34.

[80] Thangavelu T, Shanmugam P, Raj K. Modelling and control of VIENNA rectifier a single phase approach. IET Power Electron 2015;8:2471–82.

[81] Zhang Y, Jiao J, Liu J, Gao J. Direct power control of PWM rectifier with feedforward compensation of DC-bus voltage ripple under unbalanced grid conditions. IEEE Trans Ind Appl 2019;55:2890–901.

[82] Wang J, Rasekh N, Yuan X, Dagan KJ. An analytical method for fast calculation of inductor operating space for high-frequency core loss estimation in two-level and three-level PWM converters. IEEE Trans Ind Appl 2020;57:650–63.

[83] Xu X, Zheng Z, Wang K, Yang B, Li Y. A comprehensive study of common mode voltage reduction and neutral point potential balance for a back-to-back three-level NPC converter. IEEE Trans Power Electron 2019;35:7910–20.

[84] Kim SM, Won IJ, Kim J, Lee KB. DC-link ripple current reduction method for three-level inverters with optimal switching pattern. IEEE Trans Ind Electron 2018;65:9204–14.

[85] Chen M, He Y. Multiple open-circuit fault diagnosis method in NPC rectifiers using fault injection strategy. IEEE Trans Power Electron 2022;37:8554–71.

[86] Kisacikoglu MC, Ozpineci B, Tolbert LM. EV/PHEV bidirectional charger assessment for V2G reactive power operation. IEEE Trans Power Electron 2013;28:5717–27.
[87] Gupta A, Doolla S, Chatterjee K. Hybrid AC–DC microgrid: systematic evaluation of control strategies. IEEE Trans Smart Grid 2017;9:3830–43.
[88] Hoang TV, Lee H-H. Accurate power sharing with harmonic power for islanded multibus microgrids. IEEE J Emerg Sel Top Power Electron 2018;7:1286–99.
[89] Dragičević T, Lu X, Vasquez JC, Guerrero JM. DC microgrids—Part II: a review of power architectures, applications, and standardization issues. IEEE Trans Power Electron 2015;31:3528–49.
[90] Cai H, Xiang J, Wei W. Decentralized coordination control of multiple photovoltaic sources for DC bus voltage regulating and power sharing. IEEE Trans Ind Electron 2017;65:5601–10.
[91] Kakigano H, Miura Y, Ise T. Low-voltage bipolar-type dc microgrid for super high quality distribution. IEEE Trans Power Electron 2010;25:3066–75.
[92] Liu J, Hossain M, Lu J, Rafi F, Li H. A hybrid AC/DC microgrid control system based on a virtual synchronous generator for smooth transient performances. Elec Power Syst Res 2018;162:169–82.
[93] Li X, Dong C, Jiang W, Wu X. An improved coordination control for a novel hybrid AC/DC microgrid architecture with combined energy storage system. Appl Energy 2021; 292:116824.
[94] Wang J, Jin C, Wang P. A uniform control strategy for the interlinking converter in hierarchical controlled hybrid AC/DC microgrids. IEEE Trans Ind Electron 2017;65:6188–97.
[95] Huang AQ, Crow ML, Heydt GT, Zheng JP, Dale SJ. The future renewable electric energy delivery and management (FREEDM) system: the energy internet. Proc IEEE 2010;99: 133–48.
[96] Chen MJ, Poor HV. High-frequency power electronics at the grid edge: a bottom-up approach toward the smart grid. IEEE Electrif Mag 2020;8:6–17.
[97] Allende FR, Perez MA, Espinoza JR, Gajowik T, Stynski S, Malinowski M. Surveying solid-state transformer structures and controls providing highly efficient and controllable power flow in distribution grids. IEEE Ind Electron Mag 2020;14:56–70.
[98] Mishra DK, Ghadi MJ, Li L, et al. A review on solid-state transformer: a breakthrough technology for future smart distribution grids. Int J Electr Power Energy Syst 2021;133: 107255.
[99] Saleh SA, Richard C, Onge XFS, McDonald KM, Ozkop E, Chang L, et al. Solid-state transformers for distribution systems–Part I: technology and construction. IEEE Trans Ind Appl 2019;55:4524–35.
[100] Wang Y, Li Y, Cao Y, Tan Y, He L, Han J. Hybrid AC/DC microgrid architecture with comprehensive control strategy for energy management of smart building. Int J Electr Power Energy Syst 2018;101:151–61.
[101] Li Y, Han J, Cao Y, Li Y, Xiong J, Sidorov D, et al. A modular multilevel converter type solid state transformer with internal model control method. Int J Electr Power Energy Syst 2017;85:153–63.
[102] Dubey A, Santoso S. Electric vehicle charging on residential distribution systems: impacts and mitigations. IEEE Access 2015;3:1871–93.
[103] Mi Y, Chen X, Ji H, Ji L, Fu Y, Wang C, et al. The coordinated control strategy for isolated DC microgrid based on adaptive storage adjustment without communication. Appl Energy 2019;252:113465.

[104] Whaite S, Grainger B, Kwasinski A. Power quality in DC power distribution systems and microgrids. Energies 2015;8:4378–99.
[105] Majumder R, Ghosh A, Ledwich G, Zare F. Power management and power flow control with back-to-back converters in a utility connected microgrid. IEEE Trans Power Syst 2009;25:821–34.
[106] Shuai Z, Fang J, Ning F, Shen Z. Hierarchical structure and bus voltage control of DC microgrid. Renew Sustain Energy Rev 2018;82:3670–82.
[107] Khajesalehi J, Sheshyekani K, Hamzeh M, Afjei E. Maximum constant boost approach for controlling quasi-Z-source-based interlinking converters in hybrid AC–DC microgrids. IET Generation, Transmission & Distribution 2016;10:938–48.
[108] Huang S, Pillai JR, Liserre M, Bak-Jensen B. Improving photovoltaic and electric vehicle penetration in distribution grids with smart transformer. In: IEEE PES ISGT Europe 2013: IEEE; 2013. p. 1–5.
[109] Siti MW, Nicolae DV, Jimoh AA, Ukil A. Reconfiguration and load balancing in the LV and MV distribution networks for optimal performance. IEEE Trans Power Deliv 2007;22:2534–40.
[110] Devi S, Geethanjali M. Optimal location and sizing determination of distributed generation and DSTATCOM using particle swarm optimization algorithm. Int J Electr Power Energy Syst 2014;62:562–70.
[111] Nejabatkhah F, Li YW, Tian H. Power quality control of smart hybrid AC/DC microgrids: an overview. IEEE Access 2019;7:52295–318.
[112] Gorla NBY, Kolluri S, Chai M, Panda SK. A comprehensive harmonic analysis and control strategy for improved input power quality in a cascaded modular solid state transformer. IEEE Trans Power Electron 2018;34:6219–32.
[113] Phan D-M, Lee H-H. Interlinking converter to improve power quality in hybrid AC–DC microgrids with nonlinear loads. IEEE J Emerg Sel Top Power Electron 2018;7:1959–68.
[114] Lu J, Nejabatkhah F, Li Y, Wu B. DG control strategies for grid voltage unbalance compensation. In: 2014 IEEE energy conversion Congress and Exposition (ECCE): IEEE; 2014. p. 2932–9.
[115] Zhu X, Xia M, Chiang H-D. Coordinated sectional droop charging control for EV aggregator enhancing frequency stability of microgrid with high penetration of renewable energy sources. Appl Energy 2018;210:936–43.
[116] Xu Y, Zhang J, Wang W, Juneja A, Bhattacharya S. Energy router: architectures and functionalities toward energy internet. In: 2011 IEEE International Conference on smart grid communications (SmartGridComm): IEEE; 2011. p. 31–6.
[117] Chakrabortty A. Infusing autonomy in power distribution networks using smart transformers. In: 2017 IEEE Conference on control technology and applications (CCTA): IEEE; 2017. p. 1110–6.
[118] Palizban O, Kauhaniemi K. Hierarchical control structure in microgrids with distributed generation: Island and grid-connected mode. Renew Sustain Energy Rev 2015;44: 797–813.
[119] Cabrane Z, Kim J, Yoo K, Ouassaid M. HESS-based photovoltaic/batteries/supercapacitors: energy management strategy and DC bus voltage stabilization. Sol Energy 2021;216:551–63.
[120] Shi J, Gou W, Yuan H, Zhao T, Huang A. Research on voltage and power balance control for cascaded modular solid-state transformer. IEEE Trans Power Electron 2011;26: 1154–66.

[121] Roslan M, Hannan M, Ker PJ, Uddin M. Microgrid control methods toward achieving sustainable energy management. Appl Energy 2019;240:583–607.

[122] Yoo H-J, Nguyen T-T, Kim H-M. Consensus-based distributed coordination control of hybrid AC/DC microgrids. IEEE Trans Sustain Energy 2019;11:629–39.

[123] Jianfang X, Peng W, Setyawan L, Chi J, Hoong CF. Energy management system for control of hybrid AC/DC microgrids. In: 2015 IEEE 10th Conference on industrial electronics and applications (ICIEA): IEEE; 2015. p. 778–83.

[124] Diaz NL, Dragičević T, Vasquez JC, Guerrero JM. Intelligent distributed generation and storage units for DC microgrids—a new concept on cooperative control without communications beyond droop control. IEEE Trans Smart Grid 2014;5:2476–85.

[125] Qoria T, Gruson F, Colas F, Kestelyn X, Guillaud X. Current limiting algorithms and transient stability analysis of grid-forming VSCs. Elec Power Syst Res 2020;189:106726.

[126] Ahmad H, Ahmad A, Ahmad S. Efficient energy management in a microgrid. In: 2018 International Conference on power generation systems and renewable energy technologies (PGSRET). IEEE; 2018. p. 1–5.

[127] Gerber DL, Ghatpande OA, Nazir M, Heredia WGB, Feng WE, Brown R. Energy and power quality measurement for electrical distribution in AC and DC microgrid buildings. Appl Energy 2022;308:118308.

[128] Huber JE, Kolar JW. Volume/weight/cost comparison of a 1MVA 10 kV/400 V solid-state against a conventional low-frequency distribution transformer. In: 2014 IEEE energy conversion Congress and Exposition (ECCE): IEEE; 2014. p. 4545–52.

Chapter 4

Exploiting electric vehicles' benefits within household nanogrids

4.1 Introduction

Reducing global carbon dioxide emissions to net zero by 2050 has become the common goal of humankind, which requires all countries to strengthen significantly and then successfully implement their energy policies and carbon reduction measures. In recent years, a consensus scheme is the rapid promotion of renewable energy sources (RESs), particularly solar and wind energy, which provides a green electricity power supply without increasing carbon dioxide emissions. In the transportation sector, electric vehicles (EVs) have received widespread attention and application. It's estimated that the stock of EVs would exceed 250 million worldwide by 2030 [1], exhibiting a significant energy-saving and carbon mitigation potential. With the increasing penetration of the RESs in distribution systems, the grid power demand from individual customers could be relieved, along with an enhanced power supply reliability facing utility outages [2]. However, the decentralization of energy generation places a new burden on the distribution grid, which must now manage bidirectional and highly intermittent power flows. A series of challenges are also encountered in RESs-integrated distribution systems. Specifically, owing to the intermittent, uncertain, and random nature of RESs, a frequent fluctuation of feeder power would happen if no balance measures were taken, easily leading to voltage violation and system instability [3].

The microgrid has been regarded as an effective solution to integrate and manage intermittent RESs. According to the official definition of a microgrid from the US Department of Energy, a microgrid is deemed as a cluster of interconnected loads and distributed energy resources with clearly defined electrical boundaries that act as a single controllable entity with respect to the grid [4]. A microgrid could connect or disconnect from the utility grid, thus enabling either a grid-connected or islanded mode operation. The microgrid is generally served to single customer sites, such as schools, hospitals, industrial

Smart Penetration of Electric Vehicles in Distribution Grids. https://doi.org/10.1016/B978-0-443-27706-1.00006-4

parks, or data centers, etc. As a localized power system, a microgrid usually comprises distributed generations (DGs), energy storages, various loads, and a corresponding energy management system. The instability and power unbalance issue caused by the integration of RESs could be alleviated through energy storage incorporation and energy management optimization [5]. While effective, the microgrid paradigm may be over-engineered for the single-household scale, where cost, simplicity, and ease of deployment are paramount.

The nanogrid, as an emerging power system, is analogous to the microgrid in terms of the system configuration (generally contains DGs, loads, batteries, and management systems), yet exhibits dedicated characteristics like smaller scale, lower power level, and system complexity. There have been a variety of definitions of nanogrid in use. In Ref. [6], the nanogrid is described as a small-scale system that is of low power and complexity. In Ref. [7], the nanogrid is defined as a single domain for voltage, price, reliability, quality, and administration. Although this definition is of a broad generalization, confusion with power supply structures for electronic products (e.g., universal serial bus, USB) would appear [8]. There also have been a series of researches adopting a simple but effective definition, viz., a nanogrid is defined as a power distribution system for a single house/small building that can be operated in a grid-connected or islanded mode [8–10]. This claim is concise and accurate along with an obvious distinction with microgrid in terms of power hierarchy and system scale. Thus, this definition is used in this manuscript. Compared to the conventional AC nanogrid, the DC nanogrid is of more electricity saving potential, since most domestic appliances and DGs are operated with DC power essentially. Besides, there are no synchronization and reactive power issues, as well as the skin effect and AC losses in DC systems [11]. The DC nanogrid, therefore, presents a natural and efficient platform for integrating native DC sources like photovoltaics (PV) and native DC loads like EVs, minimizing the need for repeated power conversion. Due to the natural advantages over its AC counterpart, there has been an increasing interest in exploring DC nanogrid from academia and industry.

Nanogrid energy management is now being considered as an effective solution to reduce electricity bills and ensure a grid-friendly operation by dispatching nanogrid distributed energy sources and controllable loads properly. For the benefits of the nanogrid and external distribution grid, there are increasing demands toward encouraging local consumption of RESs, smoothing power fluctuation in the tie line between the nanogrid and outer grid, and realizing the peak-shaving and valley-filling of the load profile. In China, the number of private charging points has reached 0.79 million. The EV smart charging would play an essential role in energy management of the nanogrid, since the EV load (typically, several kW to over 10 kW at home) occupies a large proportion of the whole load in nanogrid. This transforms the EV from a mere transportation asset into a pivotal element of household energy infrastructure. Uncoordinated charging would not only burden the distribution grid at peak periods but also bring a series of grid issues, such as increased line loss, voltage fluctuation, and flicker [12]. Conversely, if the EV charging regime is well scheduled by shifting the charging load from peak to valley periods, the negative effect on the distribution system could be mitigated. What's more, the EV could also be controlled to feedback electricity to the power grid, thus providing auxiliary services for the grid, such as active power support, voltage regulation, power backup, and so forth. This bidirectional power exchange between EVs and the power grid is generally known as the vehicle-to-grid (V2G) [13]. It has been demonstrated that the V2G technology has the potential to improve the economy, stability, and reliability of the power grid [14].

For nanogrid, the high penetration of RESs, high-power EV charging, and various electrical appliances like the air conditioner brings significant challenges to the stable, economic, and efficient operation vision. The random power variation of RESs and abrupt load increase or decline that happen over a short time would deteriorate the operation performance, like causing point of common coupling (PCC) power fluctuation, lowering RESs utilization efficiency, and generating extra power demand in peak period. Accordingly, a real-time adjustment capability for nanogrid energy management is essential to address the occurrence of unpredicted incidents.

Nanogrid embraces huge business potential, as a mass of residential customers are inherent and could be updated to receive an efficient, green, and reliable power supply. Owing to the features of small scale and low complexity, a compact, cost-effective, and easy-to-build energy management scheme is favorable for the popularization and application of nanogrids, especially in undeveloped countries and regions. This necessitates a departure from the complex, centralized control paradigms often used in microgrids and calls for an innovative architecture that is both intelligent and inherently simple.

There have been plenty of studies focusing on the design of energy management architecture and corresponding strategies for nanogrids. In general, the common energy management architecture contains centralized, decentralized, distributed, and hierarchical architecture. In the centralized architecture, a central energy management unit is configured to execute the effective power dispatching for multiple agents in nanogrid through high-bandwidth communication channels. Based on this architecture, many researchers made a great contribution to the research of the nanogrid energy management. In Ref. [15], a self-scheduling strategy is proposed for the home power management to decide the optimal operation points of shiftable loads based on a stochastic Mixed-Integer Linear Programming (MILP) model. In Ref. [16], an MILP framework-based evaluation method is proposed for the studied smart household structure considering different demand response strategies. A convex programming problem is formulated in Ref. [17] to decide the related control decisions and parameter settings for a smart home nanogrid with plug-in EVs. Although the desirable operation performance seems to be achieved by the above optimization-based strategies, these strategies generally need to acquire extensive information and lack the real-time adjustment capability to unpredicted conditions. Their performance is inherently constrained by the accuracy of forecasts and is vulnerable to sudden disturbances. Moreover, the computational complexity should be concerned with solving optimization problems. The high complexity would lead to a long computational time, thus a high-performance processor is needed for fast solving.

Compared to the optimization-based energy management strategy, the rule-based strategy is attractive for the application in nanogrid due to relatively simpler implementation with lower computation complexity and real-time adjustment capability. A double-layer rule-based home energy supervision strategy is introduced for a smart home, aiming to flatten the domestic power load curve [18]. The proposed strategy considers diverse factors, such as different electricity prices, multiple EVs, and shiftable household appliances, achieving excellent peak-shaving and valley-filling performance for the load curve. Chen et al. produced a series of energy utilization strategies when the V2G and home distributed photovoltaic (HDPV) is enabled or disabled [19]. It is demonstrated that HDPV-V2G mode can improve the benefits of HDPV. Although the above rule-based energy management strategies could be triggered for execution in real time and respond to unpredicted changes of systems quickly, the desired operation performance may not be satisfied due to lacking considering future conditions. They often settle for suboptimal solutions and may not fully exploit the economic potential of the nanogrid. Notably, both the above optimization-based and rule-based energy management strategies rely on the centralized architecture, which requires the dedicated central controller, extensive measurements, and communication networks. This kind of design would inevitably increase the system cost and complexity. And the communication delay or failure of the central controller would impair the operation performance and stability of systems [20].

For the decentralized energy management architecture, each unit in nanogrids is controlled by corresponding local controllers with only local measured information required. The autonomous coordinated operation is expected to be realized without relying on the central controller and extensive communication links [21]. In Ref. [22], a multisegment adaptive *P/f* droop strategy is proposed to achieve the decentralized control of the PV/battery hybrid unit, thereby realizing a series of operation functions like maximum PV power harvest, battery state of charge (SoC) limitation, and power balance control. A so-called multiple-voltage-region strategy is proposed in Ref. [23] to achieve the autonomous operation of DC systems. By interpreting the voltage in the common DC bus, the different modes could be triggered when reaching the preset DC bus voltage. Although the decentralized architecture is straightforward and effective to realize the real-time coordination, it is difficult to achieve desired system performance without enough global information and overall dispatching capability.

For the distributed energy management architecture, the information is exchanged among neighboring control agents via a sparse communication network, tending to reach an agreement on the control objective relying on the consensus protocol [24]. This architecture only needs neighbor-to-neighbor communication, and is of high scalability. Little research has attempted to use the distributed architecture in nanogrid domain due to the small system scale. The overhead of a communication network for consensus-building is often unjustifiable for a system with so few components. Compared to the decentralized architecture, the distributed architecture needs communication links for system synchronization, and the time delay caused by the mutual communication, as well as the measurement errors would deteriorate the system reliability and convergence performance [25].

The hierarchical energy management architecture contains multiple management levels, in which corresponding control agents handle different operation tasks. The hierarchical architecture is a compatible scheme to take the issues concerned by the centralized, decentralized, and distributed architecture into account, namely, decreasing the computation and communication burden of central management units and improving system operation performance. The conventional hierarchical architecture usually has three control levels, and it is designed following the microgrid prototype [26]. Yet, this frame may not be a wise solution for nanogrid applications when the deviations of operation objectives, control complexities, and cost requirements of the microgrid and nanogrid are concerned. A detailed matching degree assessment of the conventional hierarchical architecture with DC microgrid and DC nanogrid is shown in Table 4.1. This table lists the features of the conventional hierarchical architecture in terms of control functions in different levels, architecture complexity, and architecture cost.

For control functions of the conventional hierarchical architecture (as shown Fig. 4.1a), a critical objective of the primary level control for microgrid is to realize the autonomous power sharing among multiple DGs like gas turbines, thus decreasing the long-distance communication and control demands, and avoiding the overload of some DGs. Yet, due to the small system scale, the single DG located in a limited area is the common configuration of nanogrids, leading to the mismatch with the initial objective of the primary level control. For the secondary level in microgrid, the restoration for the voltage deviation caused by the primary control could be implemented. Additionally, due to the possible existence of multiple power components, the multi-agents coordination would be achieved. However, in nanogrid, due to the disabled functions in the primary level control and fewer components integration, there would be no similar control requests. For the tertiary level, the system energy management is the goal of the microgrid, which is matched

TABLE 4.1 Matching degree assessment of conventional hierarchical architecture with DC microgrid and nanogrid.

Conventional hierarchical Architecture features		DC microgrid features	Match or not	DC nanogrid features	Match or not
Control Level	***Primary level*** • Power/SoC/current Sharing • Current and voltage control	Multiple DGs conducted; for power sharing	Match	Single DG in MPPT mode or other control modes	Mismatch
	Secondary level • Voltage/frequency restoration • Multiple agents coordination	Voltage restoration; Multiple agent control	Match	\	Mismatch
	Tertiary level • Optimal energy management • System supervision	System energy management; Optimal power flow	Match	System energy management	Match
Architecture complexity • Multiple control levels and agents • Relatively higher control complexity • Complex communication links		Various control functions	Match	Relatively few control functions	Mismatch
Architecture cost • Relatively higher cost due to complex architecture		Insensitive to cost because of relatively larger system scale and investment	Match	Sensitive to cost due to small system scale and residential application	Mismatch

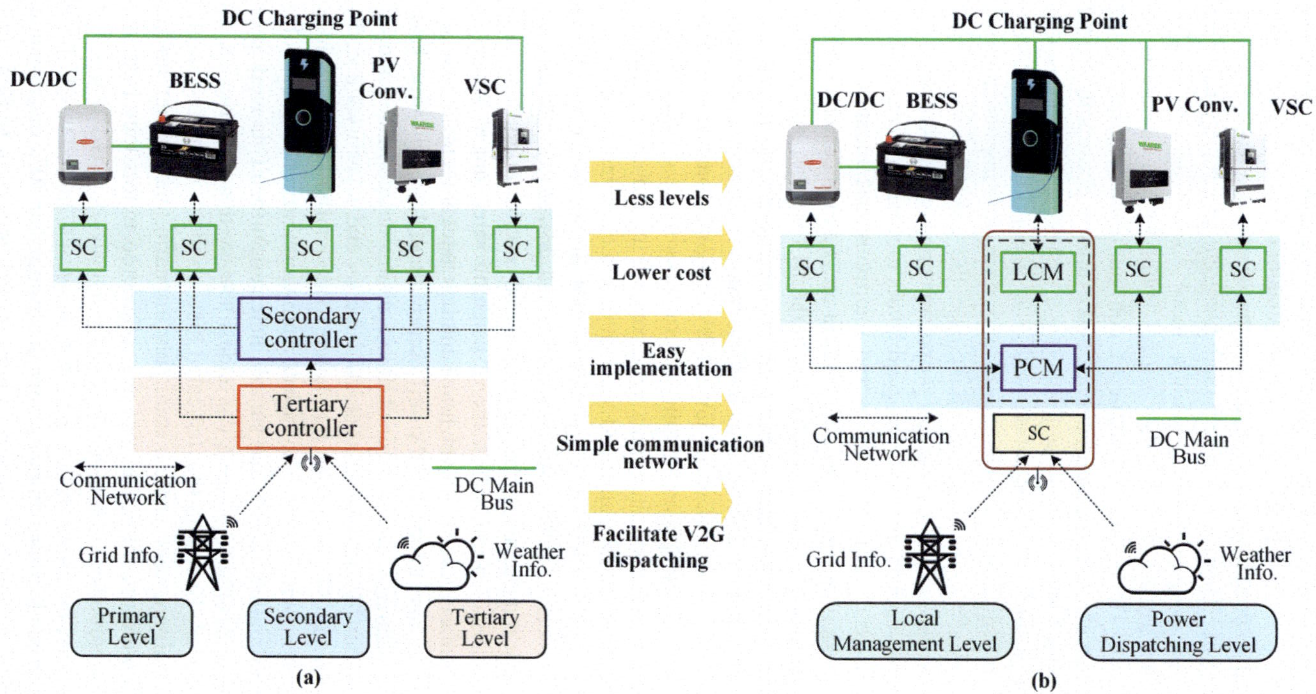

FIGURE 4.1 Diagram of system management architectures. (a) Conventional hierarchical energy management architecture. (b) The proposed pseudo hierarchical energy management architecture.

with the objective of the nanogrid. In addition, in Fig. 4.1a, the direct supervision for components through dedicated communication links could also be concerned in the tertiary level of microgrids. In conclusion, for the architecture complexity, the control functions of the microgrid are more complex than that of the nanogrid. Thus, the conventional hierarchical architecture is more compatible for microgrid via providing more control agents, levels, and communication availability. Another critical factor for assessment is the architecture cost. The relatively high-cost configuration containing multiple management levels, strong control agents, and extensive communication links, is reasonable for the microgrid due to the larger system scale and investments. However, it would be mismatched with the nanogrid, in which the requirements of low cost, compactness, and ease for implementation are preferred.

There are also a few researchers proposing improved hierarchical architectures with diverse control functions and layers to realize the desired control objectives. In Ref. [27], a three-layered energy management hierarchy is introduced to facilitate the V2G operation in an islanded commercial hybrid AC/DC microgrid. The first layer deals with the DC bus voltage regulation by coordinating DC type chargers, and the second and third layer is to realize the regulation of frequency and AC bus RMS voltage. However, this chapter has not dealt with the satisfaction of EV users' charging demand, whereas this point is vital for the widespread application of V2G. In Ref. [18], double-layer home energy supervision strategies are proposed to flatten the power load curves by coordinated load shift and V2G operation. The first layer decides the power and priority of shiftable appliances, and the second layer is used to regulate the power of EVs. Although a flattened power load curve could be achieved, this management mechanism could not respond to the short-timescale load fluctuation within the dispatching intervals. Mahmoud et al. presented a hierarchical two-layer home management system to reduce daily electricity costs and maximize PV self-consumption [28]. The first layer determines the reference settings for shiftable domestic appliances and energy storage using an MILP optimization. And the second layer uses a dedicated controller operated in 1 minute interval to conduct the storage power dispatching to compensate for the power fluctuation at PCC caused by the forecast uncertainties of RES and sample time resolution. This work provides a constructive method to address the undesirable system performance by the online regulation. However, the dedicated measurement, communication link, and control unit are necessitated, inevitably increasing the system complexity and cost. In addition, the EV is also not considered in this literature.

The autonomous operation of EV clusters is a straightforward and efficient solution to provide auxiliary services for the utility grid. In Ref. [29], the proposed sectional droop charging control strategy for EVs can provide frequency regulation for the utility grid in an autonomous way. In Refs. [30,31], an autonomous operation mechanism is proposed for EVs to provide frequency support for the utility grid with a high-penetration wind generation integrated. However, few studies have investigated the possibility of the autonomous V2G operation providing power support for nanogrids.

In this chapter, a pseudo hierarchical energy management architecture and corresponding operation strategy are proposed for DC nanogrids. The notion of "pseudo" is introduced to express that the proposed architecture has similarities with the conventional hierarchical architecture in terms of multiple levels configuration, while it does not belong to the conventional paradigm owing to varied partial control functions and hardware configuration. The major contributions of this study are concluded as:

1. The pseudo hierarchical energy management architecture built upon the smart EV charging point is firstly proposed for DC nanogrid. This architecture incorporates the upper-level central controller with the local power role creatively and comprises the local decentralized management level for the short-timescale power coordination and power dispatching level for the long-timescale energy management. This design features compactness, low cost, and simple system configuration and implementation, which favors the promotion and application in nanogrids. It also facilitates the V2G dispatching only relying on the local controller, thus avoiding the need for dedicated upper-level controllers and corresponding communication links.
2. A state-triggered droop strategy in the local management level is proposed to realize the autonomous and coordinated operation of nanogrids without digital communication links, enhancing the system resiliency and reliability. The autonomous V2G operation is also implemented in nanogrids with the capability to compensate for the unpredicted power unbalance of systems in real time, thus improving the local consumption of RESs and alleviating the peak power demand.
3. A multimode power dispatching strategy in the power dispatching level is introduced to realize the efficient power management of the nanogrid. This strategy is of low computational complexity. Based on the fundamental power coordination executed in the local management level, the dispatching strategy considering comprehensive six operation modes could realize the economic nanogrid operation, the peak-shaving and valley-filling of the nanogrid load curve, and fulfill the expected EV charging demand, accompanied with smooth mode transfer.
4. A detailed simulation model of the DC nanogrid, as well as the corresponding hardware-in-loop (HIL) experiment platform, is built to verify the effectiveness of the proposed pseudo hierarchical energy management architecture and operation strategy.

The remainder of this chapter is organized as follows. It begins with the architecture and overall configuration of the household DC nanogrid system. The subsequent part presents the modeling of the key system components, including the EV battery, the battery energy storage system (BESS), the PV generation, and the load. This chapter then describes the operation strategy of the nanogrid from both the local management level and the power dispatching level. This is followed by simulation studies under multiple operating scenarios, covering both peak and shoulder periods as well as valley periods, together with a comparative analysis. Experimental verification is then provided to further demonstrate the system performance under different operating conditions. This chapter concludes with final remarks.

4.2 System architecture

4.2.1 Typical structure of DC nanogrid system

Recently, China sets to mandate solar on at least 20% of residential roofs in pilot counties, signifying a bright market prospect of nanogrid applications. The schematic diagram of a typical residential DC nanogrid is shown in Fig. 4.2. From the point of the electrical energy flow, this system adopts the unipolar structure that contains two-wire lines to construct the DC main bus to integrate various components. This structure is a typical nanogrid scheme, which exhibits simple implementation with a low cost [32]. The BESS is linked to the DC main bus via a corresponding DC/DC converter for the bus voltage stabilization and power control. The smart DC charging point, which could execute bi-directional power flow control, is also connected to the DC main bus. Additionally, the DC main bus is coupled with PV panels through a corresponding PV converter, as well as other DC loads like air conditioners.

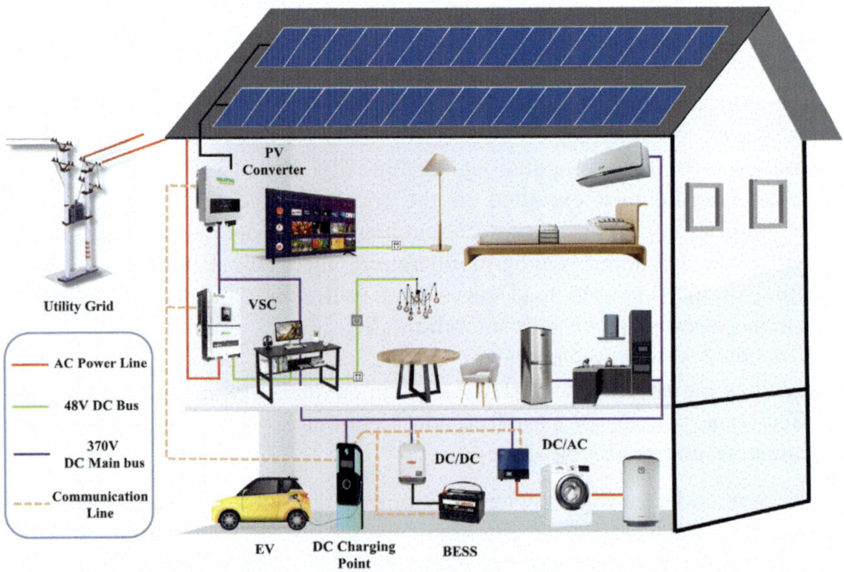

FIGURE 4.2 Schematic diagram of typical residential DC nanogrid.

There are also partial AC loads like the washer that are powered by the DC/AC converter. The voltage source converter (VSC) is tied with the utility grid by DC/AC power conversion. The flexible power exchange between the nanogrid and utility grid is supposed to be achieved.

In terms of the DC standards for residential buildings, some low power and voltage standards like the USB (typically 5 V) have been applied widely. And the single pair Ethernet standard that only uses one pair of wires to transmit data with supporting up to 52 W of DC loads embraces a bright prospect in residential applications. Yet, as for more power and higher voltage level in households, a unified DC standard is needed seriously. So far, the selection of high DC voltage levels is referred to previous experience and existing standards in other domains [32]. Referred to data center applications, the upper bound of the DC voltage in the home distribution is commonly recommended for 400 V [33]. Concerned with the DC voltage variation induced from the nanogrid operation strategy mentioned below and power loss mitigation, 370 V is adopted as the nominal DC bus voltage [34]. Notably, there also exist other lower-voltage DC buses, typically 48V for electronic devices such as lamps and desktop computers.

4.2.2 Pseudo hierarchical energy management architecture

As illustrated in Section 4.1, the significant distinction between nanogrid and microgrid lies in the system scale. For example, microgrid corresponds to larger systems like commercial buildings, industrial parks, schools, hospitals, and so forth, while nanogrids are confined to much smaller systems viz., single houses and small buildings. Unfortunately, this system-scale issue has not been paid sufficient attention yet when considering how to design the energy management architecture and corresponding operation schemes for nanogrids. In most existing research, the sophisticated architecture or operation strategy fitting for microgrids is simply applied to nanogrid in a straightforward way. Although the functions can be achieved, the undesirable performance or so-called mismatch problems (listed in Table 4.1) would inevitably present considerable challenges in practical applications. For example, it's hard to anticipate that a house owner affords a very high cost in dedicated server/controllers with extensive communication capability for the energy management in houses. What is more, especially in China, it is not suitable to set up a control center with a special area in a residential house considering very limited room space.

The compactness, easy-to-build, relatively low cost, and high system performance are basic purposes behind the nanogrid. In this context, we proposed a pseudo hierarchical energy management architecture for nanogrid as illustrated in Fig. 4.1b, which is compatible with the energy-efficient and cost-effective vision of nanogrid. The conventional hierarchical architecture is depicted in Fig. 4.1a. The discussion for the conventional architecture has been completed in Section 4.1. The factors depicted between Figs. 4.1a and b drive the transition of the conventional hierarchical architecture.

A highlighted characteristic of the pseudo hierarchical architecture is the DC charging point deemed as the central unit for nanogrid management. That's to say, the local smart controller (SC) of charging point replaces the role of the external central management unit. In the view of the information flow, the smart DC charging point communicates with essential components in the nanogrid, namely, BESS and corresponding DC/DC, PV converter, and VSC. Particularly, there are two control levels, that is, local management and power dispatching level, in this architecture. The SC of the charging point comprises the local control module (LCM) and power control module (PCM) used for handling different timescale tasks. The local management level deals with the small timescale operation issues of the grid, such as the transient voltage and current control, and autonomous power coordination. And corresponding local control strategies are implemented by the SC or LCM of controlled objects with only local measured information required without mutual communication. In the power dispatching level, the PCM of SC of the charging point is used for the regulated power dispatching of nanogrids in every 15 minutes according to the adopted upper-level energy management strategy. Therefore, only low-bandwidth communication channels are needed between PCM and other SCs or LCM, hence decreasing communication and computation burden. Moreover, the SC of the charging point obtains external useful information like electricity prices and weather forecasts through wireless or other communication mediums, thus providing references for dynamic power dispatching. A detailed comparison between the proposed pseudo hierarchical architecture and conventional hierarchical architecture is shown in Table 4.2. The fewer control levels, agents, and communication links contribute to a smaller occupied area and a lower cost, which are superior for small-scale and cost-sensitive systems.

The reasons for the incorporation of the upper-level central controller with the EV charging point are summarized as below:

TABLE 4.2 **Pseudo versus conventional hierarchical architecture.**

Architecture	Layers	Control agents	Comm. Link	Occupied area	Cost
Conventional hierarchical architecture	3	2 or more	Complex	Large	High
Pseudo hierarchical architecture	2	0	Simple	Small	Low

First, in near future, the EV charging point would be a smart node with flexible information and power interaction capability in the Energy Internet vision, rather than just a passive load. There embrace huge coordinated effects with the utility grid if large-scale charging points could realize the information interaction, namely forming the internet of smart charging points just introduced in Ref. [35]. In this context, the EV charging point conducting energy management for nanogrids facilitates the implementation of V2G services. This is because the V2G effect (desired net power at PCC) is influenced by the power of other devices like roof-top PV and residential appliances in nanogrid, thus if the EV charging point could get the essential information like PV output, it could adjust the V2G operation performance directly with more efficiency and simplicity.

Second, the EV charging point as the central management unit facilitates the energy management for the nanogrid via the direct power dispatch for the EV's charging/discharging behavior. This is because the EV load occupies a large proportion of the whole load in nanogrid, thereby playing an essential role in the overall management of nanogrid [2].

Third, this scheme holds advantages in low cost, low system complexity, and easy-to-build feature, enabling a simple and compact management frame. It is more straightforward and reliable for the charging point to make and execute the dispatching decision locally without relying on external communication links and controllers considering the fast response requirement and complexity of V2G services.

4.3 System configuration and modeling

4.3.1 DC nanogrid essential configuration

The time of use (TOU) electricity price has been widely regarded as an effective way to realize the demand responses in many regions [36]. Shenzhen is one of the representative cities in China. The characteristics of general household appliances used in Shenzhen, summer, are shown in Table 4.3, with details available in Ref. [37]. The predicted conventional load curve in the studied home and TOU pricing curve of Shenzhen are depicted in Fig. 4.3. The valley, shoulder, and peak periods are contained in the TOU pricing scheme in Shenzhen. Specifically, in the valley period 0:00 to 8:00, the power consumption like EV charging is encouraged. Yet, it's hoped to alleviate the load pressure by load shifting or load shedding in peak period 14:00 to 17:00 and 19:00 to 22:00. The nanogrid configuration should take different electricity prices into account for a more economic and efficient operation. The related design principles are listed as below:

1. The PV output is supposed to cover the load consumption during the day, and the BESS is regarded as an energy backup device for power balance. The output proportion of PV panels with 1-kW peak power characteristic in

TABLE 4.3 Characteristics of the household appliances.

Appliance	Operation periods	Power rating(W)	Appliance	Operation periods	Power rating(W)
Bedroom air conditioner	9 hours 45 minutes	2200×2	Hair dryer	15 minutes	1800
Lighting	5 hours 30 minutes	100	Desktop computer	4 hours15 minutes	150
TV	4 hours 15 minutes	80	Dishwasher	1 hour 15 minutes	2000
Electric oven	1 hour 15 minutes	2400	Laundry machine	2 hours	1400
Microwave oven	30 minutes	1200	Central air conditioner	4 hours 45 minutes	4000
Kitchen ventilator	1 hour	230	Heater	2 hours	2000
Refrigerator	24 hours	150	Rice cooker	1 hour 30 minutes	700
Electric iron	15 minutes	1500	Telephone	24 hours	5
Electric kettle	30 minutes	1500	Other fixed loads	24 hours	50
Vacuum cleaner	1 hour	2000			

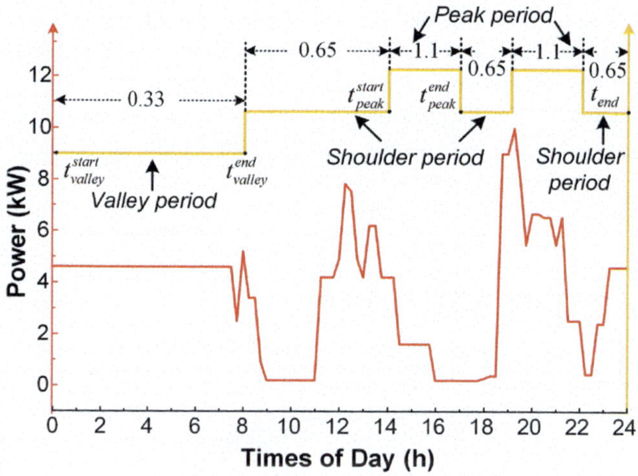

FIGURE 4.3 Predicted conventional load of the studied home and TOU electricity prices in Shenzhen.

Shenzhen is shown in Fig. 4.4, with details available in Ref. [35]. According to the average load consumption (30 kWh) during the day (8:00 to 19:00) in Fig. 4.3, 8 kW PV capacity could meet the load demand. We prefer 10 kW PV capacity (=10 kW) for reserving a certain margin.

2. The BESS is responsible for the power supply at night (19:00 to 24:00, except the valley period). The sum of load consumption in this period is 24 kWh. For the safety and battery use life considered, the allowed SoC range is set to 10%–90%. Additionally, considering the battery efficiency (assumed 0.95) and capacity reservation, the ultimate BESS capacity is designed to be 33 kWh.

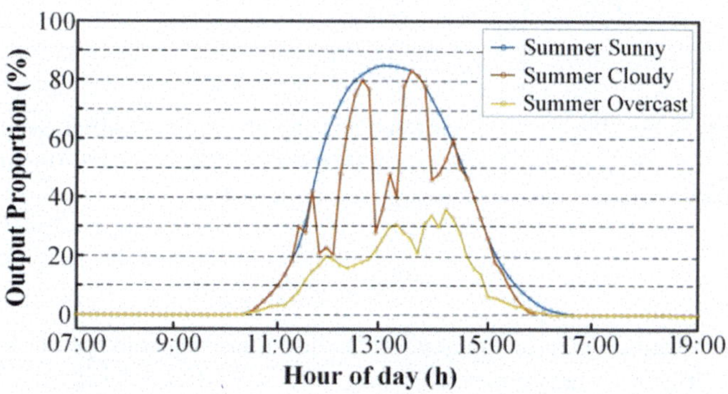

FIGURE 4.4 PV output proportion in Shenzhen during summer.

3. The EV load is not included in the above load calculations due to its random feature, high power, and bidirectional power flow capability (if V2G enabled). It's desirable for EVs to be charged in the valley and shoulder period as far as possible, and even deliver power to the nanogrid through proper power scheduling. This design would decrease the PV and storage capacity in nanogrid from an economic consideration and leave a certain margin for the V2G operation.

4.3.2 EV battery and BESS modeling

In the scope of electrochemical storage, the Li-ion battery has been widely applied in various fields like electric transportation and renewable energy generation deriving from significant energy and power density, long lifecycle, and lightweight nature [6]. The EV battery and BESS in this chapter adopt Li-ion battery, and corresponding physical behavior is described by a modified Shepherd curve-fitting model wherein a polarization voltage is taken into account in charging/discharging expression to reflect the influence of SoC variation on battery performance [38]. The battery terminal voltage is decreased as:

$$U_{Bct} = E_0 - Pol_{rec} \cdot i^* - K \cdot \frac{Q}{Q - \int_0^t idt} \cdot \int_0^t idt + A \cdot \exp\left(- B \cdot \int_0^t idt \right) \tag{4.1}$$

where Q is the total battery capacity, i^* is the battery current after filtering, i is the real battery current, and signify the open-circuit voltage and terminal voltage, respectively. K means the polarization voltage, A and B denote the exponential voltage and exponential capacity, respectively. Pol_{res} denotes the polarization resistance that varies in different battery states. The equation is expressed as:

$$Pol_{rec} = \begin{cases} K \dfrac{Q}{Q - \int_0^t idt} & i > 0, \text{discharging} \\ K \dfrac{Q}{\int_0^t idt + 0.1Q} & i < 0, \text{charging} \end{cases} \tag{4.2}$$

The SoC is another critical parameter that reflects the available capacity of batteries while cycling [9]. SoC_0 refers to the initial SoC value. By method of the Coulomb counting, the SoC value is given in Ref. [39]:

$$SoC = SoC_0 - \frac{1}{Q} \int_0^t idt \tag{4.3}$$

Battery lifetime can be optimized by maintaining appropriate operating conditions. Excess charging/discharging rate would accelerate the capacity degradation, increase the internal resistance, and produce more heat while cycling, thus imposing a negative effect on battery lifespan and stable operation [40]. According to Ref. [41], 1/3 C is adopted for the charging/discharging power limit of BESS and EV battery (when plugged in the charging point). Therefore, for BESS, the maximum discharging/charging power is set to 10 kW. For EV used is a Qing EV BYD with 60 kWh battery capacity and 500 V battery nominal voltage. The maximum power of the DC charging point is 15 kW, which is within the EV battery power limitation.

4.3.3 PV and load modeling

As the detailed PV and load model is not the focus of this study, thus they are treated as two equivalent current sources. As discussed before, the PV and load power are obtained from the real data. The corresponding current references are expressed in Eqs. (4.4) and (4.5):

$$I_{PV}^{ref} = \frac{P_{PV}^{ref}}{U_{bus}} \eta_{pv} \eta_{Cov} \tag{4.4}$$

$$I_{\text{load}}^{ref} = \frac{P_{\text{load}}^{ref}}{U_{bic}} \eta_{Cov} \tag{4.5}$$

where I_{PV}^{ref} and I_{load}^{ref} are the reference current of PV output and load, η_{pv} respectively. Denotes the PV efficiency. η_{Cov} denotes the efficiency of converters. $P_{\text{load}}^{ref} / P_{PV}^{ref}$ means the load/PV power reference.

4.4 Operation strategy

The overall nanogrid operation strategy frame for the pseudo hierarchical energy management architecture is shown in Fig. 4.5, where the main circuit topology on the top, the local control strategies for the nanogrid operation in the middle, and the block diagram of the proposed power dispatching strategy at the bottom are drawn. There are reference signals, feedback signals, and gating signals exchanged between layers in different timescales. Notably, the value of $\boldsymbol{M}_{VSC}$, $\boldsymbol{M}_{DC/DC}$, and $\boldsymbol{M}_{DAB}$ determines different control modes of related components.

As the circuit is shown in Fig. 4.5e, the BESS is connected to the DC main bus through a conventional bidirectional DC–DC converter. This DC–DC could sustain the voltage of the DC main bus by autonomous power balancing or absorb/export a certain amount of power from/to the bus according to the reference signal. Concerning the power topology of the DC charging point, the bidirectional power flow capability, galvanic isolation for safety concerns, and high power density are preferred [42]. Thus, the dual active bridge (DAB) is

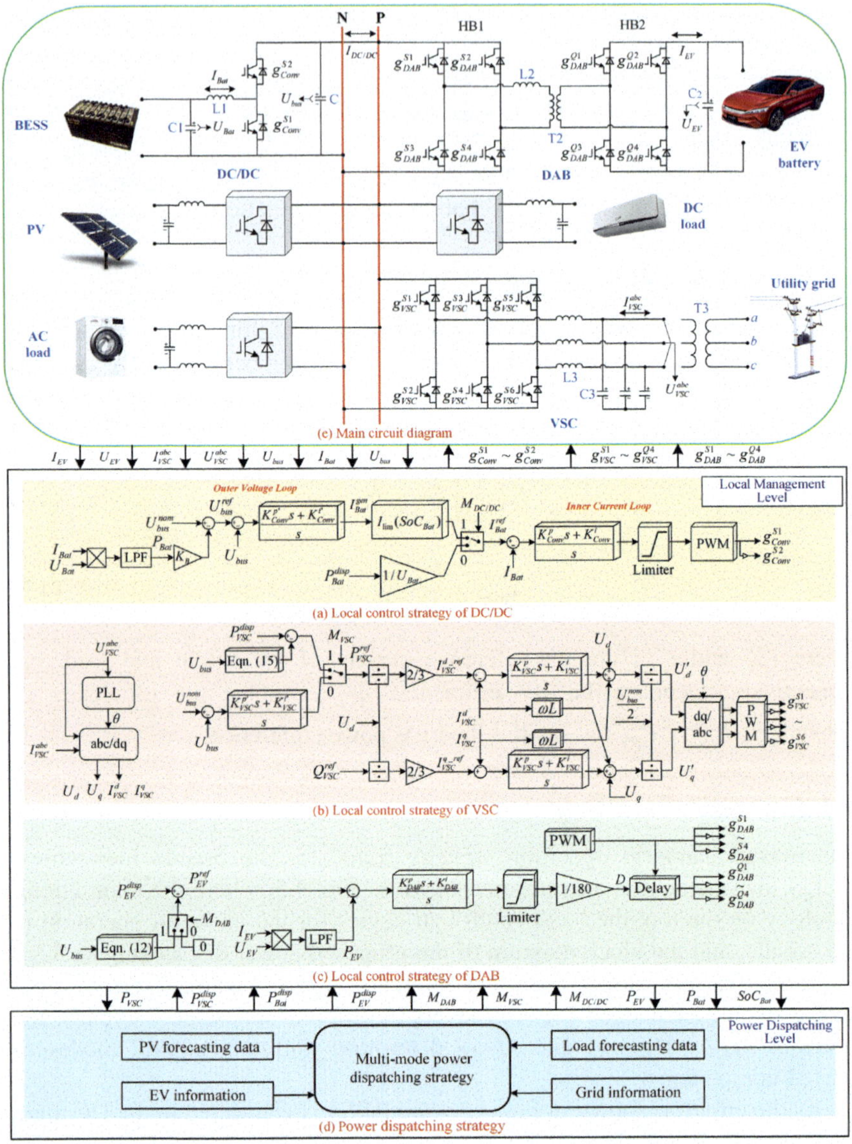

FIGURE 4.5 The nanogrid operation strategy frame for the pseudo hierarchical energy management architecture.

adopted due to high-frequency isolation oriented to higher power density merit and transition efficiency. A two-level three-phase VSC is used to interface the utility grid. The 170 V/380 V transformer is utilized for the galvanic isolation, harmonic elimination, and voltage match.

4.4.1 In local management level

For nanogrid, the local management level aims to generate the needed voltage/current for the power supply and realize the real-time power balance and coordination. In this level, the critical components for the fundamental coordination contain the BESS, EV, and VSC. Owing to the varied operation objectives, there are two scenarios to be concerned for the strategy design, namely, the shoulder and peak period, and the valley period.

During the shoulder and peak periods, a decentralized state-triggered control strategy is introduced at the local management level to promote the economic operation of the nanogrid, improve the local consumption of RESs, and reduce the peak load. Essentially, the decentralized control utilizes the global information of systems for automatic coordination like the frequency and voltage, which can be controlled in relation with the power output of energy resources and be shared simultaneously among independent devices without communication required. Accordingly, by dividing the DC bus voltage appropriately, the BESS, EV (if plugged in), and VSC could participate in or exit the coordinated operation actively in a certain order and adjust corresponding power output when the preset bus voltage is reached. According to Fig. 4.6, the operation states of essential components in the nanogrid are classified by multiple bus voltage levels. (U_{L1},U_{H1})/(U_{H1},U_{H2}) and (U_{L2},U_{L1})/(U_{H2},U_{H3}) and (U_{L3},U_{L2}) represents the BESS/EV/VSC operation range that corresponding power output varies with the changes of DC bus voltage, respectively.

4.4.1.1 BESS operation

The BESS is used to compensate the fundamental power unbalance (PV power minus the load power) via the autonomous charging/discharging behavior in nanogrid under certain range. As shown in Fig. 4.6a, the operation point of the BESS(P_{Bat}, U_{bus}^{ref}) will vary along with the red line according to the current load and PV power. Notably, it's feasible that the operation point locates in the solid red line or the dotted red line, this selection is determined by the maximum charging/discharging power limits $P_{C1}^{adj} \big/ P_{D1}^{adj\_max}$. The discharging/charging behavior of the BESS signifies a positive/negative reference power flow direction. Three working states of BESS as per different voltage ranges are summarized as below:

- State 1: $U_{L1} \leq U_{bus} \leq U_{H1}$.

In this range, the BESS would be charged/discharged based on its adaptive droop characteristic. The applied adaptive U–P droop control strategy for the BESS DC/DC converter is expressed as follows:

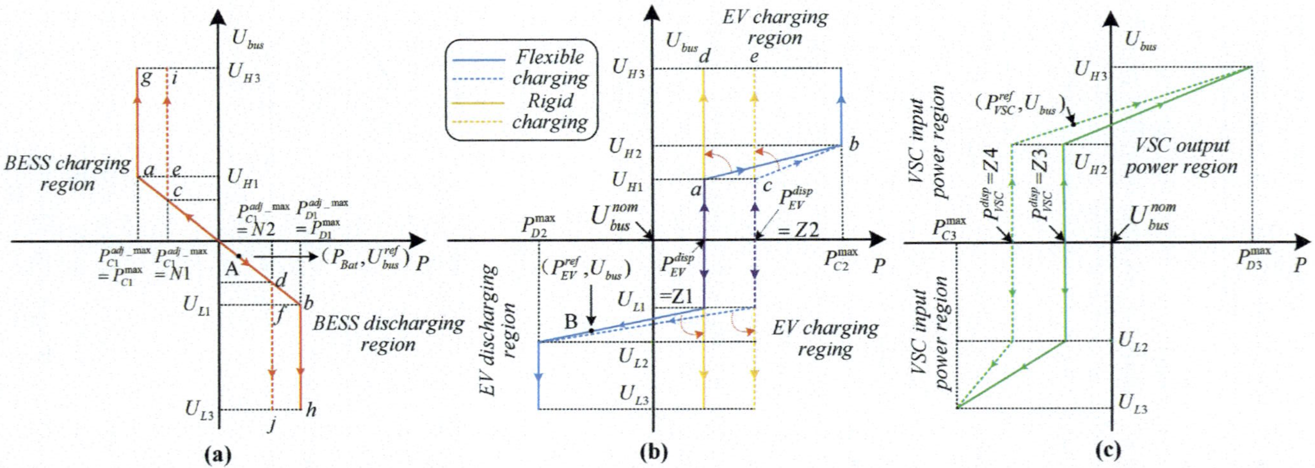

FIGURE 4.6 Operation state division based on DC bus voltage: (a) BESS operation curve; (b) EV charging/discharging power curve; (c) VSC operation curve.

$$U_{bus}^{ref} = U_{bus}^{nom} - k_B P_{Bat}, k_B = \frac{U_{H1} - U_{L1}}{P_{D1}^{max} - P_{C1}^{max}} \tag{4.6}$$

Obviously, the DC bus voltage reference will change linearly as the discharging/charging power P_{Bat} varies depicted as the line segment *ab*. In this state, the value of P_{Bat} depends on the current power unbalance value in nanogrid. Notably, the BESS operates in point a or b indicating that the BESS reaches the maximum charging or discharging power boundary P_{C1}^{max}/ P_{D1}^{max}. Meanwhile, to avoid a high depth of charging or discharging, $P_{C1}^{adj\_max}$/ $P_{D1}^{adj\_max}$ would be adjusted to other values like N1/N2 when the BESS SoC approaches the upper/lower bounds. This variation also leads to the operation line *ab* updating to the broken line $cd + ce + df$. The operation point in the vertical line *ce* and *df* indicates that the BESS power reaches the current maximum value and does not change with the U_{bus} variation. The value selection of $P_{D1}^{adj\_max} / P_{C1}^{adj\_max}$ is following the below Eqs. (4.7) and (4.8).

$$P_{D1}^{adj\_max} = \begin{cases} P_{D1}^{max}, SoC_{Bat} \geq SoC_{Bat}^{low} \\ P_{D1}^{max}\left(1 - \dfrac{SoC_{Bat}^{low} - SoC_{Bat}}{SoC_{Bat}^{low} - SoC_{Bat}^{min}}\right), SoC_{Bat}^{min} \leq SoC_{Bat} < SoC_{Bat}^{low} \\ 0, SoC_{Bat} < SoC_{Bat}^{min} \end{cases} \tag{4.7}$$

$$P_{c1}^{adj\_max} = \begin{cases} P_{c1}^{max}, SoC_{Bat} \leq SoC_{Bat}^{high} \\ P_{c1}^{max}\dfrac{SoC_{Bat}^{max} - SoC_{Bat}}{SoC_{Bat}^{max} - SoC_{Bat}^{high}}, SoC_{Bat}^{high} \leq SoC_{Bat} < SoC_{Bat}^{max} \\ 0, SoC_{Bat} \geq SoC_{Bat}^{max} \end{cases} \tag{4.8}$$

According to Ref. [43], the maximum SoC value of the BESS is set to 0.9, and the minimum SoC value is set to 0.1. As described in the above equations, the allowable charging/discharging power bound starts to linearly decrease when the SoC value reaches corresponding alarm values/(0.85/0.15). By this adaptive droop method, the autonomous BESS power adjustment concerning the current SoC value could be realized, thus mitigating the possible intense fluctuation of electrical parameters due to the abrupt power limitation when reaching the SoC boundaries.

- State 2: $U_{H1} \leq U_{bus} \leq U_{H3}$ and State 3: $U_{L3} \leq U_{bus} \leq U_{L1}$.

In these two states, the BESS has the maximum charging/discharging power output equaling the value of $P_{C1}^{adj\_max} \Big/ P_{D1}^{adj\_max}$. Moreover, accompanied by the value adjustment of $P_{C1}^{adj\_max} \Big/ P_{D1}^{adj\_max}$, the operation line also changes just like the line segment *ag* and *bh* changing to *ei* and *fj*.

The average model of the bidirectional DC/DC for the BESS is presented in the following equations:

$$\begin{aligned} L1\frac{dI_{\text{Batt}}}{dt} &= U_{\text{Batt}} - (1-\delta_B)U_{bus} \\ C\frac{dU_{bus}}{dt} &= (1-\delta_B)I_{\text{Batt}} - I_{DC/DC} \end{aligned} \tag{4.9}$$

The adopted dual-loop control strategy is illustrated in Fig. 4.5a, in which $M_{DC/DC}$ equals 1 in this scenario, meaning that the upper control loop is executed, and DC/DC works on the *U–P* droop control mode. U_{bus}^{ref} is generated according to Eq.(4.6). Then this voltage reference signal passes through an outer PI controller to generate the initial current reference I_{Bat}^{gen}. Processed by the current limiting block shown as the Eq. $P_{D1}^{adj\_max}, P_{C1}^{adj\_max}$ (4.10) (refers to Eqs. (4.7 and 4.8), the ultimate current reference of BESS is obtained. Furthermore, I_{Bat}^{ref} goes through an inner current loop to generate PWM drive signals.

$$I_{\text{lim}}(SoC_{\text{Bct}}) = \begin{cases} P_{D1}^{\alpha_{\text{cf}}\ \_\text{max}} / U_{\text{Bct}} \\ P_{C1}^{\alpha_{\text{cf}}\ \_\text{max}} / U_{\text{Bct}} \end{cases} \tag{4.10}$$

4.4.1.2 EV charging/discharging operation

As shown in Fig. 4.6b, the EV operation point (P_{EV}^{ref}, U_{bus}) will vary along the purple line, blue line, or yellow line according to the current DC bus voltage that is associated with the current power difference. The blue line segments in voltage range (U_{H1}, U_{H3}) or (U_{L1}, U_{L3}) signify that the EV battery works on the flexible charging mode (P_{EV}^{ref} varies with the U_{bus}), whereas the yellow line signifies the EV battery operates in the rigid charging mode (P_{EV}^{ref} keeps constant). The EV charging signifies the positive power flow direction. The applied *P–U* droop characteristic of the flexible charging mode can be represented as Eqs. (4.11 and 4.12):

$$P_{EV}^{ref} = P_{EV}^{\text{disp}} - P_{sub} \tag{4.11}$$

$$P_{sub} = \begin{cases} P_{EV}^{\mathrm{disp}} - P_{C2}^{\mathrm{max}} \ , U_{bus} > U_{H2} \\ \dfrac{P_{C2}^{max} - P_{C3}^{\mathrm{disp}}}{U_{H2} - U_{H1}} (U_{H1} - U_{bus}) \quad , U_{H1} < U_{bus} \le U_{H2} \\ 0 \ , U_{L1} \le U_{bus} \le U_{H1} \\ \dfrac{P_{EV}^{\mathrm{disp}} - P_{D2}^{\mathrm{max}}}{U_{L1} - U_{L2}} (U_{L1} - U_{bus}) \quad , U_{L2} \le U_{bus} < U_{L1} \\ P_{EV}^{\mathrm{disp}} - P_{D2}^{\mathrm{max}} \ , U_{bus} < U_{L2} \end{cases} \tag{4.12}$$

According to the Eq.(4.12), there are five working states of the EV charging point as per five voltage ranges:

- State 1: $U_{L1} \le U_{bus} \le U_{H1}$.

In this range, the power reference for the EV charging remains constant, which equals the dispatching reference P_{EV}^{disp} from the power dispatching level. Different dispatching references would determine the style of the operation line. For example, in the case of $P_{EV}^{\mathrm{disp}} = Z1$, the operation line is depicted as the straight line, yet when $P_{EV}^{disp} = Z2$, the operation line is shown as the dotted line.

- State 2: $U_{H1} \le U_{bus} \le U_{H2}$ and State 3: $U_{L2} \le U_{bus} \le U_{L1}$.

The DC bus voltage in state 2 or state 3 indicates that the BESS cannot fully cover/consume the power deficit/surplus of the nanogrid, thus resulting in the bus voltage exceeding the region (U_{L1}, U_{H1}). In the flexible charging mode, the EV would conduct the power balance autonomously for extra power deficit or surplus, hence increasing the local consumption of PV, as well as decreasing the imported/exported power from/to the utility grid. With the value change of the power signal P_{EV}^{disp} (e.g., Z1 to Z2) from the dispatching level, the operation curve also varies, just like the line segments *ab* and *cb*.

To guarantee the charging demand of EVs, the flexible charging mode can be converted to the rigid charging mode in some periods (shown as the line segment *ad* and *ce*), thus avoiding the possible power decline due to the droop characteristic. In the rigid charging mode, the P_{EV}^{disp} keeps constant with the pP_{EV}^{disp}.

- State 4: $U_{H2} \le U_{bus} \le U_{H3}$ and State 5: $U_{L3} \le U_{bus} \le U_{L2}$.

In these two periods, the P_{EV}^{ref} keeps constant. In the flexible charging mode, the autonomous power adjustment of the EV charging point has reached the boundary, thus P_{EV}^{ref} equals P_{D2}^{max} in state 4, and in state 5, the P_{EV}^{ref} equals P_{C2}^{max}. In the rigid charging mode, $P_{EV}^{ref} = P_{EV}^{\mathrm{disp}}$ is activated.

The adopted EV charger topology, that is, DAB converter, contains two H-bridges (HB1 and HB2) and a transformer in series with an inductance $L2$, as shown in Fig. 4.5e. Both HBs are operated at a 50% duty ratio, and the phase shift (α) between two HBs determines the transmission power as the voltage difference appears across $L2$ and produces inductance current. The relation between transmission power and the phase shift is given by Ref. [44]:

$$P_{DAB} = \frac{nU_{EV}U_{bus}}{f_s L2}\alpha_{DAB}(1 - \alpha_{DAB}) \tag{4.13}$$

where n is the primary to secondary turns ratio. α_{DAB} represents the phase shift between HB1 and HB2. f_s means the switching frequency.

It can be seen that the power output of the DAB can be controlled by the magnitude and direction adjustment of the phase shift. The applied P–U droop control strategy of the DAB is shown in Fig. 4.5c. In the condition of equaling 1, the control block of the flexible charging mode is activated with achieving P_{EV}^{ref} varying with the bus voltage. When M_{DAB} is set to zero, the control block of the rigid charging mode is enabled with P_{EV}^{ref} keeping constant, namely, the constant power strategy is executed. Then, the generated P_{EV}^{ref} is compared with the filtered power, and the deviation produces phase shift signal D through a PI closed loop.

4.4.1.3 VSC operation

The VSC is used to realize the ultimate power balance autonomously in nanogrid via the power exchange with the utility grid. In Fig. 4.6c, the operation point of the VSC will vary along the green line according to the current DC bus voltage and power difference. The power exported to the utility grid through the VSC signifies the positive power flow direction. The adopted P–U droop characteristic of the VSC is given by:

$$P_{VSC}^{ref} = P_{VSC}^{disp} - P'_{sub} \tag{4.14}$$

$$P'_{sub} = \begin{cases} \dfrac{P_{D3}^{max} - P_{PSC}^{disp}}{U_{H3} - U_{H2}}(U_{H2} - U_{bus}) & , U_{H2} < U_{bus} \le U_{H3} \\ 0 & , U_{L2} \le U_{bus} \le U_{H2} \\ \dfrac{P_{PSC}^{disp} - P_{C3}^{max}}{U_{L2} - U_{L3}}(U_{L2} - U_{bus}) & , U_{L3} \le U_{bus} < U_{L2} \end{cases} \tag{4.15}$$

There are three working states of the VSC as per three voltage ranges:

- State 1: $U_{L2} \le U_{bus} \le U_{H2}$.

In this state, the P_{VSC}^{ref} keeps constant and equals the dispatching reference P_{VSC}^{disp} from the power dispatching level. Different P_{VSC}^{disp} would determine the form of the VSC operation line, just like the straight green line/dotted green line corresponding to $P_{VSC}^{disp} = Z3 \big/ P_{VSC}^{disp} = Z4$, respectively.

- State 2: $U_{L3} \le U_{bus} \le U_{L2}$ and State 3: $U_{H2} \le U_{bus} \le U_{H3}$.

Similar to the EV in the flexible charging mode, the VSC output power is real-time regulated as the voltage fluctuates in state 2 and state 3. The imported/exported power capability of VSC (/) aids to achieve the ultimate power balance of the nanogrid.

The average model of the VSC in *dq* reference frame is given by:

$$\begin{cases} \dfrac{dI_{VSC}^{d}}{dt} - \omega I_{VSC}^{q} = \dfrac{1}{L3}\left(-U_d + p_d U_{bus}\right) \\ \dfrac{dI_{VSC}^{q}}{dt} + \omega I_{VSC}^{d} = \dfrac{1}{L3}\left(-U_q + p_q U_{bus}\right) \end{cases} \tag{4.16}$$

where I_{VSC}^{d} and I_{VSC}^{q} are the output current in *dq* frame. ω is the grid frequency angle, U_{d} and U_{q} are the grid voltage in *dq* frame. p_{d} and p_{q} are the switching space vector in *dq* frame.

The adopted *P–U* droop control strategy of the VSC is shown in Fig. 4.5b. Especially, equals 1 in droop control mode, and the upper control loop is enabled. The generated P_{VSC}^{ref} and Q_{VSC}^{ref} produce the VSC reference current in *dq* frame. P_{VSC}^{ref} is determined by the upper dispatching level and Q_{VSC}^{ref} is set to 0 to achieve a higher power factor. Then *dq* reference $I_{VSC}^{d\_ref}$ and $I_{VSC}^{q\_ref}$ is processed by PI controllers to generate the reference voltage in *dq* frame. Particularly, the cross-coupling term ($I_{VSC}^{d}\omega L$, $I_{VSC}^{q}\omega L$) in VSC model is eliminated to decouple control parameters in *dq* axes, and the feed-forward control is executed to achieve a fast and stable response.

In the valley period, to implement the valley filling of nanogrid loads, the utility grid is regarded as the main source to power the loads and BESS. The DC bus voltage signifies the power balance condition in nanogrid. Thus, the VSC converts to the constant voltage control mode with changing to 0. As shown in Fig. 4.5b, an outer voltage control loop is enabled to generate the active power reference for the inner decoupled current loop, aiming to keep the dc bus voltage constant.

To recover the SoC value of BESS effectively, the DC/DC switches from the droop control mode to the constant power mode. Accordingly, in Fig. 4.5a, $M_{DC/DC}$ is set to 0 and P_{Bat}^{disp} generates P_{Bat}^{ref} after calculation. The ultimate driving signals are generated via the current closed-loop control.

Concerning the EV (if plugged in), autonomous power regulation is not needed in this period. Thus, the EV converts to the rigid charging mode with equaling 0 in Fig. 4.5c. The summary of local control strategies is shown in Table 4.4.

TABLE 4.4 Summary of local control strategies.

	Peak/Shoulder period	Switch state (M_x)	Valley period	Switch state (M_x)
DC/DC	*U–P* droop	1	Constant power	0
DAB	*P–U* droop	1	Constant power	0
	Constant power	0		
VSC	*P–U* droop	1	U_{bus} control	0

4.4.2 In power dispatching level

In the power dispatching level, the proposed multi-mode power dispatching strategy for nanogrid is to improve the system operation performance based on the fundamental power coordination executed in the local management level. The peak load mitigation, economic operation, and the satisfaction of EV charging demand are expected to be achieved. According to Ref. [28], the time of day is divided into 96 time slots equally, that is, $t_{slot} = 1,2, \ldots, 96$. Every time slot is 15 minutes. The timeslot $t_{slot} = 1$ denotes 0:00. The reason for the choice of 15 minutes is summarized as below. First, the selection of 15-minute time slots leads to 96 time periods in a one-day cycle. A tradeoff should be concerned between the fine control and solving efficiency. On one hand, if the time period is too long (e.g., 30 minutes or 1 hour), a coarse dispatching problem may not assure a desirable operation performance. On the other hand, if the time period is too short (1 minute or 5 minutes), the generated considerable variables would increase the difficulty in solving the power dispatch problem. What's more, the autonomous operation mechanism in the local management level enables certain adaptability of nanogrid to unpredicted changes of system parameters. Accordingly, the too short time slots are not necessary in our proposed architecture. In every 15 minutes, the PCM of SC of the charging point will update new settings for the local management level, e.g., the power references and strategy switch signal. This referred information contains the predicted and current PV and load power, EV charging demand, and grid data like current electricity price, as shown in Fig. 4.5d.

The flowchart of the multi-mode power dispatching strategy is depicted in Fig. 4.7. The strategy contains six modes, and each mode is combined with one or multiple dispatching modules. These modules are listed below: EV charging scheduling (ECS), EV valley charging, VSC energy compensation (VEC), BESS valley charging (BVC), and no energy dispatching (NED) mode. Modes 1 and 4 are designed for the valley period, while the others are dedicated for the shoulder or peak period:

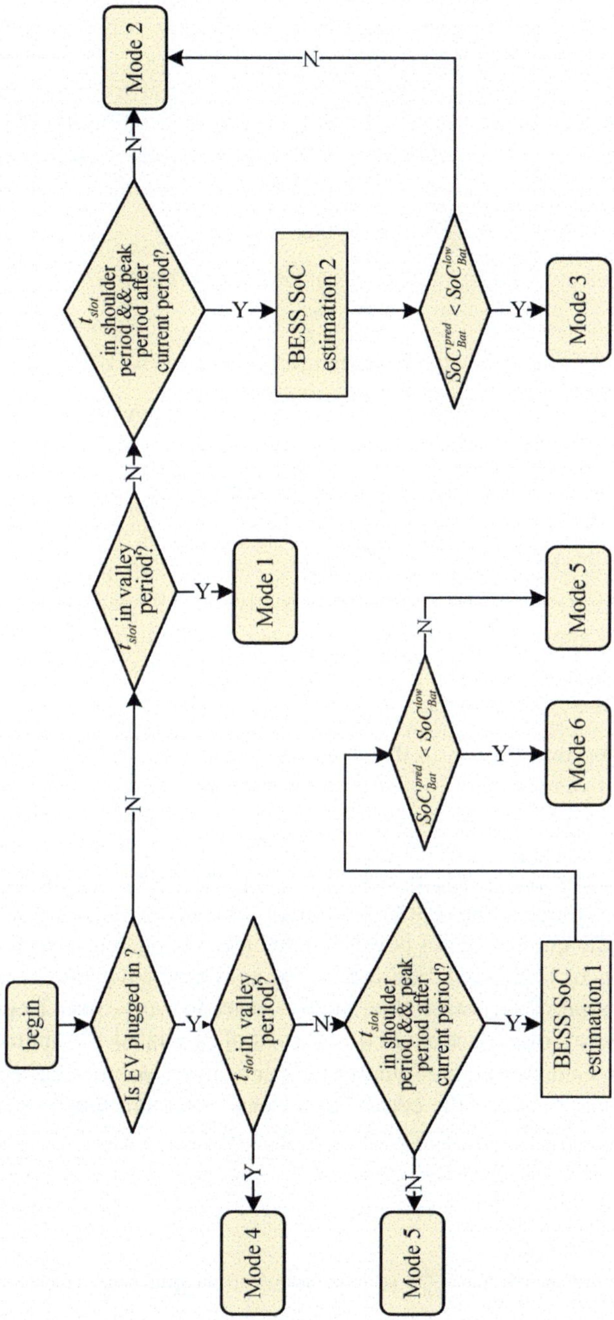

FIGURE 4.7 Flowchart illustrating the multimode power dispatching strategy.

- Mode 1 (M-1): BVC mode

In this mode, the nanogrid operates in the valley period, and the VSC stabilizes the DC bus voltage to cover all the power consumption. The BESS remains charging to receive the power reference signal from the power dispatching level. This reference signal can be decided based on the desired SoC of BESS at the end of the valley period. The related expression is represented as:

$$P_{Bat}^{\mathrm{disp}} = \frac{4\left(SoC_{Bat}^{\mathrm{valley}\_exp} - SoC_{Bat}\right)E_{Bat}}{\left(t_{\mathrm{valley}}^{end} - t_{\mathrm{slot}}\right)} \tag{4.17}$$

where $SoC_{Bat}^{\mathrm{valley}\_exp}$ denotes the expected SoC value of the BESS at the end of the valley period. t_{slot} denotes the current time slot.

- Mode 2 (M-2): NED mode

In this mode, the nanogrid conducts the autonomous power coordination only relying on the local management level, and the power dispatching reference for the VSC is zero as follows:

$$P_{VSC}^{\mathrm{disp}} = 0 \tag{4.18}$$

- Mode 3 (M-3): VEC mode

When the current time is in the shoulder period, and the next period is the peak period, it is necessary to estimate the SoC curve of the BESS in future time slots, thus making an energy reservation in advance to avoid the power imported during the peak period. If the BESS SoC would be below the low alarm value according to the predicted results, the VEC is executed for the power supply. The developed SoC estimation flowchart 2 is shown in Fig. 4.8.

In step 1, the current BESS power is obtained, and it's equal to the current power difference. Thus, the BESS SoC in the next timeslot could be estimated. The energy variation in future time slots is reset to zero. Step 2 aims to estimate the future SoC curve based on a series of predicted PV output *pre_EV (t')*, predicted load power *pre_load(t')* at future time slot t'. The sum of the power difference between the current timeslot to the end timeslot of the peak period is obtained by iterative calculation according to predicted information. Concerning step 3, the predicted SoC value of BESS at the timeslot is derived. Notably, as $SoC_{Bat}^{\mathrm{pred}}$ is just used for calculation, it may be less than, even be a negative value SoC_{Bat}^{low}.

After the value of $SoC_{Bat}^{\mathrm{pred}}$ is calculated, a comparison with SoC_{Bat}^{low} is conducted in the flowchart in Fig. 4.7. When $SoC_{Bat}^{\mathrm{pred}}$ is less than SoC_{Bat}^{low}, a

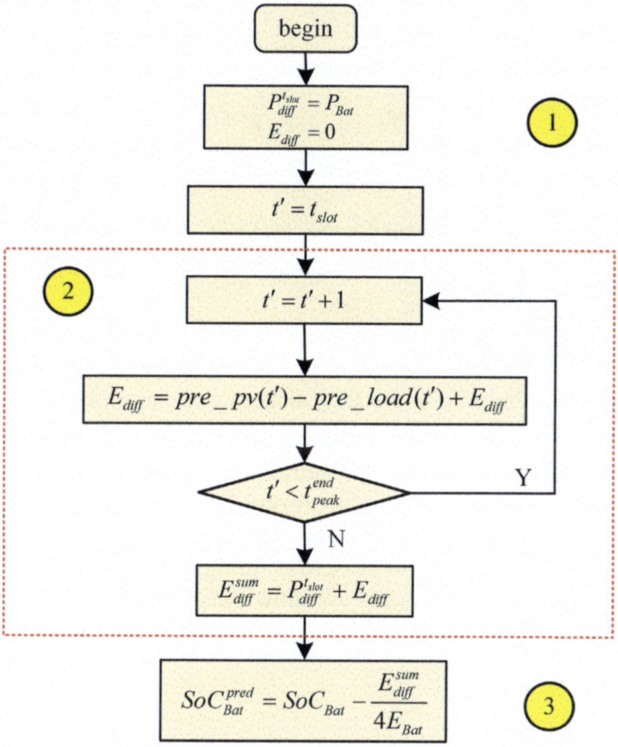

FIGURE 4.8 The developed BESS SoC estimation flowchart 2 when EV plugged out.

power compensation is executed. The dispatching reference for the VSC is illustrated as Equ. 4.19. When SoC_{Bat}^{pred} is larger than SoC_{Bat}^{low}, the BESS is predicted to have enough capacity for power supply in the next peak period. Thus, mode 2 is enabled with no power dispatching for VSC.

$$P_{VSC}^{\text{disp}} = -\frac{4\left(SoC_{Bat}^{low} - SoC_{Bat}^{\text{pred}}\right)E_{Bat}}{t_{\text{peak}}^{\text{start}} - t_{slot}} \tag{4.19}$$

- Mode 4 (M-4): BESS and EV valley charging (BEVC) mode

Similar to mode 1, the valley charging is executed for the BESS and EV. The power reference for BESS is consistent with the Eq. And the reference for the EV charging is expressed as follows:

$$
\begin{aligned}
P' &= \max\left(\frac{4\left(SoC_{El'}^{\exp} - SoC_{El'}\right)E_{El'}}{\left(t_{El'}^{\text{dep\_calj}} - t_{slot}\right)}, 0\right) \\
P_{El'}^{\text{disp}} &= \min\left(P', P_{C2}^{\max}\right) \\
t_{El'}^{dep\_\text{calj}} &= \min\left(t_{El'}^{dep}, t_{\text{valley}}^{end}\right)
\end{aligned}
\tag{4.20}
$$

where P_{EV}^{disp} is limited to the range of (0, P_{C2}^{max}). According to the departure time of the EV, the intermediate variable $t_{EV}^{dep\_adj}$ is set to limit the valley charging time.

- Mode 5 (M-5): ECS mode

In peak or shoulder period, the ECS aims to make a reasonable dispatching value P_{EV}^{disp}, to avoid the power imported in peak period and satisfy the EV user's charging demand. The overall flowchart of the proposed algorithm is shown in Fig. 4.9. Due to a large proportion of EV charging in nanogrid load, there would be a possible scenario that the whole load exceeds the BESS power limitation, thus leading to peak load. In light of this, the ECS algorithm is to find the overloaded time slots and shift the extra power demand to other low power periods. The overall flowchart of the proposed algorithm is shown in Fig. 4.9.

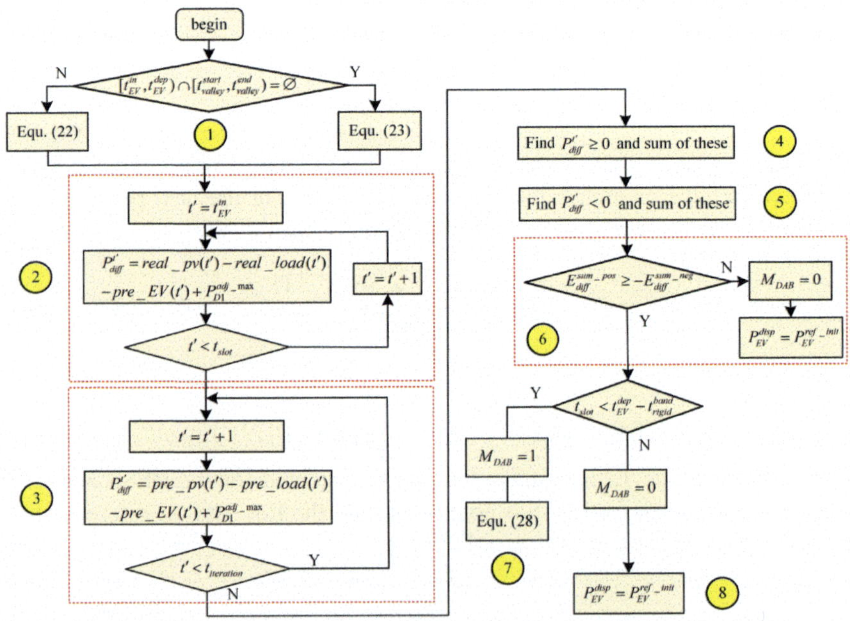

FIGURE 4.9 Overall flowchart of the EV charging scheduling algorithm.

Step 1 aims to generate the EV initial reference $P_{EV}^{ref\_init}$. When the EV plugged-in period $\left[t_{EV}^{in}, t_{EV}^{dep}\right]$ overlaps with the valley period $\left[t_{valley}^{start}, t_{valley}^{end}\right]$, the valley EV charging could be executed, and the charging power in the shoulder or peak period could be decreased. Assumed that the EV is charged with the maximum power P_{c2}^{max} in the valley period, the EV valley charging capacity $E_{EV}^{valley\_ava}$ is described as:

$$E_{EV}^{valley\_ava} = \min\left(P_{C2}^{max}\left(t_{EV}^{dep\_adj} - t_{valley}^{start}\right)/4, E_{EV}\right)$$
$$t_{EV}^{dep\_adj} = \min\left(t_{EV}^{dep}, t_{valley}^{end}\right) \tag{4.21}$$

where $t_{EV}^{dep\_adj}$ is the intermediate variable related to t_{EV}^{dep}.

When the EV valley charging would be executed, the charging reference may be zero in the peak or valley period. In this condition, the fundamental EV charging reference $P_{EV}^{ref\_init}$ is decreased as:

$$E_{EV}^{need} = E_{EV}\left(SoC_{EV}^{dep} - SoC_{EV}\right)$$
$$P_{EV}^{ref\_init} = 4^{*}\, max\left(E_{EV}^{need} - E_{EV}^{valley\_ava}, 0\right)/(t_{end} - t_{slot}) \tag{4.22}$$

When the EV plugged-in period does not overlap with the valley period, $P_{EV}^{ref\_init}$ is derived as:

$$P_{EV}^{ref\_init} = 4 * E_{EV}^{need} / \left(t_{EV}^{dep} - t_{slot}\right) \tag{4.23}$$

where $P_{EV}^{ref\_init}$ determines the value of *Pre_EV(t′)*. E_{EV}^{need} denotes the remaining EV charging demand at t_{slot}.

Step 2 is to obtain the discharging capacity of the BESS in previous time slots and the current timeslot. In this condition, if the power difference is less than zero, it is indicated that the BESS would reach the maximum power output if the EV is charged with the power $P_{EV}^{ref\_init}$, and the power supply from the utility grid is needed. Yet, the positive $P_{diff}^{t'}$ indicates that there remains discharging power capacity of the BESS in t' timeslot. In fact, the negative power $P_{diff}^{t'}$ is offset by the autonomous power decline of the EV charging executed in the local management level when the EV operates in the flexible charging mode. Moreover, the decreased EV charging power should be compensated in time slots that $P_{diff}^{t'}$ is a positive value.

Step 3 aims to acquire the discharging capacity of the BESS in future time slots by iterative calculation according to the predicted information. The value selection of the iteration parameter $t_{iteration}$ follows the Eq. (4.24). Obviously, the valley period is not included in the iteration process. The maximum value of equals the last timeslot.

$$t_{iteration} = \begin{cases} t_{EV}^{dep} & \left[t_{EV}^{in}, t_{EV}^{dep}\right) \cap \left[t_{\text{valley}}^{\text{start}}, t_{\text{valley}}^{end}\right) = \varnothing \\ t_{end} & \left[t_{EV}^{in}, t_{EV}^{dep}\right) \cap \left[t_{\text{valley}}^{\text{start}}, t_{\text{valley}}^{end}\right) \neq \varnothing \end{cases} \tag{4.24}$$

Step 4 is to find the time slots with $P_{\text{diff}}^{t'}$ large than zero and sum all of these, thus obtaining the BESS remaining capacity $E_{\text{diff}}^{sum\_pos}$, just written in Eq. (4.25).

$$E_{\text{diff}}^{sum\_pos} = \sum_{t' \in \left[t_{EV}^{in}, t_{\text{iterction}}\right]} P_{\text{diff}}^{t'}, where P_{\text{diff}}^{t'} \geq 0 \tag{4.25}$$

Step 5 is to find the time slots with $P_{\text{diff}}^{t'}$ smaller than zero, and sum all of these, thus obtaining the BESS deficit capacity $E_{\text{diff}}^{sum\_neg}$.

$$E_{\text{diff}}^{\text{sunn}\_neg} = \sum_{t' \in \left[t_{EV}^{in}, t_{\text{iteration}}\right]} P_{\text{diff}}^{t'}, where P_{\text{diff}}^{t'} < 0 \tag{4.26}$$

In step 6, if $E_{\text{diff}}^{sum\_pos} \geqslant -E_{\text{diff}}^{sum\_neg}$, it indicates that the BESS remaining energy capacity is enough for the power dispatching. Conversely, if $E_{\text{diff}}^{sum\_pos} < -E_{\text{diff}}^{sum\_neg}$, it signifies that there has to import power from the unity grid anyway. In the latter condition, P_{EV}^{disp} is generated following the Eq (4.27) and the EV is controlled to work in the rigid charging mode with $M_{DAB} = 0$.

$$P_{EV}^{\text{disp}} = P_{EV}^{ref}\_^{\text{init}} \tag{4.27}$$

The fixed period for the rigid charging is set to guarantee the user's demand, as the flexible charging mode may cause the variation of charging power, thus affecting the achievement of the charging goal. Step 7 is enabled in case of the remaining charging time surpasses (set to two time slots), the flexible charging mode is to be executed. Accordingly, is set to one. The charging scheduling could be conducted. The dispatching reference for the EV is generated according to the Eq.(4.28):

$$P_{EV}^{\text{disp}} = \begin{cases} P_{EV}^{ref^{\text{init}}} & P_{\text{diff}}^{t_{\text{zlot}}} \leq 0 \\ P_{EV}^{ref^{\text{init}}} + 4^* \dfrac{P_{\text{diff}}^{t_{\text{zlot}}}}{E_{\text{diff}}^{sum^{pos}}} \left(-E_{\text{diff}}^{sum^{neg}}\right) & P_{\text{diff}}^{t_{\text{zlot}}} > 0 \end{cases} \tag{4.28}$$

The $p_{\text{diff}}^{t_{\text{slot}}}$ denotes the power difference at the current timeslot, which could be derived at step 2. It can be seen that when $p_{\text{diff}}^{t_{\text{slot}}} \leq 0$ is set up, P_{EV}^{disp} is set to $P_{EV}^{ref\_\text{init}}$ as there is no BESS power capacity for the EV to increase the charging power. When $p_{diff}^{t_{slot}}$ the is large than zero, the reference power re-allocation to EV is carried out based on a fair proportional division for the whole deficit capacity $E_{\mathrm{diff}}^{sum\_neg}$ according to the value of $p_{\mathrm{diff}}^{t_{\mathrm{slot}}}$.

Step 8 is designed to guarantee the EV charging demand in the last $t_{\mathrm{rigid}}^{\mathrm{band}}$ before EV leaves, thus avoiding the power variation due to the autonomous power coordination. Thereby, the EV stays in rigid charging mode with $M_{DAB} = 0$. And P_{EV}^{disp} is set to $P_{EV}^{ref\_\mathrm{init}}$ as shown in Eq.(4.29)

$$P_{EV}^{\mathrm{disp}} = P_{EV}^{ref}\_^{\mathrm{init}} \tag{4.29}$$

- Mode 6 (M-6): EV charging scheduling + VSC energy compensation (ECS + VEC) mode

Similar to mode 3, the BESS SoC estimation is to be executed when the peak period comes in after the current shoulder period. This design aims to avoid the BESS SoC from dropping below SoC_{Bat}^{low}, thus increasing the power imported in the peak period. The corresponding algorithm is shown in Fig. 4.10. Step 1 aims to acquire the power difference at the current timeslot.

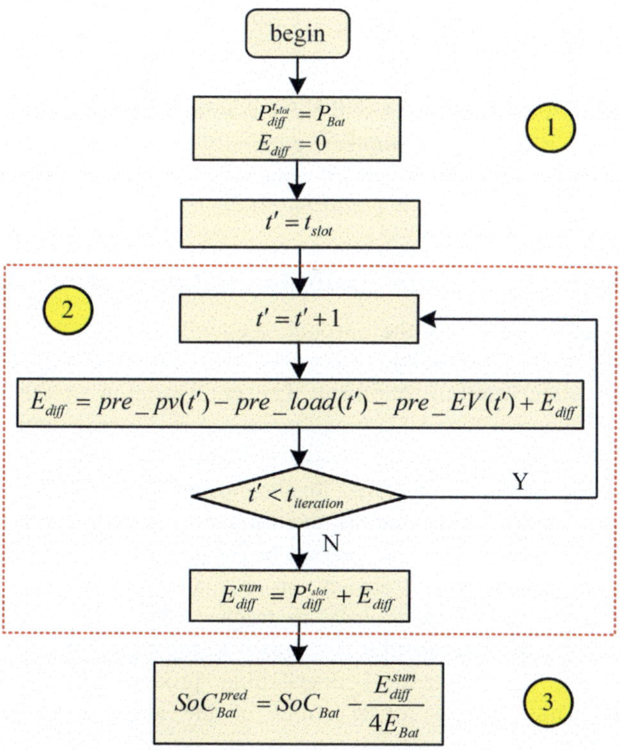

FIGURE 4.10 The developed BESS SoC estimation flowchart 1 when EV plugged in.

In step 2, the energy variation E_{diff} including the EV charging power in future time slots is calculated. The value selection of iteration times $t_{\text{iteration}}$ follows the Eq.(4.30)

$$t_{\text{iteration}} = \begin{cases} t_{EV}^{dep} & t_{EV}^{dep} \in [t_{\text{slot}}, t_{end}] \\ t_{end} & t_{EV}^{dep} \notin [t_{\text{slot}}, t_{end}] \end{cases} \tag{4.30}$$

In step 3, the predicted BESS SoC value SoC_{Bat}^{pred} is obtained. As the flowchart in Fig. 4.7, when SoC_{Bat}^{pred} is less than C_{Bat}^{low} , the power compensation is executed in advance. Corresponding power dispatching reference for the VSC is consistent with Eq.(4.17). When SoC_{Bat}^{pred} is larger than SoC_{Bat}^{low}, mode 5 is enabled with no power dispatching for VSC.

As for the ECS, the power dispatching reference for the EV is determined by current operation conditions as elaborated in Mode 5.

The computational complexity of the proposed operation strategy is related to the time and resources needed to solve a given problem [45], and it is commonly assessed by the big O function O(f(n)). In the flowchart, the computational complexity of simple calculations and assignments is O, signifying a very low and constant complexity. Concerning the flowcharts depicted in Fig. 4.8, Fig. 4.9, and Fig. 4.10, the for-loop block is contained, and the computational complexity is $O\left(t_{\text{peak}}^{end} - t_{slot}\right)$, $O\left(t_{\text{iteration}} - t_{EV}^{in}\right)$, and $O\left(t_{\text{iteration}} - t_{\text{slot}}\right)$, respectively. These belong to the so-called linear type of the big O function (O(n)). It can be seen that n equaling $t_{\text{peak}}^{end} - t_{\text{slot}}$, $t_{\text{iteration}} - t_{EV}^{in}$, or $t_{\text{iteration}} - t_{\text{slot}}$ is not a great number (generally a few tens) that depends on the current timeslot, EV departure timeslot, etc. According to Ref. [39], the solving of the kind of problems could be efficient and easily implemented.

4.5 Simulation analysis of multiple operation scenarios

To verify the effectiveness of the introduced energy management architecture and operation strategy, a simulation model corresponding to Fig. 4.5e is built in the MATLAB®/Simulink platform. The configuration of the simulation experiment is shown in Fig. 4.11. It can be seen that the main circuit topology is built as a detailed switching model to approximate the real system, whereas the PV source and load are preferred for the current source model to decrease the model complexity and accelerate the simulation speed. The local control interval is 50 us, yet the dispatch interval in the upper level is 15 minutes. To display the short time-scale transient process and long time-scale system operation performance simultaneously, an accelerated simulation is conducted, viz., 0.25 seconds simulation time is mapped to the time slot of 15 minutes. That is to say, the upper-level dispatching process is triggered every 0.25 seconds. Thus, for the observations of the long-timescale operation

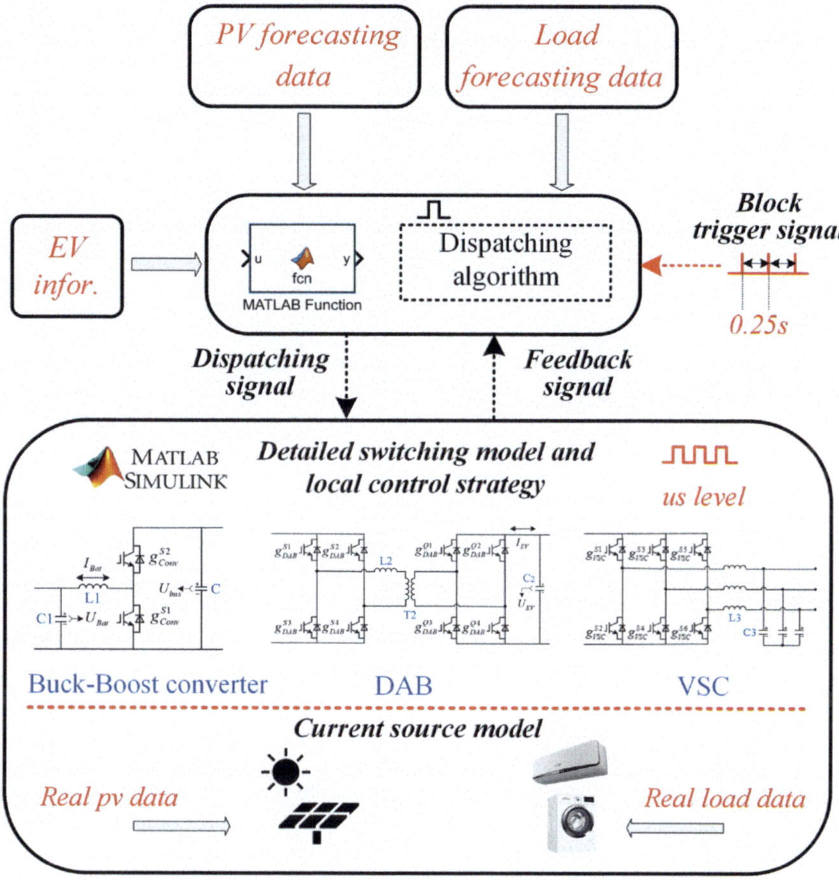

FIGURE 4.11 Simulation model setup.

performance, the stable states of related parameters in corresponding time intervals should be concerned. The parameters designed for the simulation and HIL experiment in Section 4.6 are listed in Table 4.5. Notably, the common parameters are used for both the simulation and HIL experiments.

As for the PV and load data preparation, according to the real characteristics of PV and household appliances presented in appendix, the predicted and real PV and conventional load profiles are depicted in Fig. 4.12 with 15 minutes resolution. Compared to the predicted PV output, the real PV output fluctuates frequently due to a series of climate factors, like a sudden change of solar irradiance. Likewise, the real load consumption would differ from the predicted due to unexpected power demand or load cut-off. Typical scenarios of EV arrival/departure are analyzed under considering different charging/discharging schemes of EVs. The schemes contain the EV conventional

TABLE 4.5 Simulation and experiment parameters.

Item	Variables	Value	Variables	Value
Common parm.	U_{L1}	360 V	U_{H1}	380 V
	U_{L2}	355 V	U_{H2}	385 V
	U_{L3}	340 V	U_{H3}	400 V
	P_{C1}^{max}	−10 kW	P_{D1}^{max}	10 kW
	P_{C2}^{max}	15 kW	P_{C1}^{max}	−15 kW
	P_{C3}^{max}	−25 kW	P_{C1}^{max}	25 kW
	U_{bus}^{nom}	370 V	E_{EV}	60 kWh
	K_B	1	L_1	100e-5 H
	C_1	3e-3 F	L_2	4.8e-4 H
	C_2	50e-3 F	L_3	4e-3 H
	C_3	50e-6 F	U_{VSC}^{abc}	98 V
Sim. parm.	K_{DAB}^{p}/K_{DAB}^{i}	5/0.07	K_{VSC}^{p}/K_{VSC}^{i}	2/0.04
	$K_{Conv}^{p'}/K_{Conv}^{i'}$	8/0.3	$K_{Conv}^{p}/K_{Conv}^{i}$	0.075/0.005
			$K_{VSC}^{p'}/K_{VSC}^{i'}$	0.75/0.0006
HIL parm.	K_{DAB}^{p}/K_{DAB}^{i}	0.1/0.0001	K_{VSC}^{p}/K_{VSC}^{i}	3/0.001
	$K_{Conv}^{p'}/K_{Conv}^{i'}$	0.1/0.005	$K_{Conv}^{p}/K_{Conv}^{i}$	0.05/0.0005
			$K_{VSC}^{p'}/K_{VSC}^{i'}$	0.2/0.001

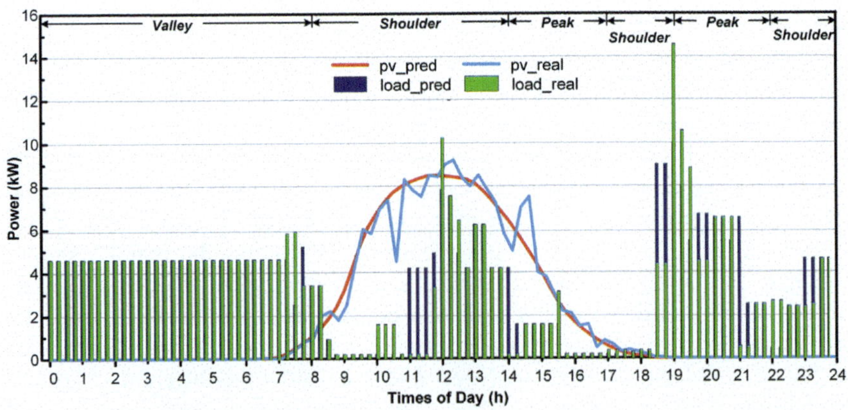

FIGURE 4.12 Real and predicted PV and conventional load curve.

charging (ECC) mode, conventional autonomous V2G operation mode (ECC + droop), and the proposed autonomous V2G operation mode (ECS + droop).

4.5.1 In peak and shoulder periods

This case investigates the operation performance of nanogrid with V2G incorporated in the peak and shoulder period under the proposed architecture and operation strategy. Additionally, the comparison of the system performance under the conventional operation strategy is also elaborated. For the conventional operation strategy of nanogrid, the BESS and PV cover the power consumption preferentially. Then the VSC is used to cover the power deficit/surplus when the power unbalances still exist [19]. Moreover, the EV remains in the ECC mode, which refers to the charging power reference equaling the needed energy dividing the remaining charging time, as described in Eq.

In this scenario, the EV is plugged in the charging point in the afternoon and plugged out before 24:00. As shown in Fig. 4.13a, at AM 8:00 (t = 0 seconds), the valley period ends, and the BESS also reaches the expected SoC value 20%. Then, due to the PV power surplus, the BESS sustains

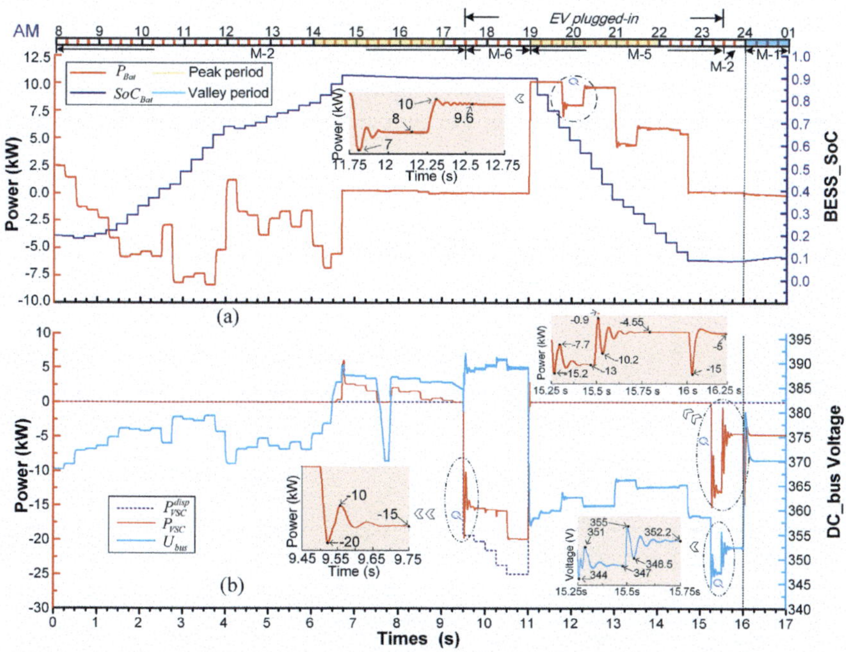

FIGURE 4.13 System performance in the peak and shoulder period under the proposed architecture and operation strategy. (a) BESS SoC and power curve; (b) VSC dispatching reference, real power, and DC bus voltage.

in the charging mode for a long time, leading to the ever-increasing SoC curve. Meanwhile, as shown in Fig. 4.13b, the DC bus voltage varies with the power change of the BESS according to the droop characteristic. Especially, the charging scenario contributes to the increase of DC bus voltage, yet the discharging leads to the voltage drop. Thanks to the autonomous power balance conducted by the BESS, the VSC stays in idle mode with zero output. Accompanied with SoC reaching the upper limit 90% at 14:30 (t = 6.5 seconds), the ultimate power balance is executed by the VSC to export excess power to the utility grid. For the previous period, the nanogrid works on the NED mode (mode 2), realizing the autonomous power balance. At 15:30 (t = 7.5 seconds), due to the abrupt load increase (3 kW as shown in Fig. 4.12), the VSC output power declines to zero accompanied with resuming to 370 V. At 17:30 (t = 9.5 seconds), the EV is plugged into the charging point with, and the preset leave time is 23:30 (t = 15.5 seconds) with target SoC 0.85. Due to the urgent charging demand, the BESS SoC would fall below 0.15 according to the SoC estimation. Therefore, the nanogrid enters in ECS + VEC mode (mode 6). Owing to BESS SoC reaching the upper bound, the grid power dispatched by the VSC is used for the charging of EV with 15-kW maximum power as shown in Fig. 4.14. As shown in the left zoom-in area of Fig. 4.14, the transient process of DAB lasts within 200 ms. In Fig. 4.13b, since is larger than the EV charging power, all loads are supposed to be supplied by the VSC. At 19:00 (t = 11 seconds), the peak period comes in, and the ECS mode (mode 5) is triggered. Meanwhile, the regular load climbs up to over 14 kW, which exceeds the 10-kW BESS power bound significantly. For now, the autonomous V2G service is carried out with power output stabilizing at 5 kW as shown in

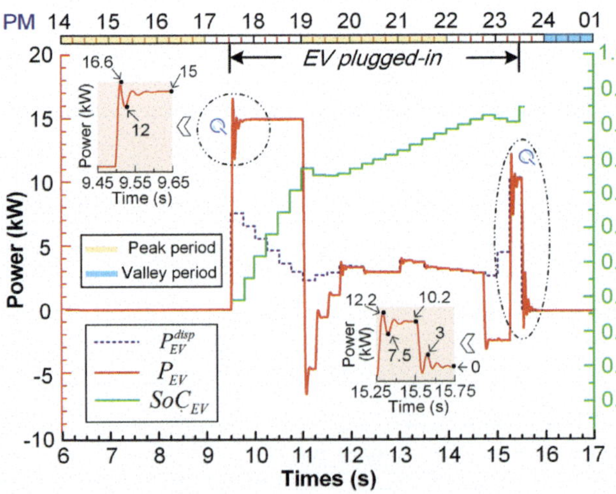

FIGURE 4.14 EV dispatching reference, real power, and SoC curve in the peak and shoulder period under the proposed architecture and operation strategy.

Fig. 4.14, thus realizing the power balance in nanogrid. In the whole peak period, thanks to the autonomous V2G operation, the imported power from the utility grid remains zero even faced with a sudden load increase. It can be seen in the zoom-in area in Fig. 4.13a, combined with the EV output in the same period, the coordination among BESS, EV, and regular load could maintain a stabilized system operation. After 22:00 (t = 14 seconds), the timeslot enters in the shoulder period. And the SoC of BESS drops to the lower bound at 22:45 (t = 14.75 seconds). Thus, in Fig. 4.14, the EV discharging meets the load consumption with 3-kW power output. At the last charging timeslot, to meet the EV SoC demand, the EV operates in rigid charging mode with 10-kW charging power. And in the zoom-in curve at the right of Fig. 4.13b, the ultimate VSC power is stabilized at −13 kW after several cycle oscillations. In addition, the DC bus voltage remains in 347 V at a steady state. Through the rigid charging, the EV SoC could get the target value. When it comes to 24:00, the nanogrid enters BVC mode (mode 1) with the VSC stabilizing the DC bus voltage and BESS operating on the power control state. As shown in Fig. 4.13b, in the transient process caused by the strategy switch, the peak power of VSC reaches −15 kW, and the power in steady state remains −5 kW for BESS charging and load supply. Meanwhile, the DC bus voltage is stabilized at 370 V after reaching 380 V peak voltage.

In the whole period, the DC bus voltage could be kept within the designed range, and the autonomous V2G operation, VEC, and smooth mode transition have been verified.

Under the conventional operation strategy, the corresponding EV charging power and SoC curve is depicted in Fig. 4.15. And the system performance is shown in Fig. 4.16. It can be seen that although the EV charging demand could

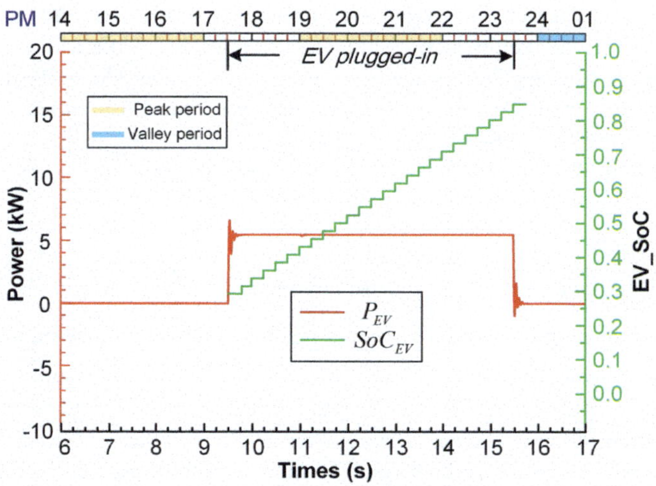

FIGURE 4.15 EV real power and SoC curve under the conventional charging mode.

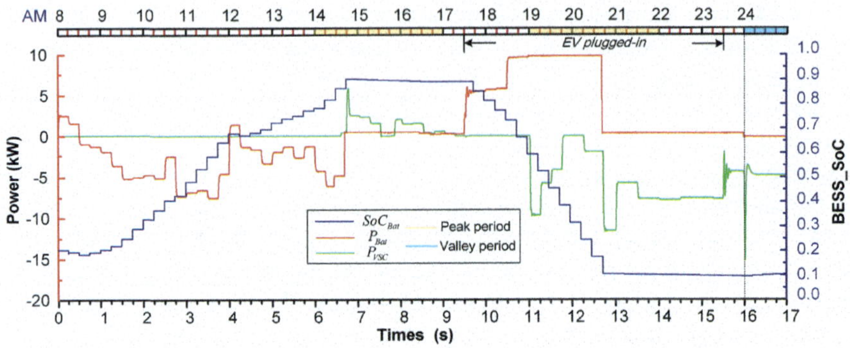

FIGURE 4.16 System performance under the conventional operation strategy.

be satisfied, the maximum imported power reaches about 12 kW at 20:45 (t = 12.75 seconds) in the peak period (19:00–22:00) when the VEC and autonomous V2G operation is disabled. And the imported energy from the utility grid during this peak period reaches 16 kWh. However, as shown in Fig. 4.13b, the maximum imported power and total energy from the utility grid is zero under the proposed architecture and operation strategy. The above results are summarized in Table 4.6. The proposed operation strategy is verified to have real-time excellent peak-shaving capability in the peak period with unpredicted load variations owing to the adopted VEC and autonomous V2G operation mechanism. The power fluctuation at PCC during the peak period is also smoothed.

4.5.2 In valley periods

This case investigates the operation performance of nanogrid with V2G incorporated in the valley period contained under the proposed architecture and operation strategy. In this scenario, the EV leaves at AM 8:30 of the next day. The other conditions keep the same with Section 4.5.1, thus resulting in the same system response before EV plugged in at 17:30, as sho wn in Figs. 4.17 and 4.18. In Fig. 4.17, since EV has enough time for charging in the

TABLE 4.6 Operation strategy comparison.

Operation strategy	Max. Imported power[a] (kW)	Imported energy[b]
Conventional strategy	12	16 kWh
Proposed strategy	0	0 kWh

[a]Refers to the peak period 19:00–22:00.
[b]Refers to the peak period 19:00–22:00.

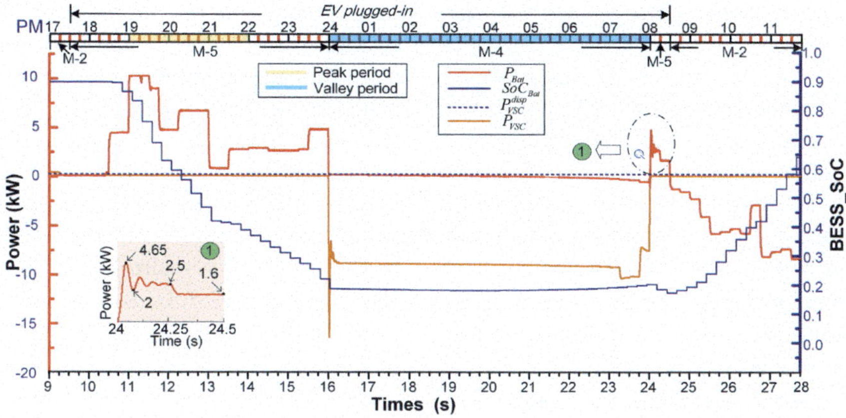

FIGURE 4.17 BESS power and SoC curve, and VSC dispatching reference and real power.

valley period, the EV is not charged in the peak and shoulder period. Meanwhile, the autonomous power compensation is conducted based on the droop characteristic. Particularly, at 19:00 (t = 11 seconds), the EV feeds the extra energy supply with about 5 kW power out. Due to the EV discharging, the EV SoC drops to 0.28. The nanogrid operates in the ECS mode (mode 5) before 24:00 with the load consumption supplied by BESS and EV. At 24:00, the nanogrid switches to BEVC mode (mode 4) with local control strategies changed. Thus, the EV charging power is stabilized at 4.2 kW in 200 ms, as depicted zoom-in area in Fig. 4.18. In the valley period, the VSC supplies all the power needed. The BESS SoC could meet the target value of 0.2, and the EV SoC gets the goal of 0.85. At 8:00, the operation mode goes back to ECS mode. As shown in the zoom-in area of Fig. 4.17, during the transition process

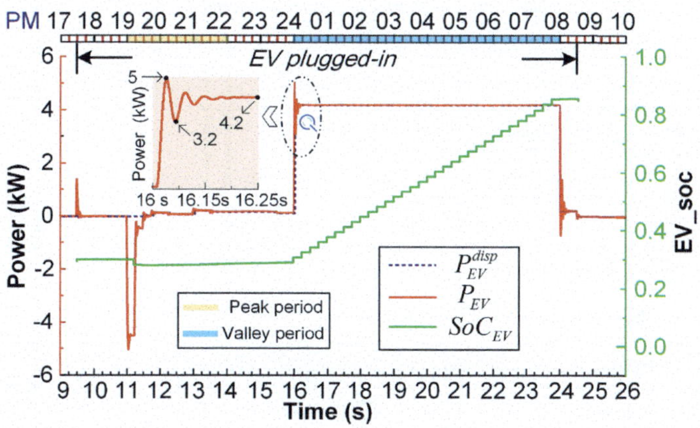

FIGURE 4.18 Comparison of the VSC output under different EV charging strategies.

of local control strategy switches, namely, the constant voltage control to *P–U* droop control of the VSC and the power control to *U–P* droop control of the BESS, the nanogrid remains in the stabilized operation. At 8:00 (t = 24 seconds), the BESS power keeps in 2.5 kW ultimately according to the current power need. The following nanogrid performance is the same as in Section 4.5.1.

The case explores a more common scenario that EV leaves in the next morning, thus emulating the daily travel demand. The results demonstrate that the shift of the EV charging load used for valley-filling is realized, and the autonomous V2G operation, as well as the targeted EV SoC, can also be achieved.

4.5.3 Comparison analysis

This case demonstrates the superiority of the proposed hierarchical V2G operation strategy, namely, the upper-level EV dispatching and local-level autonomous operation strategy (ECS + droop). A detailed comparison with the conventional autonomous V2G operation strategy (ECC + droop) is elaborated.

It is set that the time of EV plugged in is 13:00 with initial SoC equaling 0.3, and the leave time is 20:15 with equaling 0.85. The other system parameters like PV output and conventional load keep the same with Section 4.5.1. As shown in Fig. 4.19, due to the EV charging behavior, the surplus PV power is consumed locally with zero output power of VSC compared to the same period in Fig. 4.13b. As shown in Fig. 4.20, the three charging strategies, namely, the proposed ECS + droop strategy, the ECC, and the ECC + droop

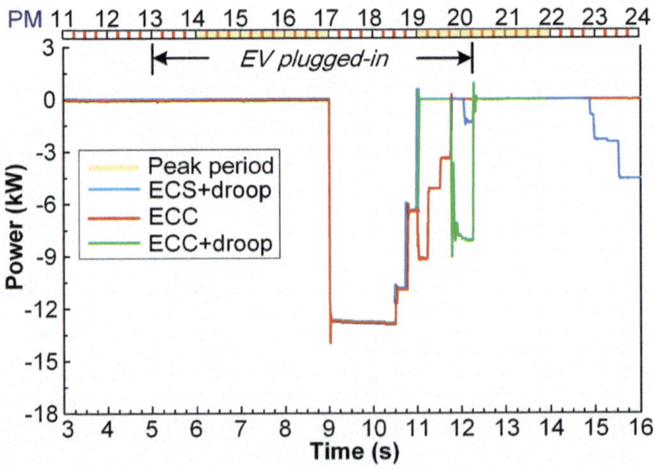

FIGURE 4.19 Comparison of the VSC output under different EV charging strategies.

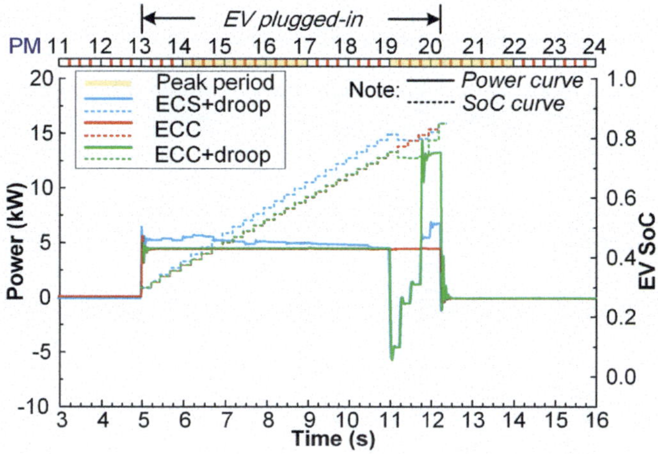

FIGURE 4.20 Comparison of three EV charging strategies.

strategy [29], could meet the EV charging demand. As the red line is shown in Fig. 4.20, the ECC strategy could not respond to the load surge at 19:00 (t = 11 seconds). And the power is always stabilized at 4.3 kW. Thus, the VSC compensates for the power deficit with 9 kW power output in Fig. 4.19. The other two strategies could realize autonomous power response. While the proposed strategy could make the best of BESS remaining capacity for high-power charging in advance, thus leading to the corresponding charging curve higher than other methods' curve, just like the blue line in Fig. 4.20. Therefore, from 13:00 to 19:00, the maximum charging power could reach about 6 kW. During 17:00–19:00, the VSC power compensation is conducted for BESS energy supplement with 13 kW maximum power. At 20:00 (t = 12 seconds), the current load is 4.535 kW. With the ECS + droop method, the EV charging power is only about 6.5 kW, whereas the charging power reaches up to 13 kW under the ECC + droop strategy. Meanwhile, the VSC power approximates −8 kW whereas it is only −1.5 kW under the proposed strategy. The imported power in the peak period is decreased. Since the imported power decreases when the ECS + droop strategy is applied, the BESS SoC would drop below 0.15 at 23:00 (t = 15 seconds). Thus, the VSC covers the power consumption under the proposed method with 3 kW power imported.

Table 4.7 shows the comparison of the three charging strategies in terms of the gird imported power, gird imported energy during the peak period (19:00–22:00), as well as the 24-hours energy cost. It can be seen that the conventional autonomous V2G strategy could not acquire prominent operation performance compared to the ECC strategy due to the inevitable rigid charging period. However, thanks to the proposed ECS + droop strategy, the maximum imported power reduces 81.2%, and the imported energy decreases 90.5% in

TABLE 4.7 V2G operation strategy comparison.

V2G operation strategy	Max. Imported power[a] (kW)	Imported energy[b]	24-h Cost
ECS + droop	1.5	0.38 kWh	¥31.5
ECC	9	4.4 kWh	¥33.53
ECC + droop [29]	8	4 kWh	¥33.07

[a]*Refers to the peak period 19:00–22:00.*
[b]*Refers to the peak period 19:00–22:00.*

the peak period compared to the ECC + droop strategy. Moreover, the 24-hours energy cost also reduces 4.7%.

This case demonstrates that the proposed V2G operation strategy could realize a desirable peak-shaving performance by regulated power dispatching and autonomous power coordination. And a more economic operation of nanogrid is also achieved.

4.6 Experimental verification

For the verification of the nanogrid operation stability and performance in comprehensive modes, the HIL experimentation platform is set up, as shown in Fig. 4.21. The system model corresponding to the circuit topology in Fig. 4.5e is built in the electromagnetic transient simulation software StarSim and operates in NI-PXIe-FPGA-7868R based real-time HIL system with the time

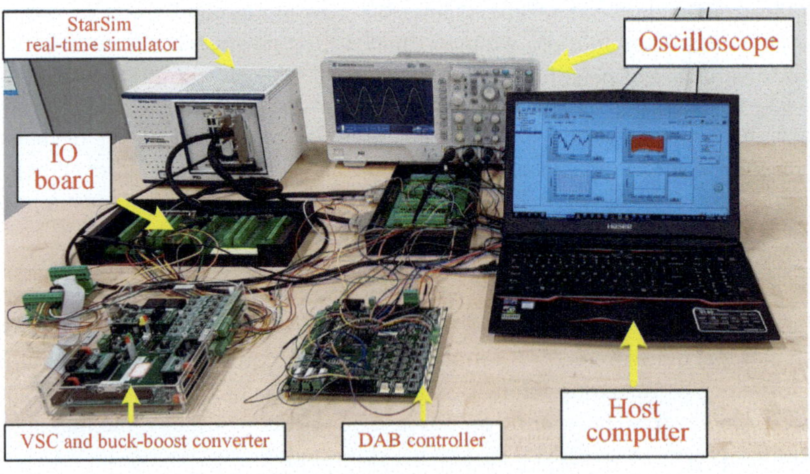

FIGURE 4.21 HIL experimentation platform.

interval of 0.5 μs [46]. The switching and sampling frequency of controllers are set to 20 kHz. The used converter controllers are based on TMS320F28335 DSP and could interact with the real-time simulator through the interface boxes with multiple Analog/Digital I/O ports. The host computer could display the experimental measurements and be used for code download and system debugging. Relevant system parameters are shown in Table 4.5.

4.6.1 System performance in peak and shoulder periods

This case investigates the system performance of comprehensive modes operation and transfer in the peak and shoulder period. Both the simulation and HIL experiment results are contained in Fig. 4.22. For the HIL experiment

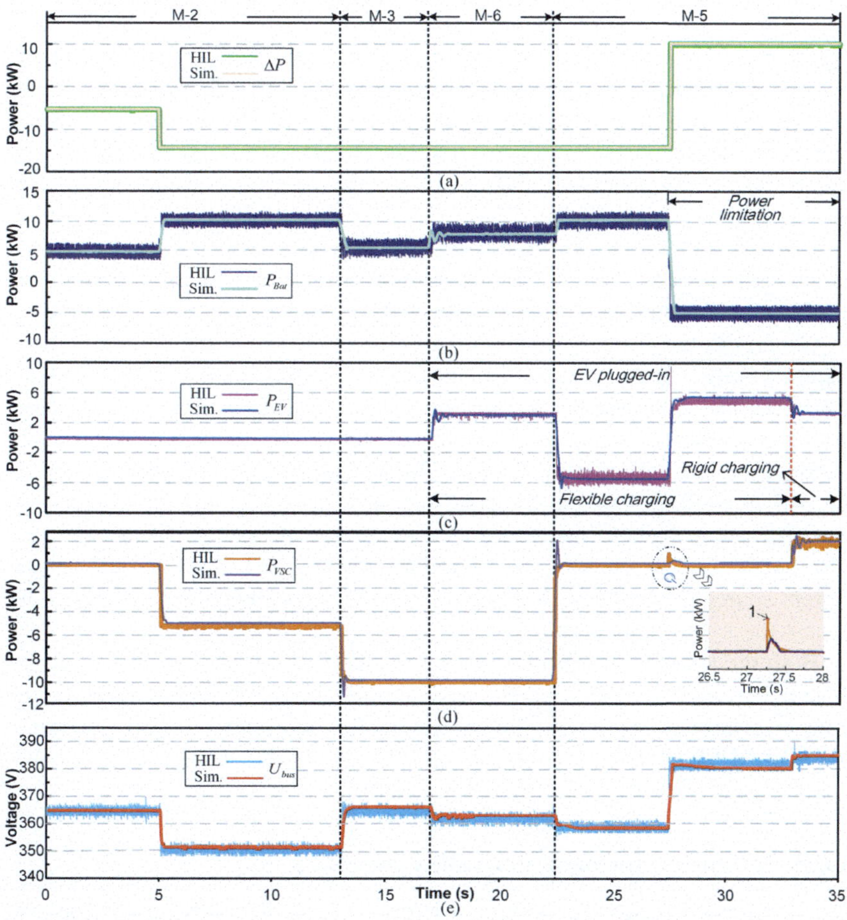

FIGURE 4.22 Simulation and HIL experiment results of mode operation and transfer in the peak and shoulder period. (a) Conventional power difference of the nanogrid; (b) BESS power; (c) EV battery power; (d) VSC output power; (e) The voltage of DC bus.

results, the nanogrid operates in NED mode (Mode 2) firstly. The conventional power difference of the nanogrid refers to the PV output power minus the conventional load power. It equals −5 kW, indicating that the PV output power could not satisfy the conventional load demand. And the deviation is covered by the BESS as shown in Fig. 4.22b. At 5 seconds, decreases to −15 kW, thus the BESS reaches the maximum power output 10 kW. Due to the V2G being disabled, the DC bus voltage drops to about 351 V (as shown in Fig. 4.22e) to trigger the power imported of the VSC. Thus, it is −5 kW as depicted in Fig. 4.22d. At 13 seconds, the VEC mode (mode 3) is enabled with the 10-kW energy compensation of the VSC. At 16.3 seconds, the EV is plugged in with 3-kW charging power in Fig. 4.22c. The ECS + VEC mode (mode 6) is triggered with both the VEC and EV power dispatching conducted. At 22.5 seconds, the nanogrid enters in ECS mode (mode 5) with resuming to zero. Meanwhile, is 10 kW, and the EV operates in the flexible charging mode to feed the 5-kW power deficiency autonomously. At 27.5 seconds, becomes 10 kW, signifying that the power surplus exists in the nanogrid. Due to the charging power limitation of the BESS SoC, is equal to −5 kW, and increases to 5 kW for charging according to the droop characteristic. Along with the switch of the EV discharging to charging process, rises up to 381 V from 359 V. Meanwhile, fluctuates at the switching process accompanied with 1-kW peak power (as shown in the zoom-in area in Fig. 4.22d) and is stabilized at zero within 500 ms. At 33.5 seconds, the EV converts to the rigid charging mode with the constant 3-kW power. Therefore, the VSC is used for the surplus power balance with 2-kW exported power.

Generally, the simulation results are broadly consistent with the experimental results in the above curves, yet exhibit less fluctuation in steady states due to the ideal sampling, filtering, and control process just like Fig. 4.22b-e. Conversely, shown as the HIL result of, due to the ripple current without filtering, the corresponding parameter fluctuates in the steady state, revealing the real conditions. For the transient process, the differences in control parameters and test platforms would lead to varied transient performance. Specifically, in Fig. 4.22a, since the PV and conventional load are built as the current source mode in the simulation and HIL experiment, the two curves are almost the same and smooth.

Both the simulation and experiment results demonstrate that the stabilized system performance of comprehensive modes operation and transfer in the peak and shoulder period is realized. And the autonomous V2G operation, adaptive BESS power limitation, flexible/rigid charging switch, and VEC are also achieved with desired power flow control.

4.6.2 System performance in valley periods

This case investigates the system performance of all modes operation and transfer in the valley period. Both the simulation and HIL experiment results are comprised in Fig. 4.23. Initially, the nanogrid operates in the NED mode (Mode 2) with 4-kW in the shoulder period in Fig. 4.23c. The corresponding is 4 kW as shown in Fig. 4.23a. In Fig. 4.22b, at 5 seconds, the current timeslot enters in the valley period, thus leading to the switch of local control strategy to realize the valley-filling of nanogrid loads. And the nanogrid works on the BVC mode (mode 1). To cover the BESS charging consumption and the conventional loads, it is stabilized at −8 kW within 500 ms ultimately after reaching the peak power −12 kW as depicted in Fig. 4.23c. At 10.5 seconds, the BEVC mode (mode 4) is enabled with the EV plugged in. The VSC imported power increases to 11 kW accompanied with 3-kW EV charging power. At 18 seconds, the ECS mode (mode 5) is enabled with the end of the valley period. Thus, goes back to zero, and the BESS switches to the voltage control mode to cover all the power consumption (7 kW). Likewise, the simulation

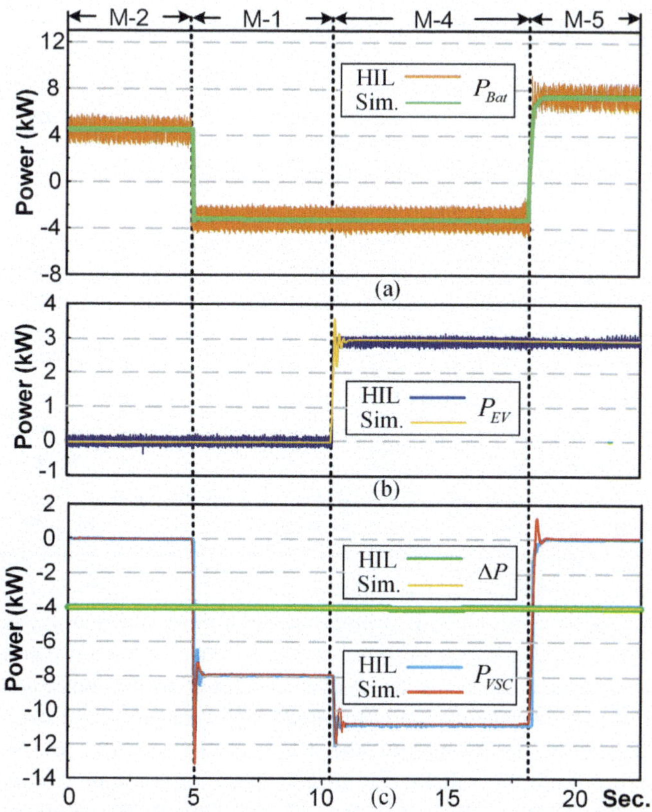

FIGURE 4.23 Simulation and HIL experiment results of mode operation and transfer in the valley period. (a) BESS power; (b) EV battery power; (c) Conventional power difference of the nanogrid and VSC output power.

TABLE 4.8 Computational time consumed by proposed dispatching strategy.

	M-1 (μs)	M-2 (μs)	M-3 (μs)	M-4 (μs)	M-5 (ms)	M-6 (ms)
Raspberry Pi	47	25	859	60	20	22

results generally keep consistent with the HIL experiment results, yet small deviations of performance are encountered in the transient process.

This case verifies the smooth transition of local control strategies and operation modes when entering in exiting the valley period. The desired power flow control is also demonstrated.

The computational performance of the proposed power dispatching strategy is tested in a low cost single-board computer (Raspberry Pi, CPU 1.5 GHz, RAM 1.0 G). Table 4.8 shows the measured computation time consumed by the six modes which are elaborated in Section 4.4.2. It can be seen that a fast solution of all modes is achieved even under the low cost and tiny computer.

4.7 Conclusion remarks

This chapter presents a pseudo hierarchical energy management architecture based on the smart EV charging point, and corresponding operation strategy. A highlight of this architecture is the incorporation of the upper-level central controller with the EV charging point, which facilitates the V2G operation and nanogrid energy management. Moreover, this architecture is free of external central controllers and extensive communication links, enabling a compact, cost-effective, and easy-to-build energy management scheme for nanogrid. The pseudo hierarchical architecture contains two management levels. In the local management level, the state-triggered droop strategy is proposed to realize the autonomous power coordination of BESS, EV, and VSC in nanogrid without communication links, thus increasing the resiliency and autonomy of systems. Moreover, the autonomous V2G operation is expected to realize the real-time adjustment to unpredicted power unbalance in nanogrid, thus improving the local consumption of RESs and decreasing the peak load. In the power dispatching level, the introduced multi-mode power dispatching strategy is of low computational complexity and could achieve the economic nanogrid operation, peak-shaving and valley filling of load curve, smooth mode transition, and the satisfaction of EV charging demand. The simulation and HIL experiment results verify the effectiveness of the proposed energy management architecture and operation strategy. Compared to the conventional autonomous V2G operation strategy, the introduced ECS + droop method within the operation strategy could achieve about 81.2% maximum power reduction and 90.5% energy demand decline in peak period, as well as 4.7% 24-hour energy cost reduction.

References

[1] Gielen D, Boshell F, Saygin D, Bazilian MD, Wagner N, Gorini R. The role of renewable energy in the global energy transformation. Energy Strategy Rev 2019;24:38–50.

[2] Salazar A, Berzoy A, Velni JM, Song W. Energy management of islanded nanogrids through nonlinear optimization using stochastic dynamic programming. IEEE Trans Industry Applic 2020;56:2129–37.

[3] Ji H, Wang C, Li P, Zhao J, Song G, Ding F, et al. An enhanced SOCP-based method for feeder load balancing using the multi-terminal soft open point in active distribution networks. Appl Energy 2017;208:986–95.

[4] Feng W, Jin M, Liu X, Bao Y, Marnay C, Yao C, et al. A review of microgrid development in the United States–A decade of progress on policies, demonstrations, controls, and software tools. Appl Energy 2018;228:1656–68.

[5] Pascual J, Arcos-Aviles D, Ursúa A, Sanchis P, Marroyo L. Energy management for an electro-thermal renewable–based residential microgrid with energy balance forecasting and demand side management. Appl Energy 2021;295:117062.

[6] Nag SS, Adda R, Ray O, Mishra SK. Current-fed switched inverter based hybrid topology for DC nanogrid application. In: IECON 2013-39th Annual Conference of the IEEE Industrial Electronics Society: IEEE; 2013. p. 7146–51.

[7] Nordman B. Nanogrids: Evolving Our Electricity Systems from the Bottom Up. Environmental Energy Technologies Division- Lawrence Berkeley National Laboratory; May 2010.

[8] Burmester D, Rayudu R, Seah W, Akinyele D. A review of nanogrid topologies and technologies. Renew Sustain Energy Rev 2017;67:760–75.

[9] Burmester D, Rayudu R, Seah WK. Distributed generation nanogrid load control system. In: 2015 IEEE PES Asia-Pacific Power and Energy Engineering Conference (APPEEC): IEEE; 2015. p. 1–5.

[10] Burmester D, Rayudu R, Seah WK. Use of maximum power point tracking signal for instantaneous management of thermostatically controlled loads in a dc nanogrid. IEEE Trans Smart Grid 2017;9:6140–8.

[11] Boroyevich D, Cvetković I, Dong D, Burgos R, Wang F, Lee F. Future electronic power distribution systems a contemplative view. In: 2010 12th International Conference on Optimization of Electrical and Electronic Equipment; 2010. p. 1369–80.

[12] Shang Y, Liu M, Shao Z, Jian L. A centralized vehicle-to-grid scheme with distributed computing capacity engaging internet of smart charging points: Case study. Int J Energy Res 2021;45:841–63.

[13] Liu C, Chau K, Wu D, Gao S. Opportunities and challenges of vehicle-to-home, vehicle-to-vehicle, and vehicle-to-grid technologies. IEEE Trans Industr Inform 2013;101:2409–27.

[14] Khemakhem S, Rekik M, Krichen L. A flexible control strategy of plug-in electric vehicles operating in seven modes for smoothing load power curves in smart grid. Energy 2017;118:197–208.

[15] Javadi MS, Gough M, Lotfi M, Esmaeel Nezhad A, Santos SF, Catalão JPS. Optimal self-scheduling of home energy management system in the presence of photovoltaic power generation and batteries. Energy 2020;210:118568.

[16] Erdinc O. Economic impacts of small-scale own generating and storage units, and electric vehicles under different demand response strategies for smart households. Appl Energy 2014;126:142–50.

[17] Wu X, Hu X, Teng Y, Qian S, Cheng R. Optimal integration of a hybrid solar-battery power source into smart home nanogrid with plug-in electric vehicle. J Power Sources 2017;363:277–83.

[18] Khemakhem S, Rekik M, Krichen L. Double layer home energy supervision strategies based on demand response and plug-in electric vehicle control for flattening power load curves in a smart grid. Energy 2019;167:312–24.

[19] Chen J, Zhang Y, Li X, Sun B, Liao Q, Tao Y, et al. Strategic integration of vehicle-to-home system with home distributed photovoltaic power generation in Shanghai. Appl Energy 2020;263:114603.

[20] Olivares DE, Mehrizi-Sani A, Etemadi AH, Cañizares CA, Iravani R, Kazerani M, et al. Trends in microgrid control. IEEE Trans Smart Grid 2014;5:1905–19.

[21] Dragičević T, Lu X, Vasquez JC, Guerrero JM. DC microgrids—Part I: A review of control strategies and stabilization techniques. IEEE Trans Power Electron 2016;31:4876–91.

[22] Mahmood H, Michaelson D, Jiang J. Decentralized power management of a PV/battery hybrid unit in a droop-controlled islanded microgrid. IEEE Trans Power Electron 2015;30:7215–29.

[23] Xiao J, Setyawan L, Wang P, Jin C. Power-capacity-based bus-voltage region partition and online droop coefficient tuning for real-time operation of DC microgrids. IEEE Trans Energy Convers 2015;30:1338–47.

[24] Saad AA, Faddel S, Mohammed O. A secured distributed control system for future interconnected smart grids. Appl Energy 2019;243:57–70.

[25] Dong C, Gao Q, Xiao Q, Yu X, Pekar L, Jia H. Time-delay stability switching boundary determination for DC microgrid clusters with the distributed control framework. Appl Energy 2018;228:189–204.

[26] Palizban O, Kauhaniemi K. Hierarchical control structure in microgrids with distributed generation: Island and grid-connected mode. Renew Sustain Energy Rev 2015;44:797–813.

[27] Rahman MS, Hossain MJ, Lu J. Coordinated control of three-phase AC and DC type EV–ESSs for efficient hybrid microgrid operations. Energy Convers Manag 2016;122: 488–503.

[28] Elkazaz M, Sumner M, Naghiyev E, Pholboon S, Davies R, Thomas D. A hierarchical two-stage energy management for a home microgrid using model predictive and real-time controllers. Appl Energy 2020;269:115118.

[29] Zhu X, Xia M, Chiang H-D. Coordinated sectional droop charging control for EV aggregator enhancing frequency stability of microgrid with high penetration of renewable energy sources. Appl Energy 2018;210:936–43.

[30] Liu H, Hu Z, Song Y, Wang J, Xie X. Vehicle-to-grid control for supplementary frequency regulation considering charging demands. IEEE Trans Power Sys 2014;30:3110–9.

[31] Vachirasricirikul S, Ngamroo I. Robust LFC in a smart grid with wind power penetration by coordinated V2G control and frequency controller. IEEE Trans Smart Grid 2014;5:371–80.

[32] Moussa S, Ghorbal MJ, Slama-Belkhodja I. DC voltage level choice in residential remote area. In: 2018 9th International renewable energy congress (IREC); 2018. p. 1–6.

[33] Siraj K, Khan HA. DC distribution for residential power networks—A framework to analyze the impact of voltage levels on energy efficiency. Energy Rep 2020;6:944–51.

[34] Noritake M, Yuasa K, Takeda T, Shimomachi K, Hara R, Kita H, et al. Experimental study of a 400 V class DC microgrid for commercial buildings. In: 2015 9th International Conference on Power Electronics and ECCE Asia (ICPE-ECCE Asia); 2015. p. 1730–5.

[35] Cousins T. Using time of use (TOU) tariffs in industrial, commercial and residential applications effectively. TLC Engineering Solutions; 2009.

[36] Boscaino V, Guerrero JM, Ciornei I, Meng L, Riva Sanseverino E, Zizzo G. Online optimization of a multi-conversion-level DC home microgrid for system efficiency enhancement. Sustain Cities Soc 2017;35:417–29.

[37] Shang Y, Liu M, Shao Z, Jian L. Internet of smart charging points with photovoltaic Integration: A high-efficiency scheme enabling optimal dispatching between electric vehicles and power grids. Appl Energy 2020;278:115640.

[38] Han Y, Yang H, Li Q, Chen W, Zare F, Guerrero JM. Mode-triggered droop method for the decentralized energy management of an islanded hybrid PV/hydrogen/battery DC microgrid. Energy 2020;199:117441.

[39] Khemakhem S, Rekik M, Krichen L. A collaborative energy management among plug-in electric vehicle, smart homes and neighbors' interaction for residential power load profile smoothing. J Build Eng 2020;27:100976.

[40] Gao Y, Jiang J, Zhang C, Zhang W, Ma Z, Jiang Y. Lithium-ion battery aging mechanisms and life model under different charging stresses. J Power Sources 2017;356:103–14.

[41] Guo Z, Qiu X, Hou G, Liaw BY, Zhang C. State of health estimation for lithium ion batteries based on charging curves. J Power Sources 2014;249:457–62.

[42] Xing W, Wang H, Lu L, Wang S, Ouyang M. An adaptive droop control for distributed battery energy storage systems in microgrids with DAB converters. Int J Electr Power Energy Syst 2021;130:106944.

[43] Yu H, Niu S, Zhang Y, Jian L. An integrated and reconfigurable hybrid AC/DC microgrid architecture with autonomous power flow control for nearly/net zero energy buildings. Appl Energy 2020;263:114610.

[44] Tu C, Xiao F, Lan Z, Guo Q, Shuai Z. Analysis and control of a novel modular-based energy router for DC microgrid cluster. IEEE J Emerg Sel Top Power Electron 2019;7:331–42.

[45] Curry DM, Dagli CH. Computational complexity measures for many-objective optimization problems. Procedia Comput Sci 2014;36:185–91.

[46] Zhang H, Liu Z, Wu S, Li Z. Input impedance modeling and verification of single-phase voltage source converters based on harmonic linearization. IEEE Trans Power Electron 2019;34:8544–54.

Chapter 5

Facilitating nearly-zero-energy-building via EV-Grid coordination

5.1 Introduction

Nowadays, the worldwide power generation is commonly dependent on fossil fuels, which could cause global warming and environmental pollution [1,2]. Buildings, as one of the major energy consumption entities, account for nearly 40% of the global energy use and over a third of the world's carbon dioxide emissions [3,4]. Particularly for China, with its urbanization rate rising sharply from 36.22% in 2000 to 59.58% at the end of 2018, the building energy consumption rises at annual rate of 5.6% and is predicted to exceed 1089 Mtce by 2020 [5,6]. These statistics highlight the critical role that buildings play in the national energy landscape, and underscore the urgency of transitioning toward sustainable, low-carbon building practices. In recent years, growing concerns over energy shortages, environmental degradation, and the need for enhanced energy efficiency have catalyzed the development of sustainable energy solutions in the building sector.

The nZEB, as a novel concept, is proposed to reduce the fossil energy consumption and carbon emission of traditional buildings. An nZEB is designed to integrate local renewable energy sources (RESs), such as solar and wind power, to meet its energy demand with minimal reliance on external grid supply [7]. The primary objective is to achieve a net balance between energy generation and consumption on an annual basis, thereby minimizing grid interaction and reducing environmental impact. By enabling local consumption of locally generated power, the intermittent nature of RESs can be better managed, reducing their negative impact on the utility grid. Additionally, this pattern of power supply, viz., local power generation and consumption, possesses more efficiency and reliability [8]. Microgrid is a small-scale power system that is usually located downstream of distribution substations [9]. It contains a variety of DGs, ESSs and loads that can be controlled in a coordinated way and can either operate in parallel with the utility grid or work on

Smart Penetration of Electric Vehicles in Distribution Grids. https://doi.org/10.1016/B978-0-443-27706-1.00005-2

the islanded mode [10]. This dual-mode functionality is particularly advantageous in scenarios requiring high reliability or during grid outages. Due to the ability of reliable integration of DGs and overall control for system components, microgrid is viewed as an effective alternative for the current power distribution system of buildings to achieve the nearly/net zero energy target by maximizing the local consumption of RESs.

The AC and DC microgrid are two main forms of the microgrid topology [11]. However, the high-efficient utilization of DGs and diversified control objectives cannot be achieved by individual AC or DC microgrid [8]. Individual AC microgrids are inefficient when interfacing with native DC sources and loads, while individual DC microgrids struggle to support legacy AC infrastructure and provide grid-forming capabilities. To address these issues, a feasible approach is to merge both the AC and DC microgrids into a single system, which is known as hybrid AC/DC microgrid [12]. In a hybrid microgrid, the AC/DC DGs and loads are directly connected to the corresponding AC/DC subgrid, leading to an improvement of overall efficiency by minimizing the power conversion losses [13]. The ILC acts as an interconnecting unit to realize desired power exchanges between two subgrids. The performance and intelligence of this ILC are therefore critical, as it serves as the central nervous system of the entire hybrid architecture. Buildings, as complicated systems that contain various types of DC and AC loads, are appropriate to employ hybrid AC/DC microgrids to achieve high-efficient and flexible operation. Specifically, the EV, as a new kind of electric load in buildings, absorbs power from utility grid for its battery charging. Moreover, it can be a "provider" to deliver power back to the grid. The bidirectional power exchange between EV and grid is termed as V2G technology [14]. With a progressive deployment of EVs in recent years, it is feasible to provide auxiliary services for utility grid through centralized V2G control. The V2G-enabled auxiliary services mainly contain frequency regulation, active and reactive power management, load balancing and shifting, grid current harmonics filtering, and so forth [15]. With respect to hybrid microgrids, these services provided by inner integrated EVs could improve microgrid economics and allow for higher penetration of RESs [16]. Thus, the building microgrid transforms from a passive energy consumer into a prosumer entity, with EVs acting as a key flexible resource to buffer renewable intermittency and optimize energy flows.

There have been many researches focused on the load power sharing among DGs in AC and DC subgrids of hybrid microgrids, in order to realize a coordinated operation without relying on communication links. This approach reduces system complexity and enhances robustness against communication failures, which is especially desirable in resource-constrained or cost-sensitive applications. As the only channel for power exchange between subgrids, the ILC should be properly controlled by corresponding strategy, which is mainly based on the so-called droop control [17]. In Ref. [18], a two-stage modified droop strategy is proposed to generate the power reference value for the ILC to realize the bidirectional power flow control. This enhances the flexibility of ILC response and allows for asymmetric power exchange as required by local load variations. Considering the limitation of tie line capacity, a generalized, autonomous droop strategy for ILC is proposed in Ref. [19]. The proposed strategy can achieve an accurate power sharing among DGs, as well as improved transient responses of DC voltage and AC frequency. To increase the tie line capacity and system redundancy, the subgrids can be interconnected through multiple parallel ILCs. In Ref. [20], a novel control strategy for ILCs in a hybrid microgrid is designed to reduce the circulating current in AC side caused by the parallel operation of ILCs. In addition, the power sharing and the enhanced DC voltage support from ILCs can be achieved. The authors from Ref. [21] design a new frequency-droop-based strategy for multiple ILCs to achieve accurate load sharing when the line impendence is considered. Such strategies help address power quality degradation and operational imbalance in multi-ILC configurations.

Although the hybrid microgrids mentioned in above articles can achieve desired power flow control, there still exist following operation limitations:

To realize power sharing, the DGs in droop control need to regulate its output according to the load conditions. Thus, they cannot always deliver current MPPT power, thereby decreasing the efficiency of RES.

Essentially, these proposed strategies belong to the decentralized control method, which uses the power lines as the only channel for communication [22]. In spite of lesser communication demand between controlled units, it is difficult for them to realize systematic coordination due to lack of global information.

An ILC that exhibits an integrated and compact structure, multiple interfaces and functions are favorable to be applied in building microgrid, which prefers small size, integration of grid components in a “plug-and-play” manner, flexible control and networking [23]. However, the single-stage converter VSCs, which is utilized as the ILC in Refs. [17–21], cannot satisfy the above requirements.

The power flow control is mainly implemented in islanded scenario of microgrid. However, the objectives of power flow control in grid-connected scenario, for example, improving the local consumption of RESs, is not considered.

In summary, the state-of-the-art reveals a triad of interconnected problems: 1) a control paradigm that sacrifices renewable efficiency for stability; 2) an ILC hardware platform that lacks the multifunctionality and integration level required for modern buildings; and 3) an operational focus that neglects the critical grid-connected mode where most nZEBs primarily function. These limitations highlight the pressing need for new control and system design strategies that can balance renewable efficiency, modular deployment, and operation in real-world building conditions.

There are two aspects, that is, control approach and microgrid architecture, that should be concentrated on to address the above issues. To realize systematic coordination, a certain level of communication between components in microgrid is crucial. Compared with the decentralized control, the centralized control approach collects necessary information through dedicated communication links, thus it can realize a coordinated control in microgrid [24]. In Ref. [25], a novel feedback-linearization control technique is proposed for the control of power-electronic converters. It has been proved that this control technique could improve the robustness of control systems under multiple operation scenarios. In Ref. [26], a centralized control approach is introduced to coordinate distributed energy resources and energy storages in islanded AC microgrids, aiming to realize equalization of SoC and a proportional power sharing among different energy storages. As is stated in the two articles, single devices or small-scale systems, which need to execute relatively less communication tasks, are considered appropriately to be controlled in a centralized way. However, as for complicated and large-scale systems, the centralized control approach will greatly increase reliance on communication networks and degrade system reliability in case of the malfunction of communication carriers [27]. This creates a single point of failure, which is unacceptable for a system designed for high reliability like an nZEB. A proper solution that contains the advantages of both the centralized and decentralized control is the hierarchical control approach, which consists of multiple control levels [28]. One or more controllers in different control levels are used to handle the control tasks that feature different time scale. Accordingly, the control and communication burden of single controllers is mitigated, and a systematic coordination could be appropriately achieved. In Ref. [29], a hierarchical control strategy is proposed to realize dynamic power sharing in microgrid and smooth the fluctuation of DC-bus voltage. Similarly, a hierarchical control strategy for isolated DC microgrid is introduced in Ref. [30]. It can achieve power sharing and power balance control among energy storage units. But both two articles focus on the individual DC microgrid. For hybrid AC/DC microgrid, just as the hierarchical control shown in Ref. [31], the DGs could operate in MPPT mode with regardless of the control strategy of ILC and load conditions, thereby improving the utilization of RES compared with working on droop control mode. However, the transition of system operation modes does not be taken into account. In Ref. [32], a uniform control strategy for ILC in hierarchical controlled hybrid AC/DC microgrid is presented to eliminate control mode switches in case of different scenarios. Thus, the reliability and resilience of the entire system could be improved. Unfortunately, this proposed strategy is only targeted at control of ILC in islanded scenario. What's more, the approach to improve local consumption of RESs is ignored by the above literature.

To facilitate the integration of DGs and loads, and allow more degree of freedom in control, there have been a relatively small body of literature concerned with the new architecture of hybrid microgrid. Such architectures aim to enhance energy autonomy, optimize local resource usage, and enable multidirectional power exchange across AC and DC domains. In Ref. [33], a hybrid microgrid architecture with grid connection through B2B converter is proposed. The multiple control modes of DGs and B2B converter facilitate the power exchange between subgrids and utility grid, enhancing system operation flexibility and reliability. However, both the AC DGs and B2B converter have to switch the control strategy when the power deficit in AC subgrid occurs with the AC DGs reaching its maximum output power. This strategy switch of DGs would increase the control complexity and degrade the power quality of the microgrid. Frequent switching may also lead to dynamic instability, harmonics, and transient overshoot, particularly under high-penetration RES conditions. Similarly, in Ref. [34], a hybrid microgrid architecture with DC connection at a B2B converter is introduced. The integration of energy storages in DC-bus enriches the operation modes of microgrid. Energy storages provide additional flexibility for load shifting, emergency backup, and renewable smoothing. However, the optimal power flow between AC and DC subgrid cannot be realized under different cases. Peng et al. [35] proposes a three-port hybrid microgrid architecture that consists of three subgrids, that is, DC subgrid, AC subgrid and distributed storage subgrid. However, the architecture is only controlled to implement the power management among three subgrids in islanded operation mode. SST is a highly compact power electronic device that integrates the transformer, the converter stage and the filter. Yu et al. [36] presents a kind of SST interfaced microgrid architecture, which enables multiple mode operations (islanded and SST enabled operation) and seamless mode switch. Nevertheless, the power dispatching between AC and DC subgrid is not considered. This limits the SST's application as a full interlinking converter in hybrid systems where bidirectional subgrid interaction is required. Similarly, a hybrid microgrid architecture for smart building based on SST is proposed in Ref. [37]. The additional functions such as dynamic compensation of reactive power and optimal power dispatching can also be provided by SST. Unfortunately, the power dispatching is only considered between the utility grid and microgrid in grid-connected mode, and the power dispatching between subgrids is not involved. This omission restricts the system's flexibility in managing local power balancing, especially when utility grid access is constrained or temporarily unavailable.

The architectural survey reveals that while innovation is ongoing, most proposals are plagued by narrow operational focus, control mode transitions that compromise power quality, or an inability to manage internal power flows optimally in all scenarios. There is a clear need for an integrated architecture that is reconfigurable, avoids control mode switching, and enables optimal power dispatch both internally and with the main grid.

In this chapter, a novel integrated and reconfigurable hybrid AC/DC microgrid architecture with its hierarchical control strategy is proposed.

Tables 5.1 and 5.2a show the comparisons of the proposed ILC with other ILCs mentioned before in terms of provided functionalities and power conversion losses, respectively. Specifically, given that different ILCs perform different directional power flow control, the corresponding number of power conversion stages is recorded in Table 5.2, and a compared evaluation of power conversion losses of ILCs is also involved. The main contributions of this chapter are as follows: 1) A new topology of ILC named SIU is developed to interface AC and DC subgrids; 2) The SIU-based hybrid microgrid architecture is proposed, which features compact structure, multiple interfaces (include dedicated interfaces for EVs), reconfiguration ability and more freedom in control; 3) Based on the established hierarchical control structure, the hierarchical control strategy of SIU is proposed, which contains local control in primary control level and power flow control in secondary control level; 4) The hierarchical control strategy can achieve optimal power dispatching with steady operation of microgrid, realizing mutual power support between subgrids, as well as improving utilization and local consumption of RESs.

The remainder of this chapter is organized as follows. It first introduces the proposed microgrid architecture for nearly zero energy buildings, together with the associated hierarchical management structure. The chapter then presents the system configuration, followed by the modeling and control considerations of the key interface and energy components. On this basis, the hierarchical management strategies are described from both the local control level and the power flow dispatching level. The subsequent part evaluates the system operation performance under different energy-flow scenarios,

TABLE 5.1 Functionalities provided by SIU in comparison with contemporary ILCs.

Functionalities	Single VSC [17–19]	Multiple VSCs [20,21]	B2B converter [34]	SST [36,37]	SIU This chapter
Storage integration	×	×	√	×	√
Dedicated interfaces for EVs	×	×	×	×	√
Reconfigurable architecture	×	×	×	×	√
Power support between subgrids	√	√	√	√	√
Increased local consumption of RESs	×	×	×	×	√

TABLE 5.2 Conversion loss of SIU in comparison with contemporary ILCs.

ILC	AC side to utility grid	DC side to utility grid	AC side to DC side	Conversion loss
Single VSC [18–20]	0	1	1	L[a]
Multiple VSCs [21,22]	0	1	1	L
B2B converter [28]	2	1	1	M[b]
SST [30,31]	3	2	1	H[c]
SIU (This chapter)	0	1	1	L

[a]*L-low.*
[b]*M-medium.*
[c]*H-high.*

including multiobjective dispatching and islanded operation. The chapter concludes with final remarks.

5.2 Nearly zero energy building

Nearly zero energy building (nZEB) has emerged as one of the most representative paradigms in the transition from conventional buildings to sustainable, low-carbon, and highly efficient energy systems [38]. In essence, an nZEB is not merely a building with low electricity consumption, but a building whose annual or long-term energy demand is extremely low and can be largely covered by renewable energy generated on-site or nearby. Compared with traditional building energy systems, the concept of nZEB emphasizes the coordinated optimization of building envelope, end-use efficiency, distributed renewable generation, energy storage, and intelligent energy management [39]. As a result, the building is no longer regarded as a passive energy consumer. Instead, it becomes an active participant in local energy balancing and, in some cases, a flexible resource capable of interacting with the surrounding grid [40]. This shift in perspective is of fundamental importance, because it means that the evaluation of building performance should no longer be limited to static indicators such as annual energy consumption alone. Greater attention should also be paid to how effectively the building can coordinate local generation, storage, and load under varying operating conditions. In other words, the concept of nZEB extends beyond energy saving in the conventional sense and moves toward active energy management at the building scale. The development of nZEBs is driven by several converging factors. First, buildings account for a substantial proportion of global final energy consumption and carbon emissions. Therefore, the decarbonization of the building sector is essential to achieving broader climate and energy-transition goals. Second, the rapid decline in the cost of photovoltaic panels, energy storage devices, and power electronic converters has made distributed clean-energy supply increasingly practical at the building scale. Third, the digitalization of energy systems, including sensing, communication, and advanced control, has enabled more refined coordination among generation, storage, loads, and grid interfaces. Under these conditions, the idea of bringing building energy demand close to net zero has become technically feasible and strategically attractive. Meanwhile, policy support, building energy codes, and carbon-neutrality targets in many countries and regions have further accelerated the promotion of nZEB-related technologies [41]. These developments have encouraged the integration of local renewable resources, electrified end-use systems, and intelligent control platforms into next-generation buildings. As a consequence, nZEBs are increasingly regarded not only as energy-efficient buildings, but also as important nodes in future smart and low-carbon energy systems.

However, the phrase "nearly zero" should be understood carefully. It does not necessarily imply that the building is fully self-sufficient at every moment, nor that it can operate without any exchange with the utility grid. In practice, the energy behavior of an nZEB is highly dynamic. Renewable generation, especially photovoltaic output, is intermittent and strongly dependent on weather and solar irradiation. Meanwhile, building loads vary with occupancy, equipment operation, air-conditioning demand, and user behavior. For this reason, even a highly efficient building with abundant on-site renewable generation may experience periods of power deficit and periods of power surplus. The essence of nZEB operation therefore lies not in eliminating all grid interaction, but in minimizing net external energy dependence over time while maintaining reliability, comfort, and operational flexibility. This dynamic nature makes energy management a central issue in nZEBs. Traditional buildings are typically supplied by the grid in a largely one-way manner. Their internal loads are served according to demand, and the control objective is often limited to local equipment operation or comfort maintenance. By contrast, an nZEB involves multiple coupled energy elements. Renewable generation must be utilized effectively rather than curtailed unnecessarily. Storage devices must be charged and discharged in a coordinated way. Different categories of loads may need to be supplied with different priorities. In addition, interaction with the utility grid must be managed in a manner that is both economically and technically appropriate. Such a coordinated operation problem is inherently more complicated than that of a conventional building, because it requires simultaneous consideration of source-side uncertainty, load-side variability, and the operational limits of storage and conversion devices. In particular, once the building contains both controllable and uncontrollable loads, as well as intermittent and dispatchable energy resources, the energy management task becomes a system-level optimization problem rather than a simple local control problem.

The problem becomes even more complex when EVs are integrated into the building energy system, since EVs introduce both additional electricity demand and additional flexibility [42]. In recent years, EVs have gradually evolved from independent transportation devices into important components of integrated energy systems. When EVs are connected to a building-scale microgrid, they can play multiple roles. At the most basic level, they represent a new electrical load whose charging demand must be accommodated. At a more advanced level, they can also act as flexible storage resources capable of shifting demand in time, absorbing surplus renewable power, and, under appropriate control, supplying power back to the building or the grid. This dual characteristic makes EVs particularly relevant to nZEBs. On the one hand, EV charging may increase the energy burden of the building if it is not well coordinated. On the other hand, if charging and discharging are scheduled intelligently, EVs can significantly enhance renewable-energy utilization, reduce grid dependence, and improve the flexibility of the entire building energy system [43]. From the perspective of building-grid interaction, EVs therefore occupy a unique position. Unlike fixed loads, their power demand can often be temporally shifted within a certain user-acceptable window. Unlike stationary storage devices, they are mobile assets whose availability is uncertain but whose energy capacity may be substantial. This combination of mobility and flexibility makes EVs both valuable and challenging for nZEB operation. The opportunity lies in the fact that EV batteries can help bridge the temporal mismatch between renewable generation and building demand. For example, surplus photovoltaic power produced at midday can be directed to EV charging rather than being exported or curtailed. During periods when local generation is insufficient, controlled discharge from EVs or coordinated charging deferral can reduce the stress on the building-side energy system. This capability is especially valuable in buildings with highly variable daily load profiles or with strong renewable penetration. The challenge, however, is that EV energy demand is inherently uncertain. Arrival time, departure time, charging duration, state of charge, and user expectations all vary from case to case. As a result, EV integration increases the stochasticity of building energy management and requires more sophisticated coordination mechanisms than those needed in conventional building energy systems.

Besides EVs, battery energy storage systems (BESSs) are another critical component in nZEB operation [44]. Although EV batteries can provide a degree of flexibility, they are not always available when the energy system requires support, since vehicle usage depends on transportation needs rather than solely on building operation. A stationary BESS, by contrast, can be designed and controlled specifically for building energy management. It can absorb excess renewable generation, provide power during deficit periods, support bus-voltage stability, smooth short-term fluctuations, and reduce the magnitude of power exchanged with the utility grid. In this sense, the BESS serves as a controllable buffer between variable generation and variable loads. When used together with EVs, it enables the building energy system to exploit both stationary and mobile storage resources, thereby improving operational resilience and dispatching flexibility. This complementary relationship is highly valuable. The BESS provides predictable and continuously available support, whereas EVs contribute additional flexible capacity whenever they are connected. The coordinated use of these two types of storage can significantly enlarge the feasible operating region of the building energy system and improve its ability to respond to fluctuating renewable generation and changing load demand. Photovoltaic generation is usually the most important renewable source in nZEBs, particularly for residential and commercial buildings with rooftop space. PV systems are attractive because of their modularity, low operating cost, and direct compatibility with building-scale deployment. In nZEB applications, PV generation is not simply an add-on source of electricity. It is a core enabler of the near-zero-energy target. The degree to which local PV output can be consumed within the building or its associated microgrid strongly affects the actual energy performance of the building. If a large share of PV power is exported while electricity must later be imported during deficit periods, the operational value of local generation is not fully realized. Consequently, an important objective in nZEB operation is to increase the self-consumption of photovoltaic energy [45]. This requires coordinated scheduling among PV generation, building loads, storage devices, EV charging demand, and grid exchange. Accordingly, the realization of high-performance nZEB operation depends not only on the presence of renewable generation and storage devices, but also on the existence of an appropriate system architecture and an effective energy management framework. This is precisely why the following sections of this chapter focus on a novel microgrid architecture and hierarchical management strategy for nZEBs, in which PV generation, EVs, BESS, and building loads can be coordinated in an integrated manner.

At this point, the need for a dedicated microgrid architecture becomes evident. The energy system of an nZEB is fundamentally different from that of a conventional building because it contains multiple distributed resources and multiple layers of controllable loads. Directly coupling all these elements through a simple radial or equipment-oriented design often leads to limited flexibility, complicated energy conversion paths, and inefficient operation. For example, many modern building-side sources and loads are inherently DC in nature, including PV panels, batteries, EV batteries, LED lighting, electronic devices, and many types of power electronics. Yet conventional building power systems are predominantly AC-based. Repeated AC/DC and DC/AC conversion not only introduces losses but also increases control complexity.

Therefore, hybrid AC/DC microgrid architectures have become increasingly attractive for nZEB applications, since they can better accommodate the coexistence of AC and DC resources while reducing unnecessary conversion stages. Within such a hybrid microgrid, the arrangement of subgrids and the design of the interlinking interface are of particular importance. The AC subgrid can continue to support conventional building loads and maintain compatibility with utility-grid interconnection, while the DC subgrid can more naturally integrate PV generation, batteries, EV chargers, and other DC-oriented components. The interlinking converter or system interface unit then becomes the key device for coordinating power exchange between the two subgrids. Its role is not only to transfer power physically, but also to support the operational objectives of the overall microgrid, such as renewable-energy prioritization, state-of-charge regulation, and grid-interaction control. This architectural logic explains why the present chapter does not discuss nZEBs in isolation, but instead places them in the broader context of a novel microgrid architecture and hierarchical energy-management strategy.

The hierarchical nature of management in nZEBs deserves special attention. Because the building energy system contains devices with different response speeds, control objectives, and physical characteristics, it is generally neither efficient nor robust to manage everything at a single level. A lower-level control layer is needed to ensure local stability, voltage regulation, current tracking, and safe operation of individual converters and subsystems. Above this, a higher-level dispatching layer is required to determine how energy should be allocated among PV generation, EV charging/discharging, the BESS, building loads, and the utility grid. These two layers are closely related but functionally distinct. Local control ensures that each device behaves properly in real time, whereas upper-level dispatching coordinates the overall energy flow according to broader operational goals. A well-designed nZEB therefore requires both reliable local control and intelligent system-level management.

From an operational perspective, the objectives of nZEB management are inherently multidimensional. Reducing purchased electricity is important, but it is not the only goal. The system may also seek to maximize photovoltaic self-consumption, limit the fluctuation of exchanged power, maintain appropriate battery state of charge, reduce converter stress, and ensure secure operation under different grid conditions. In addition, user-side considerations remain essential. Building comfort cannot be sacrificed arbitrarily, and EV users must still receive sufficient charging service. The practical operation of an nZEB is therefore a multiobjective balancing process among energy efficiency, renewable utilization, operational security, user requirements, and control feasibility. This multiobjective nature becomes more prominent when the building is subject to different external operating conditions. During some periods, such as midday with strong solar radiation and moderate building demand, the main challenge may be how to absorb surplus photovoltaic output efficiently. During evening hours, when PV generation falls and building demand remains high, the issue may shift toward deficit compensation and reduction of grid dependence. In off-peak periods, charging the BESS or EVs may become preferable if future renewable or load conditions justify it. In islanded or disturbed conditions, the priority may change again, focusing on critical-load supply and local stability. These varying conditions mean that a static energy-management rule is rarely sufficient. Instead, the control and dispatching framework of an nZEB should be able to adapt to different scenarios while preserving a coherent overall objective.

Another reason why nZEBs merit separate discussion is that they sit at the intersection of several traditionally independent domains: building energy systems, distributed generation, microgrids, storage technologies, and transportation electrification. If these domains are considered separately, many potential synergies remain underexploited. For instance, building energy studies may emphasize thermal performance and load reduction without fully considering EV flexibility. Conversely, EV integration studies may focus on charging coordination without examining how a building-side microgrid can reshape the energy context in which charging takes place. The nZEB framework provides a useful convergence point because it naturally requires an integrated view of demand reduction, local renewable supply, storage coordination, and flexible electrical loads. In this respect, EVs are not external additions to the nZEB concept. They are increasingly part of its practical realization. For residential and community-scale applications in particular, the relevance of nZEB-oriented microgrids is growing. As distributed PV, home storage, and private EV ownership expand simultaneously, the household or building-scale energy system is becoming more active and more complex. This development challenges the conventional notion that electricity consumers are passive endpoints of the grid. Instead, households and buildings can behave as prosumer-type entities that both consume and manage energy. In this setting, the energy-management strategy must go beyond simple self-consumption optimization and account for the interactions among multiple devices, multiple time scales, and multiple operating modes. The design of a suitable architecture is therefore inseparable from the design of the control strategy.

The discussion in this chapter adopts precisely this perspective. Rather than treating the nZEB as a static building-performance label, it considers the nZEB as an operationally active energy node supported by a hybrid microgrid structure. In this framework, the nearly-zero-energy objective is approached through coordinated management of photovoltaic generation, EV charging/discharging, a battery energy storage system, building loads, and utility-grid interaction. The emphasis is placed not only on reducing net external energy dependence, but also on ensuring that the internal energy flows of the microgrid are organized efficiently and robustly. This shift from "low-consumption building" to "actively managed building energy system" is central to understanding the technical direction of this chapter.

Accordingly, the next section introduces a novel microgrid architecture for nZEBs, focusing on the system configuration and the hierarchical management structure that supports coordinated operation. On this basis, the chapter then develops the local control strategies and power-flow dispatching strategies required for practical implementation. The subsequent analysis demonstrates how the proposed system performs under different energy-flow directions, multiobjective dispatching requirements, and islanded operation scenarios. Through this progression, the chapter aims to show that the realization of a nearly zero energy building depends not only on efficient components, but also on an appropriate system architecture and an effective hierarchical management framework.

5.3 Novel microgrid architecture for nZEBs

5.3.1 System configuration

The proposed hybrid AC/DC microgrid architecture for nZEB is shown in Fig. 5.1. The core component of the microgrid is the SIU, which is used to

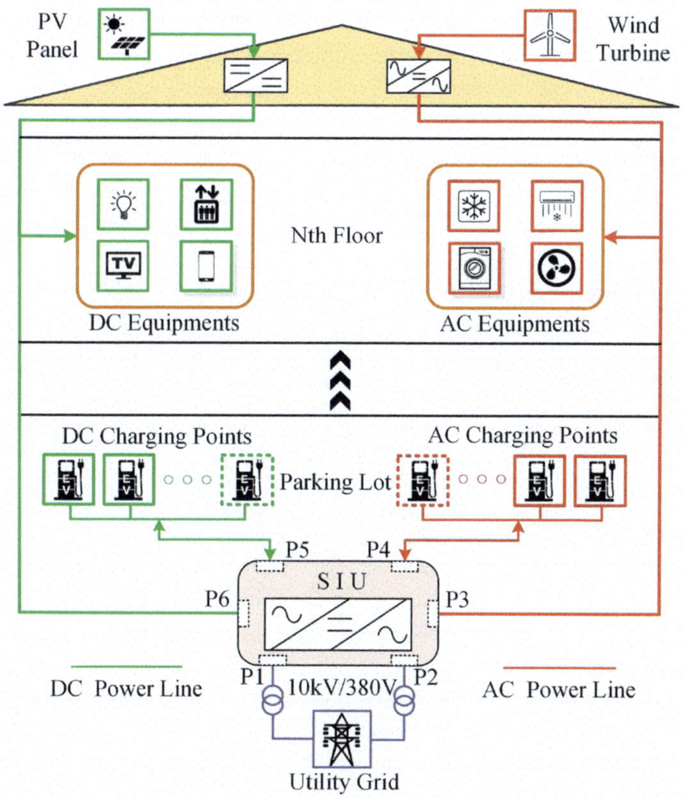

FIGURE 5.1 Proposed hybrid AC/DC microgrid architecture for nZEB.

realize expected power flow among different interfaces (P1–P6) by the power conversions (AC-DC, DC-AC and AC-DC-AC). Its subcomponents and main circuit topology will be illustrated in Section 5.4. The hybrid microgrid architecture contains two subgrids, that is, DC and AC subgrid. In DC subgrid, the DC DGs (e.g., PV panels), as well as the DC loads (e.g., lights and TVs, etc.) are coupled to the common DC feeder derived from the interface P6. In addition, the DC charging points, which provide the fast-charging service for EVs, are clustered together and connected to the SIU through the dedicated interface P5. Similarly, in AC subgrid, the AC DGs (e.g., wind turbine) and AC loads (e.g., fans and refrigerators, etc.) are integrated to the common AC bus that is connected to P3. Multiple AC charging points are interconnected and interfaced with SIU through P4. Furthermore, the dual interfaces (P1 and P2) are linked to the utility grid through the 10 kV/380V transformers. The information of each interface such as voltage, current, and frequency magnitude is collected by SIU in real time.

5.3.2 Hierarchical management structure

Based on the proposed hybrid microgrid architecture, the hierarchical control structure is established according to the time scales of control objectives. As shown in Fig. 5.2, the hierarchical control structure contains three control levels, namely, the primary control, the secondary control and the tertiary control level.

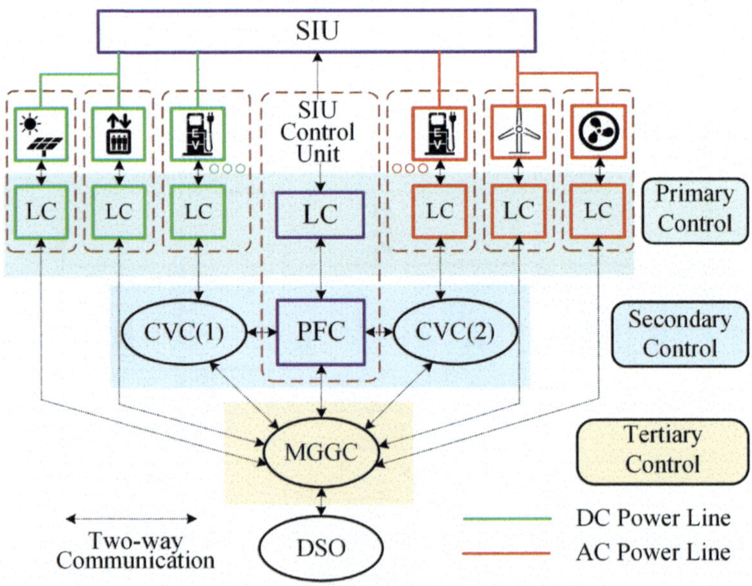

FIGURE 5.2 Hierarchical control structure of proposed hybrid microgrid.

The primary control usually features the fastest response in the three control levels. It realizes the independent local control for components in microgrid by respective LCs. The control objectives contain: 1) AC/DC voltage, frequency and power control of SIU; 2) MPPT and power control of DGs; 3) charging and discharging control for EVs; 4) load management, that is, load shedding and recovering. The controlled components can operate in assigned mode or state according to the commands from the higher control levels.

Compared with the primary control, the secondary control operates in a relatively slower time scale. In this control level, the PFC of SIU collects and processes the data from corresponding LC on a regular basis. In return, the power dispatching command is generated online and delivered back to the LC of SIU according to the power flow control strategy. The CVC (1)/CVC (2) are responsible for the centralized power management for the EVs connected to DC/AC charging points, respectively. The communication between the CVCs and PFC can be accessed to realize the coordinated operation between EVs and SIU.

In tertiary control level, the global coordinated control is to be implemented by MGCC to realize a long-term optimal operation of the microgrid. As the interface between the microgrid and the DSO, the MGCC directly supervises and manages the SIU controller, CVCs, as well as LCs of loads and DGs. Moreover, the MGCC exchanges information with the DSO to facilitate EVs to participate in electricity market. When the EVs plugged into AC charging points provide V2G service, the MGCC will submit real-time information of the power capacity of EVs and dispatchable power information of SIU to DSO. Based on the information, the DSO calculates and generates the reference signal for power demand in auxiliary service, which is sent back to MGCC. Finally, the MGCC determines the power dispatching commands for CVC (2) and PFC to implement the power control for EVs and SIU. Therefore, the coordinated V2G operation in this microgrid architecture can be achieved.

The basic configuration of SIU is depicted in Fig. 5.3. The VSC (1) and VSC (2) form a B2B structure that encompasses a common DC bus. In the DC side, except the interfaces (P5 and P6), there is a BESS that can be served as a slack terminal to make a power balance in the DC-bus by control of the bidirectional DC/DC converter. It also can be controlled to absorb/deliver a certain amount of power according to the reference. In the AC side, the states (ON or OFF) of the switches S1 and S2 remain the same and determine the operation scenarios (grid-connected/islanded operation) of the microgrid. In addition, the energy path S is responsible for the power exchange between the AC buses in the AC side of the VSC (1) and VSC (2). And its ON/OFF state is determined by S3. The converters, BESS and switches are supervised and controlled directly by the LC of SIU.

ESS could be roughly divided into mechanical, electro-chemical, chemical, electrical, thermal and hybrid types [46]. Among them, the BESS based on Li-

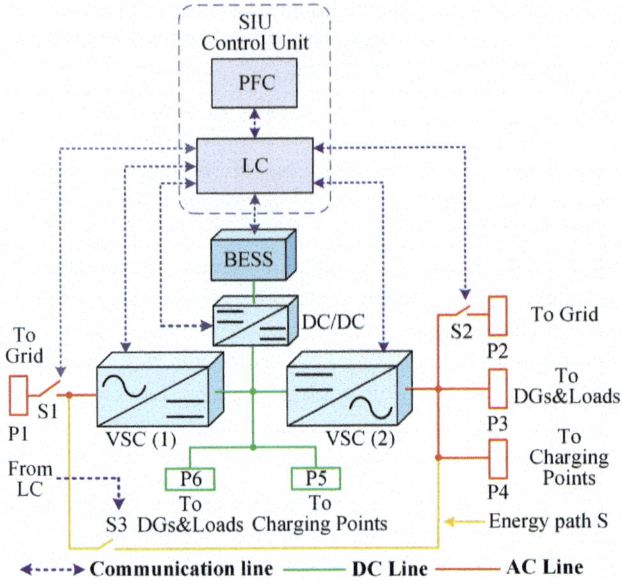

FIGURE 5.3 Basic configuration of SIU.

on battery has been proven suitable for the grid energy storage since it has the advantages of high energy density, long lifecycle and low self-discharge [47]. According to Ref. [48], the Li-on battery mode can be described like the following equations:

The SoC represents the available battery capacity that can be described as:

$$\mathrm{SOC} = 100\left(1 - \frac{1}{Q}\int_0^t idt\right) \tag{5.1}$$

When batteries stay in discharging state ($i > 0$), the terminal voltage of batteries is represented as:

$$\begin{aligned} u_{BAT} = E_0 - k \times \frac{Q}{Q - \int_0^t idt} \times i^* - k \times \frac{Q}{Q - \int_0^t idt} \times \int_0^t idt \\ + A \times exp\left(-B \times \int_0^t idt\right) \end{aligned} \tag{5.2}$$

When batteries stay in charging state ($i < 0$), the terminal voltage of batteries is represented as:

$$\begin{aligned} u_{BAT} = E_0 - k \times \frac{Q}{\int_0^t idt + 0.1Q} \times i^* - k \times \frac{Q}{Q - \int_0^t idt} \times \int_0^t idt \\ + A \times exp\left(-B \times \int_0^t idt\right) \end{aligned} \tag{5.3}$$

where Q is the maximum battery capacity, i is the battery current, i^* is the filtered battery current, E_0 and U_{Bat} denote the open circuit voltage and terminal voltage of the battery, respectively. k denotes the polarization voltage, A and B are the exponential voltage and exponential capacity, respectively.

The overcharging and overdischarging of battery will put an adverse effect on its service life. Therefore, the upper limit of SoC is set to 90% while the lower limit of SoC is set to 10%. In addition, the battery charging/discharging rate should also be concerned and remained in appropriate range for the stable operation and longer life of the BESS. Based on [49], 0.5 C is selected for the reference upper limit value of the charging/discharging current. And it can be calculated that and is around 400 kW according to the magnitude of the nominal voltage (330 V) and total capacity of the BESS (2400 Ah).

5.4 Hierarchical management strategies

5.4.1 Local control strategies

The overall local control strategies of SIU that are implemented in primary control level are presented in Fig. 5.4, where the main circuit diagram of SIU on the top, the local control block diagram of the DC/DC converter in the middle, and the control block diagram of the VSC (N) at the bottom are drawn. As illustrated in main circuit diagram, the VSCs have three-phase two-level topology with a series of IGBTs fully controlled by gating signals derived from corresponding local control strategies. The LCL filter is utilized to reduce the current harmonics around the switching frequency caused by PWM modulation [50]. For the DC/DC converter, the conventional buck/boost topology is used to achieve the bidirectional power flow control for the BESS. The detailed parameters of SIU are shown in Table 5.3. During different operation scenarios of the microgrid, there are different combinations of the switch state and local control strategies of the VSCs and DC/DC converter.

During the grid-connected operation, the switches S1 and S2 are turned ON and S3 remains in OFF state. In the DC-bus, the power generation and consumption should be balanced all the time to maintain microgrid a steady operation with limited fluctuation of the DC-bus voltage. The DC-bus voltage control can be implemented by VSC (1) or DC/DC converter based on the SoC value of BESS. The three conditions are summarized in Table 5.4. For the latter two conditions, the return difference can avoid possible iterative switch of local control strategies at threshold points.

5.4.1.1 Condition I

In this condition, the BESS is utilized for the DC-bus voltage control. As depicted in Fig. 5.4b, the DC-bus voltage control strategy of DC/DC contains two control loops, that is, the outer voltage loop and inner current loop. In outer voltage loop, the difference between V_{DC}^{ref} and V_{DC} is calculated by an

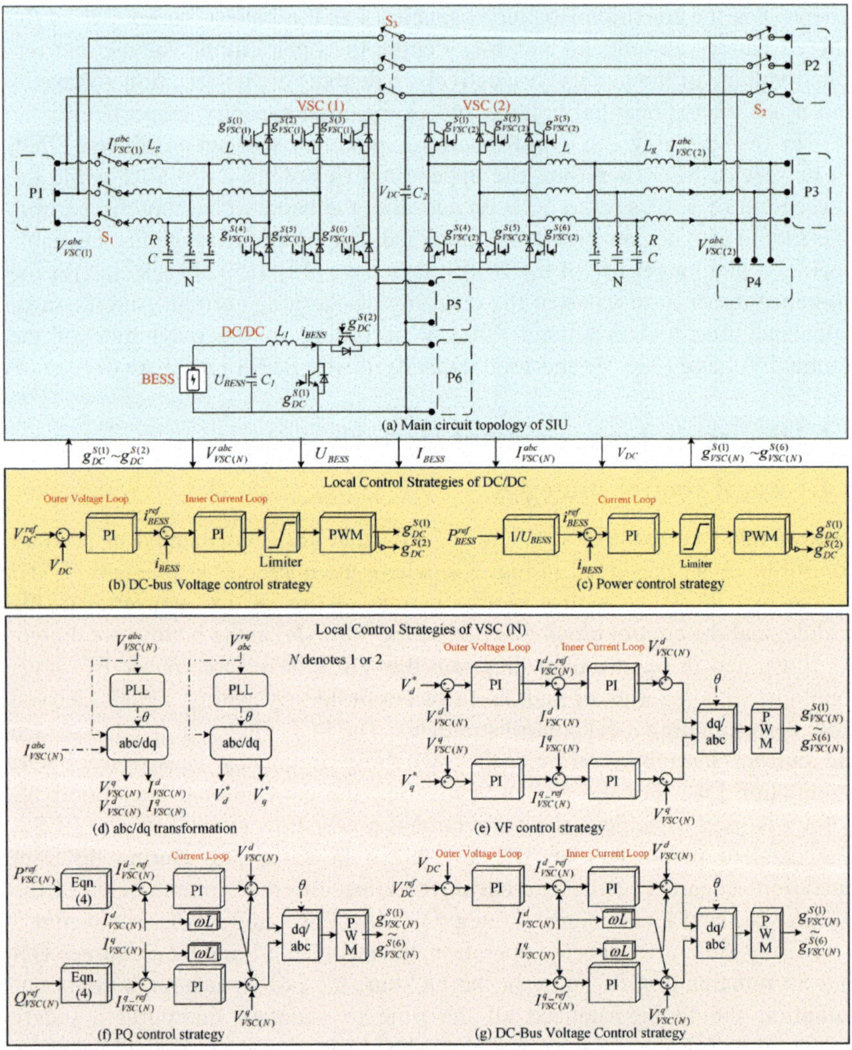

FIGURE 5.4 Main circuit topology and overall local control strategies of SIU.

outer-loop PI controller to generate i_{BESS}^{ref} for the inner current loop. Then the output signals from the inner-loop PI controller are processed by the PWM signal generator to generate driver signals $g_{DC}^{S(1)}$ and $g_{DC}^{S(2)}$ for the switch elements of DC/DC.

Since the utility grid effectively supports the AC voltage and frequency of the microgrid, the VSC (1) is able to work on the grid-feeding mode to absorb/deliver prespecified value of power from/to the utility grid. The active and

TABLE 5.3 Paremeters of SIU.

Symbol	Parameter	Value
L	Inverter side inductance of LCL filter/H	500×10^{-6}
L_g	Grid side inductance of LCL filter/H	100×10^{-6}
R	Series resistance of LCL filter/W	1
C	Filter capacitance of LCL filter/F	30×10^{-6}
C_1	Capacitance of output side of BESS/F	3×10^{-3}
C_2	Filter capacitance of DC-bus/F	1.2×10^{-2}
L_1	Filter inductance of DC/DC/H	3.04×10^{-4}
f_{VSC}	PWM carrier frequency of VSC/Hz	15,000
$f_{DC/DC}$	PWM carrier frequency of DC/DC/Hz	5000

reactive power (PQ) control strategy, as shown in Fig. 5.4f, is utilized by VSC (1) to perform the power control. The voltage feedback signals ($V^{abc}_{VSC(1)}$) and current feedback signals ($I^{abc}_{VSC(1)}$) are transformed to dq-frame by abc/dq transformations (depicted in Fig. 5.4d) to simplify control loop design. And the reference power signals ($P^{ref}_{VSC(1)}$, $Q^{ref}_{VSC(1)}$) are transformed to the reference current signals by Eq. (5.4). Since the SIU is designed to only provide active power during grid-connected operation, $Q^{ref}_{VSC(1)}$ is set to zero.

$$\begin{bmatrix} I^{d\_ref}_{VSC(1)} \\ I^{q\_ref}_{VSC(1)} \end{bmatrix} = \frac{1}{{V^{d}_{VSC(1)}}^{2} + {V^{q}_{VSC(1)}}^{2}} \begin{bmatrix} V^{d}_{VSC(1)} & V^{q}_{VSC(1)} \\ V^{q}_{VSC(1)} & -V^{d}_{VSC(1)} \end{bmatrix} \begin{bmatrix} P^{ref}_{VSC(1)} \\ Q^{ref}_{VSC(1)} \end{bmatrix} \tag{5.4}$$

TABLE 5.4 SoC condition.

Condition	The range of SoC (%)
I	10% < SoC value < 90
II	SoC value reaches 90% and afterward do not drop to 80
III	SoC value reduces to 10% and afterward do not recover to 20

The generated $I_{VSC(1)}^{d\_ref}$ and $I_{VSC(1)}^{q\_ref}$ are tracked by the measured current signals $I_{VSC(1)}^{d}$ and $I_{VSC(1)}^{q}$ through the inner current loop that are given in Eq. (5.5) It should be noted that the cross-coupling term $\omega L I_{VSC(1)}^{q}$ and $\omega L I_{VSC(1)}^{d}$ are eliminated by the feed-forward decoupling control to decouple the control variables of dq axes for better control performance [51]. Finally, the obtained voltage reference U_{d} and U_{q} are transformed back to abc coordinates by dq/abc transformation in order to generate drive signals $g_{VSC(1)}^{S(1)} \sim g_{VSC(1)}^{S(6)}$ by PWM modulation.

$$\begin{cases} U_d = V_{VSC(1)}^{d} + \omega L I_{VSC(1)}^{q} + \left(K_p + \frac{K_i}{s}\right)\left(I_{VSC(1)}^{d\_ref} - I_{VSC(1)}^{d}\right) \\ U_q = V_{VSC(1)}^{q} - \omega L I_{VSC(1)}^{d} + \left(K_p + \frac{K_i}{s}\right)\left(I_{VSC(1)}^{q\_ref} - I_{VSC(1)}^{q}\right) \end{cases} \tag{5.5}$$

5.4.1.2 *Condition II and III*

In these two conditions, the DC/DC converter works on the power control mode, in which the output power of the BESS is determined by P_{BESS}^{ref} . Thus, overcharging and overdischarging of the BESS can be avoided by setting the proper power reference value. Fig. 5.4c shows the control block diagram of the power control strategy of DC/DC. Firstly, i_{BESS}^{ref} is generated by P_{BESS}^{ref}. Then the inner current loop regulates the output current of BESS to track i_{BESS}^{ref}.

The VSC (1) switches to control the DC-bus voltage. As illustrated in Fig. 5.4g, the DC-bus voltage control strategy is composed of the outer voltage loop and inner current loop. The outer voltage loop generates $I_{VSC(1)}^{d\_ref}$ for inner active current controller, and $I_{VSC(1)}^{q\_ref}$ for inner reactive current controller is set to zero to realize the unity power factor operation of VSC (1). Additionally, regardless of the SoC value, the VSC (2) always works on the grid-feeding mode to realize power control during the grid-connected operation.

In the islanded operation, the switches S1 and S2 are in OFF state. And the S3 is turned on, thus the power exchange between the VSC (1) and VSC (2) are available. Due to lack of the electricity support from the utility grid, the VSC (2) should work on the grid-forming mode to provide stable frequency and voltage for the AC subgrid. The adopted VF control strategy is depicted in Fig. 5.4e. In this control loop, the outer voltage loop controls the generated voltage to match the amplitude and frequency of V_{abc}^{ref}, and the inner current loop regulates the output current to balance the power deficit in AC subgrid. The VSC (1) switches to operate in the grid-feeding mode to deliver specified output power by tracking the AC voltage and frequency generated by VSC (2). Meanwhile, the BESS becomes the unique component to stabilize the DC bus voltage regardless of its current SoC value. Table 5.5 summarizes the local control strategies of the DC/DC and VSCs.

5.4.2 Power flow dispatching strategies

The power flow control in secondary control level is conducted by the PFC of SIU to realize an appropriate power transfer between the subgrids. For the power flow control strategy, the power references (P^{ref}_{BESS}, $P^{ref}_{VSC(1)}$, $P^{ref}_{VSC(2)}$) are synthesized for the local control loop of the converters in SIU. The strategy exhibits the following benefits: 1) it can realize mutual power support between subgrids; 2) it can be implemented only based on local information of the SIU during grid-connected operation of the microgrid; 3) the BESS can be controlled to operate with appropriate SoC and charging/discharging power during steady state. Similar to the local control strategy, the power flow control strategy of SIU is designed concerned with the following scenarios. As defined here, the power direction from DC to AC side is positive, and the reverse is negative. For BESS, positive power represents that it performs discharging action, and the reverse is charging action.

During grid-connected operation, the power flow dispatching strategy can be further described under the following conditions.

5.4.2.1 Condition I

In this condition, the DC-bus voltage is stabilized by the DC/DC and the power flow control strategy generates the power reference value for the VSCs. This strategy is developed based on the priority of RESs utilization that is represented as $\lambda^{\mathrm{current}}_{\mathrm{sub-grid}} > \lambda^{\mathrm{adjacent}}_{\mathrm{sub-grid}} > \lambda_{\mathrm{BESS}} > \lambda_{\mathrm{grid}}$, which assigns the highest priority to the output of DC/AC DGs to be consumed by the respective load in DC/AC subgrid. The second priority aims to use the surplus power in a certain subgrid transferred to balance the power deficit in adjacent subgrid. Afterward, the third priority focuses on utilization of BESS to absorb/supply the excess/

TABLE 5.5 Local control strategies of DC/DC and VSCs.

Operation scenario	SoC	DC/DC	VSC (1)	VSC (2)
Grid-connected operation	Condition I	V_{DC} control	PQ	PQ
	Condition II	Power control	V_{DC} control	PQ
	Condition III	Power control	V_{DC} control	PQ
Islanded operation	/	V_{DC} control	PQ	VF

deficient power of the microgrid. Finally, the utility grid has the lowest priority to make a final power balance in microgrid.

As illustrated in Fig. 5.5, the power flow control strategy contains eight modes under this scenario.

Mode 1 ($\Delta P_{DC} >= 0\& \ |P_{\text{BESS}}| >= P_{\text{char}}^{\text{lim}} \& \ \Delta P_{AC} >= 0$): In this mode, the power in AC and DC subgrid is surplus, revealing that the two subgrids are in light-load condition. In addition, the charging power of BESS exceeds the upper limit. According to the priority mentioned before, the VSC (2) will work on the idle mode with no power transfer executed. And the surplus power in AC subgrid is autonomously balanced by utility grid. To limit BESS charging power, the excess charging power that is beyond upper limit is fed to the utility grid through VSC (1). The power commands for VSC (1) and VSC (2) are expressed as:

$$P_{VSC(1)}^{ref} = \Delta P_{DC} - P_{\text{char}}^{\text{lim}}, P_{VSC(2)}^{ref} = 0 \tag{5.6}$$

Mode 2 ($\Delta P_{DC} >= 0\& \ |P_{Bs}| > P_m^P \& \ bP_a < 0$): In this mode, the AC subgrid operates in heavy-load condition. And the power deficiency in AC subgrid can be supplemented by DC subgrid through the power dispatching of VSC (2). Similar to mode 1, the limitation for charging power of BESS should also be considered.

$$P_{VSC(1)}^{ref} = \max\left(\Delta P_{DC} - P_{\text{char}}^{\text{lim}} + \Delta P_{AC}, 0\right), P_{VSC(2)}^{ref} = -\Delta P_{AC} \tag{5.7}$$

Mode 3 ($\Delta P_{DC} >= 0\& \ |P_{Bs}| < P_m^P \& \ \Delta P_{AC} >= 0$): Since the charging power of the BESS is within the boundary, the VSC (2) can deliver surplus

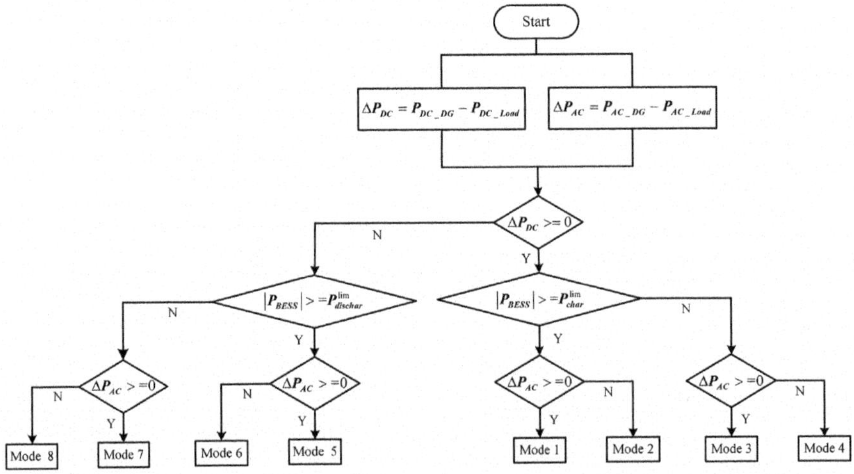

FIGURE 5.5 Proposed power flow control strategy in condition of grid-connected operation and SoC within normal range.

power from AC to DC subgrid for battery charging to increase the local consumption of DGs output. And the VSC (1) remains in idle state.

$$P^{ref}_{VSC(1)} = 0, P^{ref}_{VSC(2)} = -\min(\Delta P_{AC}, P^{\lim}_{\text{char}} - \Delta P_{DC} \tag{5.8}$$

Mode 4 ($\Delta P_{DC} >= 0\& \; |P_{\text{BESS}}| < P^{\lim}_{\text{char}}\& \; \Delta P_{AC} < 0$): The VSC (1) operates in the idle mode, and the BESS power and surplus power in DC subgrid can be dispatched to the AC subgrid for power balance under considering the limitation of discharging power of BESS.

$$P^{ref}_{VSC(1)} = 0, P^{ref}_{VSC(2)} = \min\left(-\Delta P_{AC}, \Delta P_{DC} + P^{\lim}_{\text{dischar}}\right) \tag{5.9}$$

Mode 5 ($\Delta P_{DC} < 0\& \; |P_{\text{BESS}}| >= P^{\lim}_{\text{dischar}}\& \; \Delta P_{AC} >= 0$): In this mode, the surplus power in AC subgrid can be conveyed to the DC subgrid to compensate the power deficit. Meanwhile, the battery discharging power could be limited by the power control of VSC (1).

$$P^{ref}_{VSC(1)} = \min\left(\Delta P_{DC} + P^{\lim}_{\text{dischar}} + \Delta P_{AC}, 0\right), P^{ref}_{VSC(2)} = -\Delta P_{AC} \tag{5.10}$$

Mode 6 ($\Delta P_{DC} < 0\& \; |P_{\text{BESS}}| >= P^{\lim}_{\text{dischar}}\& \; \Delta P_{AC} < 0$): In this mode, the power deficiency in AC subgrid is smoothly balanced by utility grid, and the VSC (2) enters the idle mode. The utility grid compensates the part of the discharging power of BESS that exceed $P^{\lim}_{\text{dischar}}$ by the power dispatching of VSC (1).

$$P^{ref}_{VSC(1)} = \Delta P_{DC} + P^{\lim}_{\text{dischar}}, P^{ref}_{VSC(2)} = 0 \tag{5.11}$$

Mode 7 ($\Delta P_{DC} < 0\& \; |P_{\text{BESS}}| < P^{\max}_{\text{dischar}}\& \; \Delta P_{AC} >= 0$): Since the BESS power can satisfy the load demand in DC subgrid within the boundary, the VSC (1) enters the idle mode. Meanwhile, the VSC (2) dispatches the surplus power in AC to DC subgrid.

$$P^{ref}_{VSC(1)} = 0, P^{ref}_{VSC(2)} = -\min\left(\Delta P_{AC}, P^{\lim}_{\text{char}} - \Delta P_{DC}\right) \tag{5.12}$$

Mode 8 ($\Delta P_{DC} < 0\& \; |P_{\text{BESS}}| < P^{\max}_{\text{dischar}}\& \; \Delta P_{AC} < 0$): In this mode, the BESS could meet the power demand of the DC subgrid. And the utility grid supports the AC loads. Both the VSC (1) and VSC (2) are in the idle mode.

$$P^{ref}_{VSC(1)} = 0, P^{ref}_{VSC(2)} = 0 \tag{5.13}$$

5.4.2.2 Condition II

Under this condition, the VSC (1) switches to control the DC-bus voltage. The utilization priority for RESs is the same as that of condition I. The four modes and corresponding power references for DC/DC and VSC (2) are listed in Table 5.6. In condition of $\Delta P_{DC} >= 0$, if $\Delta P_{AC} >= 0$, the DC/DC and VSC

TABLE 5.6 Power references for mode 9 to 12.

Mode	Condition	Reference power
9	$\Delta P_{DC} >= 0 \& \Delta P_{AC} >= 0$	$P_{\text{BESS}}^{ref} = 0, P_{VSC(2)}^{ref} = 0$
10	$\Delta P_{DC} \geq 0 \& \Delta P_{AC} < 0$	$P' = -\min(\Delta P_{DC} + \Delta P_{AC}, 0)$ $P_{\text{BESS}}^{ref} = \min\left(P', P_{\text{dischar}}^{\text{lim}}\right)$ $P_{VSC(2)}^{ref} = P_{\text{BESS}}^{ref} + \Delta P_{DC}$
11	$\Delta P_{DC} < 0 \& \Delta P_{AC} >= 0$	$P_{VSC(2)}^{ref} = -\min(-\Delta P_{DC} \Delta P_{AC})$ $P_{\text{BESS}}^{ref} = \max\left(P_{VSC(2)}^{ref} - \Delta P_{DC}, P_{\text{dischar}}^{lim}\right)$
12	$\Delta P_{DC} < 0 \& \Delta P_{AC} < 0$	$P_{VSC(2)}^{ref} = 0$ $P_{\text{BESS}}^{ref} = \max\left(-\Delta P_{DC}, P_{\text{dischar}}^{\text{lim}}\right)$

(2) remain idle, and the surplus power in DC subgrid is transferred to utility grid by VSC (1). If $\Delta P_{AC} < 0$, the BESS and surplus power in DC subgrid can be used to compensate the power deficit in AC subgrid. Similarly, the limitation of discharging power of BESS should be considered. In condition of $\Delta P_{DC} < 0$, if $\Delta P_{AC} >= 0$, a certain amount of the AC surplus power is conveyed to DC subgrid to cover the power deficit. If $\Delta P_{AC} < 0$, the BESS satisfies the power demand of DC subgrid and the utility grid supplies power to AC loads.

5.4.2.3 Condition III

In this scenario, the low SoC value of BESS will decrease the operational capability of the microgrid to response to the emergency, such as the unplanned islanded phenomenon. In order to recover the SoC value quickly, the usage priority for RESs is adjusted to $\lambda_{\text{sub-grid}}^{\text{current}} > \lambda_{\text{BESS}} > \lambda_{\text{sub-grid}}^{\text{adjacent}} > \lambda_{\text{grid}}$. In this way, a higher priority is assigned to BESS to utilize the surplus power in DC subgrid for charging.

Table 5.7 shows the corresponding power reference for modes under this scenario. In condition of $\Delta P_{DC} >= 0$, if $\Delta P_{AC} >= 0$, the surplus power in both subgrids is used for the BESS charging. If $\Delta P_{AC} < 0$, the power deficiency in AC subgrid is provided by utility grid, and the surplus power in DC subgrid is utilized for BESS charging with a higher priority. In condition of $\Delta P_{DC} < 0$, if $\Delta P_{AC} >= 0$, the surplus power is injected into the DC subgrid for load demand and BESS charging. If $\Delta P_{AC} < 0$, the BESS and VSC (2) remain idle, and utility grid balances the all power deficit in microgrid.

TABLE 5.7 Power reference for mode 13 to 16.

Mode	Condition	Reference power
13	$\Delta P_{DC} >= 0 \& \Delta P_{AC} >= 0$	$P' = \min\left(\Delta P_{DC}, P_{char}^{lim}\right)$ $P_{VSC(2)}^{ref} = -\min\left(P_{char}^{lim} - P', \Delta P_{DC}\right)$ $P_{BESS}^{ref} = P_{VSC(2)}^{ref} - P'$
14	$\Delta P_{DC} >= 0 \& \Delta P_{AC} < 0$	$P_{BESS}^{ref} = -\min\left(\Delta P_{DC}, P_{char}^{lim}\right)$ $P_{VSC(2)}^{ref} = \max\left(P_{BESS}^{ref} + \Delta P_{DC}, 0\right)$
15	$\Delta P_{DC} < 0 \& \Delta P_{AC} \geq 0$	$P_{VSC(2)}^{ref} = -\Delta P_{AC}$ $P' = \max(\Delta P_{DC} + \Delta P_{AC}, 0)$ $P_{BESS}^{ref} = -\min\left(P', P_{char}^{lim}\right)$
16	$\Delta P_{DC} < 0 \& \Delta P_{AC} < 0$	$P_{VSC(2)}^{ref} = 0$ $P_{BESS}^{ref} = 0$

During islanded operation, the microgrid works on mode 17. The VF control strategy is utilized by VSC (2) to automatically achieve the power balance in AC subgrid. The power balance in DC subgrid is implemented by BESS. If the BESS SoC surpasses 90%, the DGs have to decrease their output power by coordination of the MGCC and related LCs to avoid overcharging of the BESS. If the SoC is lower than 10%, the load shedding should be implemented to avoid overdischarging. Moreover, the reference upper limit for BESS power do not take effect in this scenario. This aims to improve the continuity of power supply without the utility grid.

Particularly, when $\Delta P_{AC} \geq 0$, the AC DGs can meet the AC power demand, hence the output power of the VSC (2) decreases to zero, indicating that the VSC (2) only provides the reference of voltage and frequency for the AC subgrid. Meanwhile, The VSC (1) transfers the surplus power in AC to DC subgrid through the energy path S. The power reference for VSC (1) is given by:

$$P_{VSC(1)}^{ref} = -\max(\Delta P_{AC}, 0) \tag{5.14}$$

5.5 System operation performance

To validate the proposed hybrid AC/DC microgrid architecture and the hierarchical control strategy, a simulation model corresponding to Fig. 5.1 is built in the MATLAB®/SIMULINK platform. To approximate the real system, the switching model is preferred for the SIU modeling. To decrease model complexity and accelerate simulation speed, the average model is adopted for modeling of DGs. Based on the constructed control structure, the communication network of the microgrid is established in this model, in order to realize system coordination. It should be noted that the communication delay between different components is ignored in modeling. However, in real applications, the time delay may exist due to long communication distance, external disturbance or other factors. And it would affect system operation performance [52]. Thus, the effects of time delay on system control should be considered in the design of communication network for real application. In terms of verification, selected several typical cases are tested to demonstrate the performance of the microgrid in a certain mode and between mode transitions. The specifications of the microgrid model are given in Table 5.8.

5.5.1 Power flow from DC to AC subgrid

In this case, the power dispatching from DC to AC subgrid is investigated when the microgrid is in grid-connected operation. The simulation results are depicted in Fig. 5.6 with three modes involved. As shown in Fig. 5.6d, the SoC

TABLE 5.8 System parameters of hybrid microgrid.

Parameter	Value
Rated DC bus voltage	700 V
Rated AC bus voltage	380 V
Rated frequency of AC subgrid	50 Hz
Maximum power from DC DGs	600 kW
Maximum power from AC DGs	500 kW
Maximum DC load (include EV load)	600 kW
Maximum AC load (include EV load)	500 kW
Power factor of AC load	0.9
Transformation ratio of T1 and T2	10 kV/380 V
Transformation capacity	1000 kVA
Upper/Lower SoC threshold of BESS	90%/10%
Reference upper limit of BESS power	400 kW
Nominal BESS capacity	2400 Ah
Nominal BESS voltage	330 V

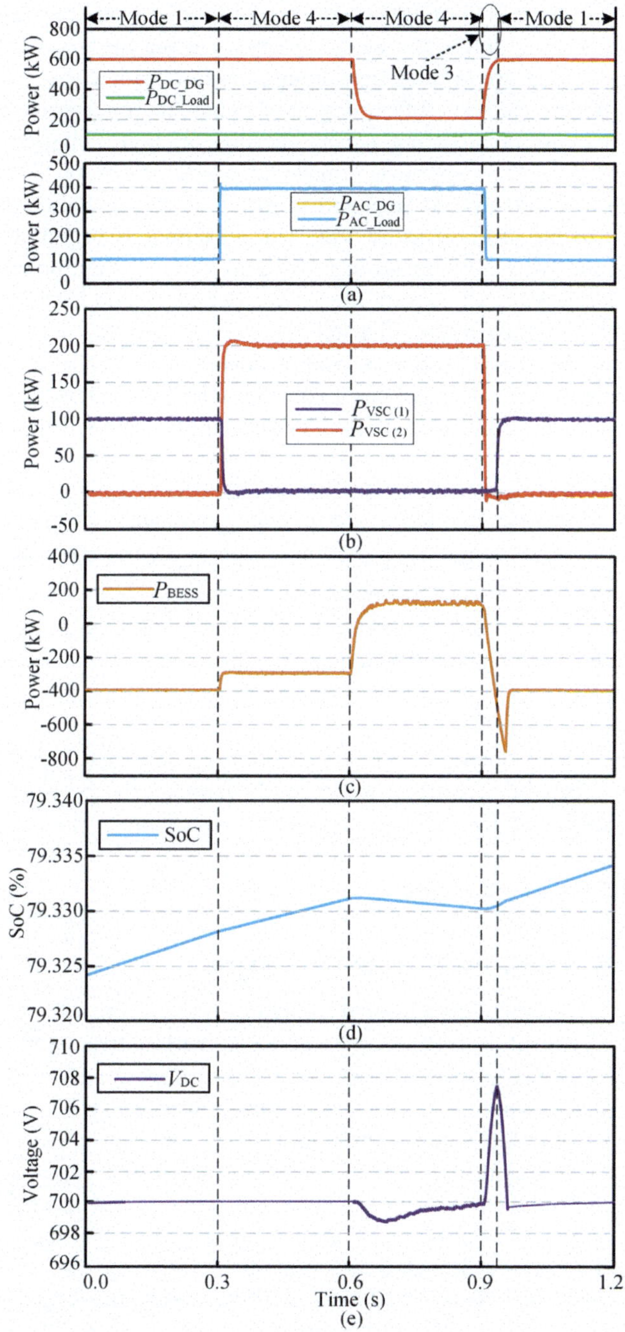

FIGURE 5.6 Simulation results for case 1: (a) Power of AC/DC DGs and loads; (b) Output power of VSC (1) and VSC (2); (c) BESS power; (d) SoC value of BESS; (e) DC-bus voltage of SIU.

of BESS is in the range of 10%–90%, therefore the local control strategies of SIU keep the same during the whole process. During the first 0.3 seconds, the dc and ac subgrid are both in light load condition with $\Delta P_{DC} = 500$ kW and $\Delta P_{AC} = 100$ kW calculated from Fig. 5.6a. Thus, the microgrid stays in mode 1. Since there is not existing power shortage in both subgrids, part of the surplus power in DC subgrid is consumed preferentially by BESS, which is charged at the (400 kW) as depicted in Fig. 5.6c; the remaining (100 kW) is transferred to the utility grid by VSC (1), as shown in Fig. 5.6b. At 0.3 seconds, with the AC load demand increasing to 400 kW, the microgrid switches to mode 4.The VSC (2) covers the power deficit (−200 kW) in AC subgrid through the executed power dispatching from DC to AC side. Afterward at 0.6 seconds, the sharp decline of output power of DC DGs causes a sudden power shortage. In response to that, the BESS smoothly switches from charging to discharging state at 100-kW discharging power. From 0.9 seconds, the power of DGs and loads are as the same as that of 0–0.3 seconds. Considering the response time of BESS, the BESS charging power takes 30 ms to reach P_{char}^{lim}. During this period, the microgrid operates in mode 3. Accordingly, VSCs remain idle and the excess power in DC subgrid causes a rapid rise in amplitude of DC bus voltage as shown in Fig. 5.6e. When $P_{\mathrm{char}}^{\mathrm{lim}}$ is reached, microgrid returns to mode 1. Afterward, with the power dispatching conducted by VSC (1), the BESS charging power is restored within $P_{\mathrm{char}}^{\mathrm{lim}}$ after reaching a peak value. Meanwhile, the DC voltage gets back to 700 V within 50 ms.

For the whole operation process, it can be concluded that the microgrid architecture can be controlled to smoothly balance the power deficit in AC subgrid with a stabilized system operation. Additionally, in steady state, the BESS charging power can be limited within $P_{\mathrm{char}}^{\mathrm{lim}}$ by the VSC (1)-enabled power delivery to utility grid.

5.5.2 Power flow from AC to DC subgrid

This case investigates the power dispatching from AC to DC subgrid in condition of microgrid in grid-tied state and SoC within 10%–90%. Initially, the microgrid operates in mode 7 with AC subgrid underloaded and DC subgrid overloaded as shown in Fig. 5.7a.The surplus power (200 kW) in AC subgrid is transferred to the DC subgrid through VSC (2) as Fig. 5.7b shows. And the power is consumed in two ways: half of it is used to compensate the power deficit in DC subgrid, and the remaining power is used to charge BESS as Fig. 5.7c shows. At 0.7 seconds, with a rapid rise in DC load demand, the generated DC power deficit is autonomously filled by the BESS with discharging power increasing to 200 kW. At 1.4 seconds, the output power of AC DGs decreases to 250 kW, causing the power delivered by VSC (2) dropping to 50 kW. Meanwhile, the output power of DC DGs declines to zero, causing

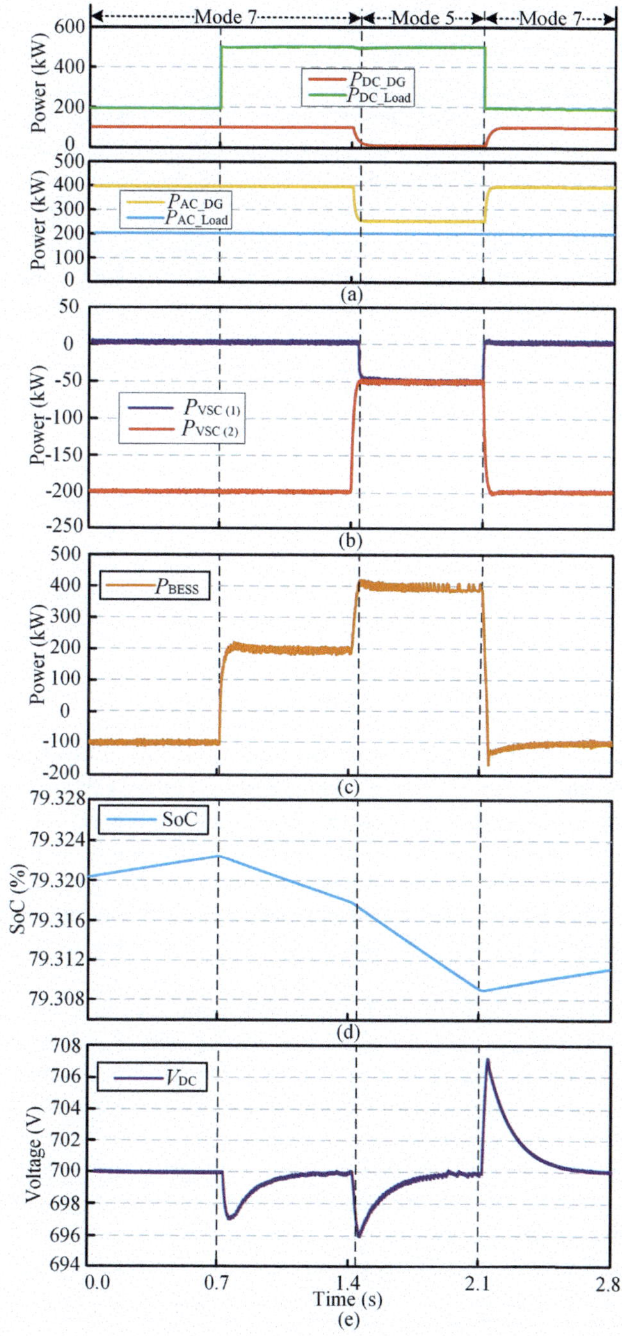

FIGURE 5.7 Simulation results for case 2: (a) Power of AC/DC DGs and loads; (b) Output power of VSC (1) and VSC(2); (c) BESS power; (d) SoC value of BESS; (e) DC-bus voltage of SIU.

the discharging power of BESS increasing continuously to balance the power deficit. Once the discharging power reaches $P^{\mathrm{lim}}_{\mathrm{dischar}}$ (400 kW), the microgrid switches to mode 5. The VSC (1) supplies the power of 50 kW from utility grid to the DC subgrid to limit the discharging power of BESS within $P^{\mathrm{lim}}_{\mathrm{dischar}}$. During the last 0.7 seconds, the microgrid returns to mode seven and its steady state is as the same as that of the first 0.7 seconds. As shown in Fig. 5.7e, the variation of DC-bus voltage does not exceed ±7 V in the whole process.

The case study reveals that the power dispatching from AC to DC subgrid could be realized autonomously with a stabilized operation, smooth mode transition and effective limitation for the discharging power of BESS.

5.5.3 Multiobjective dispatching results

5.5.3.1 DC-bus voltage regulator: from BESS to VSC (1)

In this case, the operation performance of the microgrid is investigated when the SoC value of BESS reaches the 90% (Condition II). At the beginning, the microgrid works on mode 3. Fig. 5.8b shows that the VSC (2) transfers the surplus power in AC subgrid (100 kW) to the DC subgrid. Plus ΔP_{DC} equaling to 200 kW, the BESS is charged at the power of 300 kW as shown in Fig. 5.8c. At 0.41 seconds, the SoC value of BESS reaches 90% as shown in Fig. 5.8d. To avoid overcharging of the BESS, the DC/DC converter quits stabilizing the DC-bus voltage and switches to power control mode with P_{BESS} adjusted to zero. Meanwhile, the VSC (1) switches to control the DC-bus voltage and the microgrid enters mode 9. For the mode transition, Fig. 5.8f shows that the phase voltage and current at the AC side of VSC (1) (P1) fluctuates first and then recovers to a steady and synchronized state within 50 ms. The corresponding output power of VSC (1) experiences an overshoot and stabilizes at 200 kW ultimately, implying that the power balance in DC subgrid is achieved by VSC (1). Meanwhile, after a drop, the DC-bus voltage rapidly stabilizes at 700 V within 400 ms as Fig. 5.8e shows. At 1.0 seconds, mode 11 is originated with the power deficiency (600 kW) in DC subgrid that compels the discharging power of BESS to reach the upper limit. Besides, the transferred power from AC subgrid (100 kW) and the offset power provided by VSC (1) (100 kW) compensate the rest of power shortage. At 1.5 seconds, the mode nine is activated and the operating points recovers to the steady state as the same as that of 0.5–1 seconds.

The simulation result indicates that the microgrid can recover to steady state quickly even faced with the switch of local control strategies of SIU when the SoC value of BESS reaches 90%. Additionally, the desired power dispatching and limits for the BESS power are also achieved.

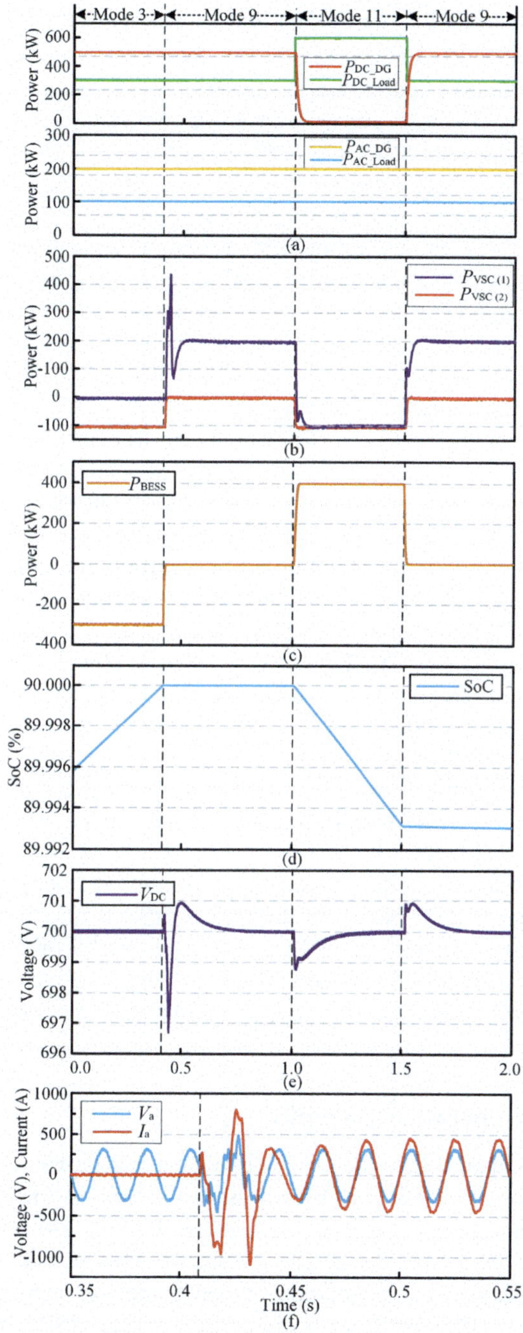

FIGURE 5.8 Simulation results for case 3: (a) Power of AC/DC DGs and loads; (b) Output power of VSC (1) and VSC (2); (c) BESS power; (d) SoC value of BESS; (e) DC-bus voltage of SIU; (f) Voltage and current profile at P1.

5.5.3.2 DC-bus voltage regulator: from VSC (1) to BESS

This case elaborates the performance of the microgrid when the SoC value of BESS is restored to 20% (condition III). From 0 to 0.5 seconds, the VSC (1) is responsible for the DC-bus voltage control. Fig. 5.9a shows that the microgrid operates in mode 15 with $\Delta P_{DC} = -300$ kW and $\Delta P_{AC} = 100$ kW. In addition to the 100-kW dispatched power by VSC (2) from AC to DC subgrid, the VSC (1) transfers the power of 200 kW from utility grid to DC subgrid to stabilize the DC-bus voltage as shown in Fig. 5.9b. During this period, Fig. 5.9c shows that the BESS remains in idle state to avoid the overdischarging. At 0.5 seconds, with the output power of DC DGs increasing to 600 kW and AC DGs decreasing to 0 kW, the microgrid switches to mode 14. As illustrated in Fig. 5.9c, the whole of the surplus power (300 kW) in DC subgrid is used for the BESS charging despite of existing power shortage in AC subgrid. Thus, the higher usage priority of RES for BESS charging in current condition is verified. At 1.55 seconds, the SoC value of BESS reaches 20%, thus the DC/DC converter takes over the responsibility of stabilizing the DC-bus voltage. And the VSC (1) returns to the power control mode. Thus, the mode four is activated. As depicted in Fig. 5.9f, the voltage and current at P1enters steady state within 100 ms after several periodic fluctuations during the mode transition. Moreover, Fig. 5.9b shows that the VSC (1) returns to idle state after a spike of its output power. And the VSC (2) delivers 100-kW power to the AC subgrid. Meanwhile the charging power of BESS decreases to 200 kW. This verifies that the RESs used for power support between subgrids have higher priority after the SoC value returns to 20%.

The simulation results demonstrate that the microgrid remains in a stabilized operation after mode transitions and the envisioned power flow is realized among the subgrids.

With respect to the certain instants of the above four cases, the local consumption rate of RES based on the typical hybrid microgrid architecture and control strategy mentioned in Ref. [18] is shown in the right column of Table 5.9. By comparison, it is obvious that the proposed microgrid architecture and control strategy in this chapter can increase the local consumption of RESs.

5.5.4 Islanded operation scenario

This case investigates the performance of the microgrid during islanded operation. Initially, as shown in Fig. 5.10a, the AC and DC subgrid both have the 100-kW power deficit, which is filled by BESS with 200-kW output power shown in Fig. 5.10c. At 0.5 seconds, an autonomous power balance is performed by VSC (2) to satisfy the increased load demand in AC subgrid shown in Fig. 5.10b. At 1.0 seconds, with the surplus power (100 kW) emerging in AC subgrid, the VSC (1) is controlled to transfer the surplus power to DC

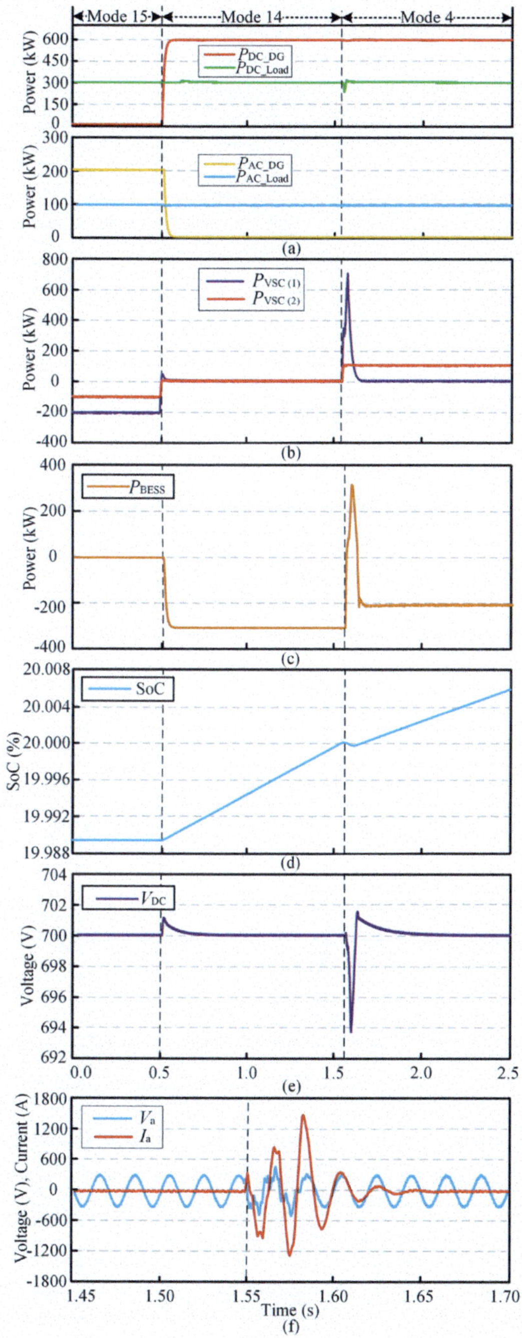

FIGURE 5.9 Simulation results for case 4: (a) Power of AC/DC DGs and loads; (b) Output power of VSC (1) and VSC (2); (c) BESS power; (d) SoC value of BESS; (e) DC-bus voltage of SIU; (f) Voltage and current profile at P1.

TABLE 5.9 Comparison of local consumption rate of RES.

Scenario	The time in scenario (s)	The local consumption rate of RES	
		This chapter (%)	Ref. [19] (%)
Case 1	0.2	75	25
Case 2	0.4	100	80
Case 3	0.2	100	57
Case 4	1	100	67

subgrid through the energy path S. And the output power of VSC (2) declines to zero, indicating that the VSC (2) only provides the reference of the voltage and frequency for the AC subgrid. In addition, the BESS power decreases to zero gradually after 1 seconds, showing that the total load demand of microgrid is solely satisfied by the DGs. Since no grid support is given during islanded operation of the microgrid, it is particularly crucial to keep the AC voltage and frequency stable in the AC subgrid. Based on the IEC standard 61000-2-2, the limitation for frequency deviation is set to 2%, viz. 1 Hz

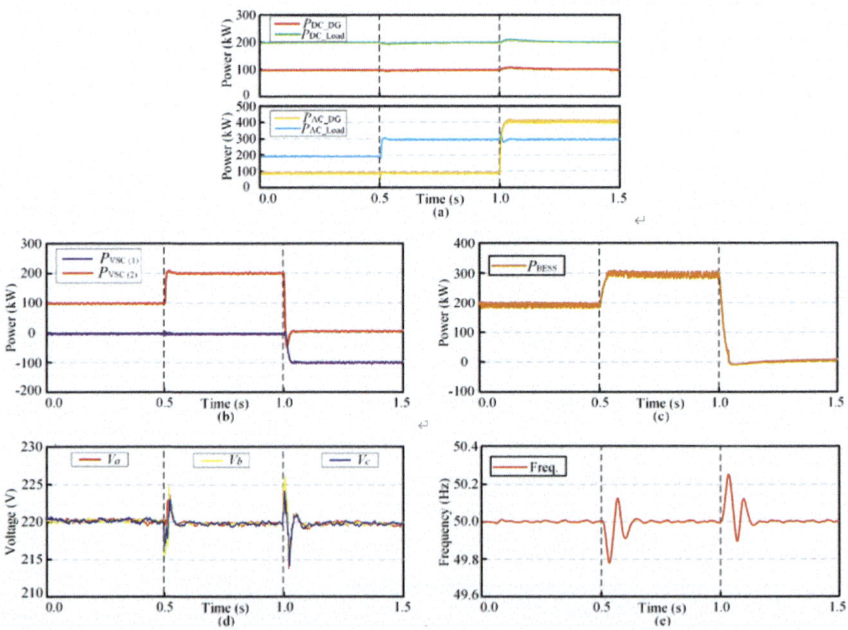

FIGURE 5.10 Simulation results for case 5: (a) Power of AC/DC DGs and loads; (b) Output power of VSC (1) and VSC (2); (c) BESS power; (d) AC RMS voltage; (e) AC frequency.

(absolute value). And according to the IEC standard 60038, the boundary for AC voltage deviation is set to 10%, namely 22 V (absolute value). The frequency and RMS voltage of phase a, b, c is presented in Figs. 5.10d and e. It can be observed that the maximum AC voltage and frequency deviation is 0.247 Hz and 7 V respectively, which are within the acceptable range.

This case reveals that the topology reconfiguration and power flow control during islanded operation improve the utilization efficiency of RES. In addition, the steady AC voltage and frequency are provided by SIU under the variance of DG output and load power.

5.6 Concluding remarks

In this chapter, a novel hybrid AC/DC microgrid architecture with a hierarchical control strategy is proposed to achieve nearly/net-zero-energy-targeted buildings. The proposed microgrid architecture is designed based on the two-stage ILC named SIU, which is composed of a B2B converter, a BESS, a DC–DC converter, as well as multiple ports and switches. The architecture is characterized by compact structure with multiple interfaces, reconfiguration ability and control flexibility. To reduce reliance on communication and achieve systematic coordination, a three-level control hierarchy of the proposed architecture is developed. Specifically, the dedicated interfaces and cluster controllers for EVs in secondary control level are designed to facilitate the centralized V2G control in microgrid. The operation scenarios of microgrid (grid-connected or islanded operation) and BESS SoC value determine the combination of the switch state and local control strategies of SIU in primary control level. In secondary control level, the power flow control strategy of SIU is developed to realize the desired power dispatching in microgrid. It is designed based on the utilization priority for RESs and contains 17 modes under considering the SoC and power boundaries of BESS, as well as the DG output and load power. The simulation results reveal that the proposed microgrid architecture and hierarchical control strategy can achieve a stabilized and flexible system operation under different scenarios. And the mutual power support between subgrids, accompanied with increased utilization and local consumption of RESs are realized.

The proposed microgrid architecture and control method provide a new vision for the design of nZEBs and could be applied in other scenarios, such as seaports and ships. Based on the current fundamental work, a detailed stability analysis of the microgrid system and experimental verification will be carried out in the future study. In addition, the interconnected-microgrids system is a promising solution to construct the energy of internet. Due to the merit of high integration of the SIU, it's meaningful to study the microgrid clusters that consist of the proposed microgrid architecture.

References

[1] Rahman MS, Hossain MJ, Lu J. Utilization of parked EV-ESS for power management in a grid-tied hybrid AC/DC microgrid. In: 2015 Australasian Universities Power Engineering Conference (AUPEC); 2015. p. 1–6.

[2] Strasser T, Andren F, Kathan J, Cecati C, Buccella C, Siano P, et al. A review of architectures and concepts for intelligence in future electric energy systems. IEEE Trans Ind Electron 2015;62(4):2424–38.

[3] Thomas D, Deblecker O, Ioakimidis CS. Optimal operation of an energy management system for a grid-connected smart building considering photovoltaics' uncertainty and stochastic electric vehicles' driving schedule. Appl Energy 2018;210:1188–206.

[4] Hannan MA, Faisal M, Ker PJ, Mun LH, Parvin K, Mahlia TMI, et al. A review of internet of energy based building energy management systems: Issues and recommendations. IEEE Access 2018;6:38997–9014.

[5] Li C, Ding Z, Zhao D, Yi J, Zhang G. Building energy consumption prediction: An extreme deep learning approach. Energies 2017;10:1525.

[6] Zhong H, Wang J, Jia H, Mu Y, Lv S. Vector field-based support vector regression for building energy consumption prediction. Appl Energy 2019;242:403–14.

[7] Robledo CB, Oldenbroek V, Abbruzzese F, Wijk AJM. Integrating a hydrogen fuel cell electric vehicle with vehicle-to-grid technology, photovoltaic power and a residential building. Appl Energy 2018;215:615–29.

[8] Gupta A, Doolla S, Chatterjee K. Hybrid AC-DC microgrid: Systematic evaluation of control strategies. IEEE Trans Smart Grid 2018;9(4):3830–43.

[9] Lede AMR, Molina MG, Martinez M, Mercado PE. Microgrid architectures for distributed generation: A brief review. In: 2017 IEEE PES Innovative Smart Grid Technologies Conference - Latin America (ISGT Latin America); 2017. p. 1–6.

[10] Bullich E, Díaz-González F, Aragüés-Peñalba M, Girbau-Llistuella F, Olivella-Rosell P, Sumper A. Microgrid clustering architectures. Appl Energy 2018;212:340–61.

[11] Nejabatkhah F, Li YW. Overview of power management strategies of hybrid AC/DC microgrid. IEEE Trans Power Electron 2015;30(12):7072–89.

[12] Hussain A, Bui V-H, Kim H-M. Microgrids as a resilience resource and strategies used by microgrids for enhancing resilience. Appl Energy 2019;240:56–72.

[13] Jianfang X, Peng W, Setyawan L, Chi J, Hoong CF. Energy management system for control of hybrid AC/DC microgrids. In: 2015 IEEE 10th Conference on Industrial Electronics and Applications (ICIEA); 2015. p. 778–83.

[14] Peng C, Zou J, Lian L, Li L. An optimal dispatching strategy for V2G aggregator participating in supplementary frequency regulation considering EV driving demand and aggregator's benefits. Appl Energy 2017;190:591–9.

[15] Jian L, Zheng Y, Xiao X, Chan CC. Optimal scheduling for vehicle-to-grid operation with stochastic connection of plug-in electric vehicles to smart grid. Appl Energy 2015;146:150–61.

[16] Zhang M, Chen J. The energy management and optimized operation of electric vehicles based on microgrid. IEEE Trans Power Deliv 2014;29(3):1427–35.

[17] Maryama V, Zeni V, Pica CQ, Ortmann MS, Heldwein ML. Unified hybrid (AC/DC) active distribution networks droop-based load-sharing strategy. In: IEEE PES innovative smart grid technologies, Europe; 2014. p. 1–6.

[18] Eghtedarpour N, Farjah E. Power control and management in a hybrid AC/DC microgrid. IEEE Trans Smart Grid 2014;5(3):1494–505.

[19] Malik S, Sun Y, Huang W, Ai X, Shuai Z. A generalized droop strategy for interlinking converter in a standalone hybrid microgrid. Appl Energy 2018;226:1056–63.

[20] Xia Y, Peng Y, Yang P, Yu M, Wei W. Distributed coordination control for multiple bidirectional power converters in a hybrid AC/DC microgrid. IEEE Trans Power Electron 2017;32(6):4949–59.

[21] Peyghami S, Mokhtari H, Blaabjerg F. Autonomous operation of a hybrid AC/DC microgrid with multiple interlinking converters. IEEE Trans Smart Grid 2018;9(6):6480–8.

[22] Dragičević T, Lu X, Vasquez JC, Guerrero JM. DC microgrids—Part I: A review of control strategies and stabilization techniques. IEEE Trans Power Electron 2016;31(7):4876–91.

[23] Burgio A, Menniti D, Sorrentino N, Pinnarelli A, Motta M. A compact nanogrid for home applications with a behavior-tree-based central controller. Appl Energy 2018;225:14–26.

[24] Olivares DE, Mehrizi-Sani A, Etemadi AH, Cañizares CA, Iravani R, Kazerani M, et al. Trends in microgrid control. IEEE Trans Smart Grid 2014;5(4):1905–19.

[25] Perez R, Silva C, Yazdani A. Control of an islanded power-electronic converter as an oscillator. In: 2016 IEEE energy conversion congress and exposition (ECCE); 2016. p. 1–6.

[26] Díaz NL, Luna AC, Vasquez JC, Guerrero JM. Centralized control architecture for coordination of distributed renewable generation and energy storage in islanded AC microgrids. IEEE Trans Power Electron 2017;32:5202–13.

[27] Lin P, Jin C, Xiao J, Li X, Shi D, Tang Y, et al. A distributed control architecture for global system economic operation in autonomous hybrid AC/DC microgrids. IEEE Trans Smart Grid 2019;10(3):2603–17.

[28] Shuai Z, Fang J, Ning F, Shen Z. Hierarchical structure and bus voltage control of DC microgrid. Renew Sustain Energy Rev 2018;82:3670–82.

[29] Yuan M, Fu Y, Mi Y, Li Z, Wang C. Hierarchical control of DC microgrid with dynamical load power sharing. Appl Energy 2019;239:1–11.

[30] Mi Y, Chen X, Ji H, Ji L, Fu Y, Wang C, Wang J. The coordinated control strategy for isolated DC microgrid based on adaptive storage adjustment without communication. Appl Energy 2019;252:113465.

[31] Lu X, Guerrero JM, Sun K, Vasquez JC, Teodorescu R, Huang L. Hierarchical control of parallel AC-DC converter interfaces for hybrid microgrids. IEEE Trans Smart Grid 2014;5:683–92.

[32] Wang J, Jin C, Wang P. A uniform control strategy for the interlinking converter in hierarchical controlled hybrid AC/DC microgrids. IEEE Trans Ind Electron 2018;65:6188–97.

[33] Majumder R, Ghosh A, Ledwich G, Zare F. Power management and power flow control with back-to-back converters in a utility connected microgrid. IEEE Trans Power Syst 2010;25 (2):821–34.

[34] Majumder R. A hybrid microgrid with DC connection at back to back converters. IEEE Trans Smart Grid 2014;5(1):251–9.

[35] Wang P, Jin C, Zhu D, Tang Y, Loh PC, Choo FH. Distributed control for autonomous operation of a three-port AC/DC/DS hybrid microgrid. IEEE Trans Ind Electron 2015;62 (2):1279–90.

[36] Yu X, She X, Ni X, Huang AQ. System integration and hierarchical power management strategy for a solid-state transformer interfaced microgrid system. IEEE Trans Power Electron 2014;29(8):4414–25.

[37] Wang Y, Li Y, Cao Y, Tan Y, He L, Han J. Hybrid AC/DC microgrid architecture with comprehensive control strategy for energy management of smart building. Int J Electr Power Energy Syst 2018;101:151–61.

[38] Becchio C, Corgnati SP, Delmastro C, Fabi V, Lombardi P. The role of nearly-zero energy buildings in the transition towards post-carbon cities. Sustain Cities Soc 2016;27:324–37.

[39] Liu Z, Zhou Q, Tian Z, He B, Jin G. A comprehensive analysis on definitions, development, and policies of nearly zero energy buildings in China. Renew Sustain Energy Rev 2019;114:109314.

[40] Karlessi T, et al. The concept of smart and NZEB buildings and the integrated design approach. Procedia Eng 2017;180:1316–25.

[41] Tori F, Bustamante W, Vera S. Analysis of net zero energy buildings public policies at the residential building sector: A comparison between Chile and selected countries. Energy Policy 2022;161:112707.

[42] Lu M, Sun Y, Dou W, Ma Z. Towards sustainable and resilient net/nearly zero energy buildings: A flexible energy framework integrating renewable energy, storage, and electric vehicles with grid-friendly interactions under uncertainties. Appl Energy 2026;402:127002.

[43] Hou Y, Zhu N, Zhang Z, Morovat N, Zhou W. Performance assessment of grid-connected PV-battery-EV integration in rural residential net-zero energy building. J Build Eng 2026;118:114916.

[44] Georgiou GS, Christodoulides P, Kalogirou SA. Optimizing the energy storage schedule of a battery in a PV grid-connected nZEB using linear programming. Energy 2020;208:118177.

[45] Zhan S, Dong B, Chong A. Improving energy flexibility and PV self-consumption for a tropical net zero energy office building. Energy Build 2023;278:112606.

[46] Hannan MA, Hoque MM, Mohamed A, Ayob A. Review of energy storage systems for electric vehicle applications: Issues and challenges. Renew Sustain Energy Rev 2017;69:771–89.

[47] Raijmakers LHJ, Danilov DL, Eichel RA, Notten PHL. A review on various temperature-indication methods for Li-ion batteries. Appl Energy 2019;240:918–45.

[48] Tremblay O, Dessaint LA. Experimental validation of a battery dynamic model. World Electr Vehic J 2009;3:1–10.

[49] Guo Z, Qiu X, Guangdong H, Liaw BY, Zhang C. State of health estimation for lithium ion batteries based on charging curves. J Power Sources 2014;249:457–62.

[50] Hanif M, Khadkikar V, Xiao W, Kirtley J. Two degrees of freedom active damping technique for filter-based grid connected PV systems. IEEE Trans Ind Electron 2014;61(6):2795–803.

[51] Gaubert J-P, Champenois G. An improved maximum power point tracking for photovoltaic grid-connected nverter based on voltage-oriented control. IEEE Trans Ind Electron 2011;58(1):66–75.

[52] Liu S, Wang X, Liu PX. Impact of communication delays on secondary frequency control in an islanded microgrid. IEEE Trans Ind Electron 2015;62:2021–31.

Chapter 6

Addressing difficult-to-charge issue in old residential communities

6.1 Introduction

Electric vehicles (EVs) are widely recognized as a promising solution to mitigate carbon emissions in densely populated areas as the energy and transportation industry shifts toward decarbonization [1]. Their integration not only reduces local air pollution but also aligns with global climate goals, making them a focal point of sustainable urban mobility strategies. In China, the EV industry has experienced remarkable growth in recent years, primarily due to the implementation of incentive policies and the continuous advancement of core technologies. Supportive measures such as purchase subsidies, tax exemptions, and license plate privileges have significantly accelerated consumer adoption. According to statistics released by the Chinese Transportation Bureau, China's EV stock has reached an impressive 13 million by the end of 2022, maintaining its position as the world's leader. As EVs assume an increasingly crucial role in the transportation sector, the planning and deployment of charging infrastructures have garnered extensive attention and research from various stakeholders, including government departments, grid operators, and automakers [2]. Ensuring the scalability, accessibility, and resilience of charging systems has become a pivotal challenge in the ongoing evolution of EV ecosystems.

However, the rapid proliferation of EVs is not uniformly feasible across all urban landscapes. A critical and challenging frontier for widespread adoption lies within the vast number of old residential communities, where physical and electrical constraints create a "last-mile" problem for charging infrastructure. These areas typically lack designated parking spaces, upgraded electrical wiring, and spatial allowances for installing modern charging equipment, severely restricting equitable access to EV charging services. Compared with newly planned residential areas, old communities are often characterized by legacy infrastructure, fragmented property management, and limited room for

Smart Penetration of Electric Vehicles in Distribution Grids. https://doi.org/10.1016/B978-0-443-27706-1.00007-6

coordinated renovation. As a consequence, the charging issue in such communities is not simply a technical problem of power supply, but also a combined problem involving land use, construction feasibility, operational convenience, and long-term system sustainability.

In terms of the power levels of charging infrastructures, EV charging can be divided into AC slow and DC fast charging modes [3]. The slow charging mode features a relatively low charging rate (typically, 3.3 kW or 7 kW in China), thus needing a rather long charging serving time. For instance, with a BYD Yuan EV equipped with a 45 kWh battery, it will take about 7 h to charge it from empty to full using a 7-kW charging power. The charging circuit of slow charging mode adopts the on-board charger, which favors the deployment in a car owing to its compact size, lightweight, and low cost [4]. Due to a long charging period, the slow charging mode is the mainstream for charging EVs at residential garages, workplaces, or roadside parking [5]. It is especially suited for overnight charging scenarios, where long dwell times align well with slow energy replenishment. As for the DC fast charging mode, the enabled charging power varies from tens of kilowatts to hundreds of kilowatts, thus decreasing the charging period dramatically compared with the AC counterpart. For example, the superchargers exclusive to Tesla vehicles could provide the fastest charging rate of 120 kW, shortening the refueling period from 200 miles down to below 30 min [6]. Unlike the on-board charger, the used off-board charger for fast charging is deployed in the parking spot with fewer concerns in weight and size. This allows for the integration of more advanced thermal management and power electronics, thereby supporting high power delivery and operational stability. The fast charging mode is intended to satisfy the urgent charging demand in a short period, thus usually applied in shopping complexes, hospitals, or at highways [7]. Its strategic deployment is vital for reducing range anxiety and enabling intercity travel, which are key factors in mass EV adoption. In June 19, 2023, according to a guideline released by the General Office of the State Council, China will further boost the building of a high-quality charging infrastructure network to meet the rapid development of the country's new energy vehicles (NEVs). The guideline emphasizes coordinated planning, intelligent management, and the integration of renewable energy sources to enhance the sustainability and efficiency of the charging network. From the perspective of residential application, however, neither charging mode can be adopted mechanically. Slow charging is highly dependent on the availability of fixed parking spaces and long residence times, whereas fast charging requires stronger power support, higher equipment investment, and more careful coordination with local grid constraints. Therefore, the selection of charging mode in old residential communities must be made in conjunction with spatial conditions, user charging habits, and the actual supply capability of the local distribution grid.

The availability of charging facilities is ensured not only through existing charging infrastructure deployment plans but also by administrative directives mandating allocation reservations for the future. Specifically, all newly constructed neighborhoods are required to have or reserve a 100% redundancy of installation conditions for chargers, with at least 10% of this ratio dedicated to public places [8]. This policy framework demonstrates the government's proactive stance in embedding charging accessibility into urban development from the outset. However, concerning old residential communities, the extensive deployment of charging infrastructures is not a simple task considering its realities. As per the General Office of the State Council of China, old residential communities can be described as residential neighborhoods constructed during an earlier era (e.g., 20 years ago), exhibiting inadequate maintenance, lack of proper management, limited municipal and community service facilities, and where residents express a strong desire for renewal and revitalization. Specifically, due to the delayed design planning and the fast-growing number of vehicles, the number of reserved parking spaces in old residential communities often cannot meet the rapidly growing demand for parking. To address the parking issue, gasoline vehicle owners have adopted a series of flexible parking methods, such as roadside parking or parking in nondesignated areas (e.g., green areas). These adaptations, though functional for internal combustion engine vehicles, expose significant infrastructure and regulatory gaps when transitioning to an electrified vehicle ecosystem. In contrast, EVs require dedicated charging points, rendering such ad hoc parking methods unsuitable and further exacerbating the parking dilemma. The fundamental incompatibility between the spatial randomness of flexible parking and the fixed-location nature of conventional charging piles constitutes a major logistical challenge in the electrification of these communities.In this context, relying solely on the conventional approach—namely, large-scale deployment of AC slow chargers—is infeasible in old residential areas due to the limited availability of fixed parking spaces. Furthermore, narrow installation corridors, complex renovation processes, and high retrofitting costs significantly hinder comprehensive infrastructure deployment. Thus, given the constraints on physical space and investment, and the necessity to serve a growing EV population, a centralized, space-efficient, and cost-effective charging station model is more suitable for these settings. Such centralized solutions not only optimize land use but also allow for more coordinated grid connections, maintenance, and future upgrades. The stereo garage, a high-capacity, space-saving parking system, has been considered a viable option to alleviate the parking issue; however, it is typically deployed in commercial contexts due to its high capital investment and stringent safety management requirements [9]. These facts indicate that the charging problem in old residential communities is structurally different from that in newly built neighborhoods. It cannot be solved simply by replicating the infrastructure standards of new developments. Instead, it calls for a renovation-oriented solution that can work within the constraints of existing parking patterns, existing feeders, and existing community spaces.

Besides the related construction issues of charging stations, the ever-increasing EV penetration would burden the distribution grids significantly. Since the charging load shares a large proportion of the household load level, the limited line/transformer capacity and the space scarcity for line corridors would hamper the deployment of charging facilities in sparse private parking spaces [10]. In addition, for residential EV owners, their charging demand often synchronizes, as used EVs are typically plugged into charging points upon returning home from their last daily travel [11]. This synchronization leads to a natural overlap between the traditional peak load and the generated charging loads during evening rush hours, typically occurring from 17:00 to 21:00, thus resulting in undesirable consequences for distribution grids. Some of the negative impacts include network congestion, voltage deviations, and increased power losses [12]. These issues are even more pronounced in old residential communities, where outdated electrical infrastructure and delayed grid modernization efforts further limit the ability to accommodate additional EV loads. As a result, such neighborhoods are more prone to power quality degradation, reduced operational efficiency, and limited EV adoption potential. The conventional response—upgrading grid infrastructure by replacing distribution lines and transformers with higher-capacity alternatives—is technically sound but economically and logistically impractical at scale. Especially in densely built-up urban areas, retrofitting requires extensive coordination, incurs high capital expenditure, and often faces resistance due to disruption of daily life. Therefore, a more feasible and strategic approach is needed to balance the competing demands of electrification, cost-efficiency, and infrastructure limitations.

To enhance the power transfer capacity and system operation performance of distribution grids, new grid architecture has gained widespread attention from industrial and academic communities. Soft open point (SOP) or the so-called flexible interconnect device (FID) is an emerging power electronic device that can be used in distribution grids. It could realize the flexible interconnection between individual feeders by replacing the conventional tie switch [13]. In Ref. [14], for the FID reconstructed 10 kV distribution grids, an improved system performance with power loss reduction, feeder load balancing, and increased distributed generators (DGs) penetration is to be achieved through the proper power dispatching of FID. The paper [15] reveals the significant economic merit of FID deployment in multiple microgrid planning compared with the conventional scheme of storage installation. In Ref. [16], a bi-level coordinated planning model of DGs and SOPs is introduced, considering a complete information dynamic game. The simulation results reveal that the balance of interests of different stakeholders, including DG investors, distribution companies, and electricity consumers, could be achieved along with enhancing the operation state of distribution grids. This stakeholder-aware perspective is critical for practical implementation, especially in urban areas with diverse and conflicting priorities. Haidar S. et al.

proposed an efficient strategy for the real-time operation of FID-interconnected microgrids in normal, islanding, and fault conditions [17]. A stabilized and coordinated system operation could be implemented. Despite these comprehensive studies on advanced grid architectures, one critical gap remains: the integration of electric vehicles (EVs) and charging stations is typically overlooked. Given the unique spatiotemporal characteristics of EV loads, such integration could significantly impact grid planning and operation outcomes. Moreover, the majority of the studies target medium-voltage distribution networks, which differ significantly from low-voltage networks, particularly in old residential communities, in terms of system topology, node density, and operational constraints. Such discrepancies limit the direct applicability of existing medium-voltage-oriented solutions to low-voltage scenarios.

The Solid-State Transformer (SST) is another emerging and highly integrated multistage power electronic device, showing great promise in enhancing grid capabilities through embedded intelligence and control flexibility [18]. In Ref. [19], an SST-based fast charging station architecture is introduced. By replacing power frequency transformers, the newly built architecture is of less space occupation and power losses. In Ref. [20], concerning the SST-based charging station, a rule-based energy management strategy is proposed to ensure the effectiveness of provided auxiliary services and the quality of charging services for EV owners. It can be seen that, different from the SOP (FID), the SST is used to build a new power system rather than conducting system reconstruction relying on existing infrastructures. Thus, the SST does not favor grid reconstruction in old residential communities. Lu Z. et al. introduced a novel reconstruction scheme for distribution grids, viz., converting the existing AC medium voltage line to DC through configuring a voltage source converter (VSC) [21]. The released transfer capacity of DC lines and the node voltage adjustment by the power transfer of VSCs enhance the EV penetration prominently. Yet, the low-voltage distribution grid scenario is not considered. In Ref. [22], a hybrid microgrid reconstruction strategy tailored for old residential areas is introduced. By implementing DC renovation on existing LV feeders, the scheme significantly enhances line transfer capacity and facilitates higher EV penetration. Nonetheless, the solution requires the deployment of VSCs at each building along the DC feeder, which increases system complexity and capital expenditure considerably. Furthermore, the issue of physical space occupation by charging stations is not addressed, representing a critical oversight in the context of dense urban environments where spatial constraints are paramount. Taken together, these studies show that improving charging availability in constrained residential areas requires more than higher power ratings or more advanced converters. The architecture must also be compact, renovation-friendly, and economically acceptable, while still being capable of coordinated operation under uncertain EV charging behavior.

It has been proven that the random nature of the charging schedule and EV owners' behavior is inconvenient to maintain a controlled and stable grid. In other words, if the charging regimes of EV clusters can be well arranged, for example, declining the charging demand in peak load period, the adverse effects on grid operation derived from EV charging could be mitigated, and more EVs' integration would be satisfied [23]. The coordinated power interaction between EVs and the power grid is generally referred to as vehicle-to-grid (V2G) technology [24]. The V2G technology can provide multiple auxiliary services for power grids, such as peak shaving and valley filling for load curves, voltage regulation, and frequency support [25]. These services contribute not only to operational efficiency but also to the economic dispatch and resilience of modern distribution networks. There has been a series of research focusing on the design of V2G dispatching strategies. In Ref. [26], a peak-shaving and valley-filling EV scheduling algorithm is developed considering six typical V2G scenarios. A desirable flattened load curve and node voltage levels are achieved in the targeted cases. Authors in Ref. [27] developed a mathematical model based on real-world data of power consumption and parking lot occupation. By the executed V2G dispatching, the peak power consumption of buildings is reduced by 20%. The above research puts emphasis on the V2G strategy design, and the considered grid architecture belongs to the typical AC grid architecture rather than the novel grid architecture. In Ref. [28], an SST-based photovoltaic-assisted charging station is introduced. Accompanied by the introduced autonomous energy management strategy, the frequency regulation command could be tracked timely, and the EV charging demand can be guaranteed. Unfortunately, the strategy fell short in addressing broader grid operational objectives, such as transformer load balancing and voltage profile restoration, which are particularly critical in stressed or capacity-constrained urban grids.

In this chapter, a cost-effective and high-efficient shared fast charging scheme is introduced for old residential communities to address the existing difficult-to-charge issue. The charging scheme comprises two aspects: charging mechanism and charging station architecture. The first one refers to the solution for organizing the regime of EV charging, such as charging upon arrival mode. The other one refers to the solution for establishing the electrical power system of charging stations. Once these two aspects are implemented, the corresponding fundamental charging services can be provided. The proposed approach aims to balance limited physical space, constrained power supply, and growing charging demand through a highly integrated design. Considering the limited construction spaces and power supply availability in old residential communities, the space-saving and high-efficient charging mechanism is preferred. Regarding the grid level, the introduced smart interlinking unit (SIU) is the critical component of the reconstructed grid architecture for charging stations, exhibiting multiple AC/DC interfaces, a dedicated DC power loop, and flexible control capability. The major contributions of this chapter are listed below:

- A cost-effective and high-efficient shared fast charging scheme is proposed for old residential communities. With this scheme, more EV accommodation is to be achieved in communities, along with less system investment and occupied areas compared with the conventional AC charging scheme.
- The SIU-based charging station architecture favors grid reconstruction by reusing existing power infrastructures. By replacing the typical long-distance AC power transfer, the SIU-enabled DC power transfer reduces line costs and improves average transfer efficiency by 1.3% compared with the AC counterpart.
- A hierarchical coordinated operation strategy is introduced for the reconstructed architecture. The comprised upper-level real-time V2G dispatching strategy considers the added fast charging load and uncertainty of EV charging behavior. An efficient solving and optimized load peak-shaving and valley-filling performance are to be realized.
- A lower-level coordination strategy is proposed based on the architecture features to enhance the short-timescale grid operation performance. The desired transformer load balance and DC voltage restoration could be implemented simultaneously.

The remainder of this chapter is organized as follows. It first examines the charging challenges in old residential communities. It then introduces the shared fast charging mechanism and the new charging station architecture, followed by the hierarchical management structure and the corresponding coordinated operation strategy. Two case studies are subsequently presented to validate the effectiveness of the proposed method. The chapter closes with concluding remarks.

6.2 Charging difficulties in old residential communities

6.2.1 Limited charging parking spaces

Old residential communities belong to the most basic units of life in cities, but are also the most vulnerable areas of cities [29]. Over a long period, the lack of regular maintenance and renovation has given rise to a series of issues, such as poor outdoor environment, aging of fundamental facilities, and crowded public spaces. Especially for the EVs accommodation in old residential communities, there are many difficulties to solve. Specifically, as shown in Fig. 6.1a, due to limited public spaces and lagging planning, there usually are not enough fixed parking spaces to accommodate ever-increasing vehicles, resulting in obstacles in building private or public charging facilities. Additionally, in Fig. 6.1b, the outdated grid infrastructure usually features limited power capacity and narrow power supply corridors, hampering the widespread deployment of charging facilities. Accordingly, as shown in Fig. 6.1c, many residents have to charge their EVs through private temporary wiring due to a lack of complete charging facilities. This kind of behavior implies a significant security and fire

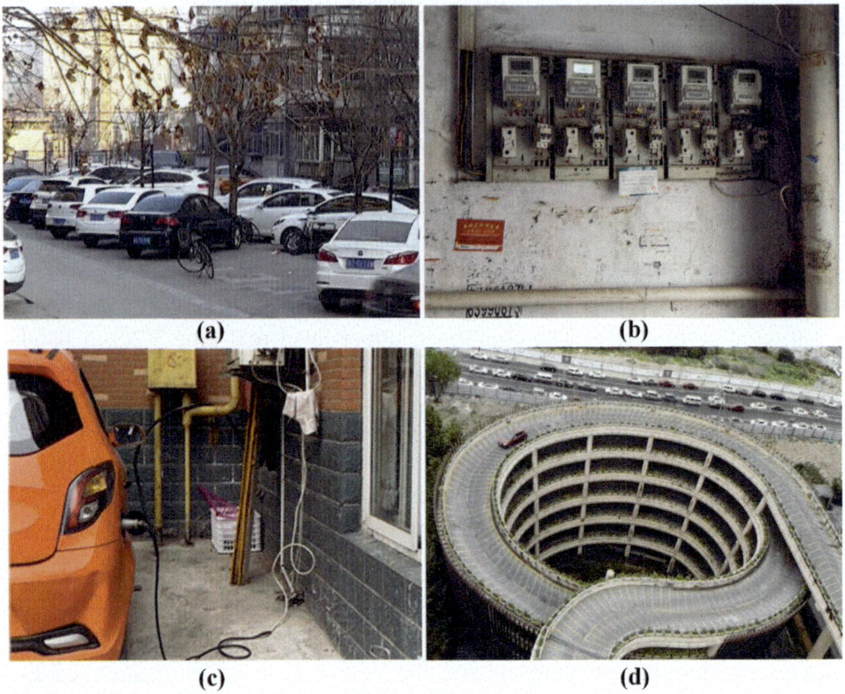

FIGURE 6.1 Related to EV charging in old residential communities. (a) Narrow public areas and limited fixed parking spaces; (b) aging of grid infrastructure; (c) Risky charging behavior through private temporary wiring; (d) stereo garage to save occupied spaces in commercial fields.

risk. According to statistics from the China State Council, there are more than 17,000 old residential communities such as these in China, covering over 100 million residents. Necessary policies and measures should be taken to address the EV charging issues of residents in old residential communities. Just like the stereo garage used in commercial fields as Fig. 6.1(d), it would be favorable to design a centralized public charging station featuring space saving and high utilization efficiency when considering the scarce space and electrical power resources in old residential communities.

6.2.2 Grid capacity and power quality issues

Old residential community is the basic component of life in the city and also belongs to the most vulnerable areas of the city. Owing to lacking daily maintenance and renewal in a long time, old residential communities are subject to varied kinds of issues, such as crowded narrow roads, poor environmental quality, limited public spaces, and old fundamental facilities. In terms of distribution grids, as illustrated in Fig. 6.2, the messy overhead lines (left up) and aging of distribution grid facilities (right up) are often

FIGURE 6.2 Real pictures: Messy overhead lines (*left up*); aging of distribution grid facilities (*right up*); limited charging point deployment (*left down*); EV charging through private temporary wiring (*right down*).

encountered in old residential communities. According to reports, in an old residential community in Beijing, the rated power capacity of each household is 2kVA, leading to limited charging point deployment (left down). In addition, due to lacking enough charging facilities, many residents charge their EVs through private temporary wiring (right down). This behavior implies a huge security and fire risk. All these factors convey that satisfying the increasing demand for EV charging in old residential communities is a systematic task and full of challenges.

When encountering large-scale EV integration, the distribution grids of old residential communities are also easily suffered from a series of power quality issues, as shown in Fig. 6.3. Initially, due to the delayed grid planning and limited current capacity of transfer lines, feeder congestion would occur if EV penetration level increases to a certain level. This point hinders EVs integration significantly. In addition, since the EV charger belongs to a nonlinear power electronic device, the generated harmonic distortion would be more prominent. As for the extensive single-phase EV/Grid connections, the aggravation of the voltage imbalances between phases needs more concern. Moreover, the increased energy losses and possible voltage violations need to be mitigated as EV penetration increases. Thus, in a long-term perspective, to further exploit the potential of EVs will require more renovations of

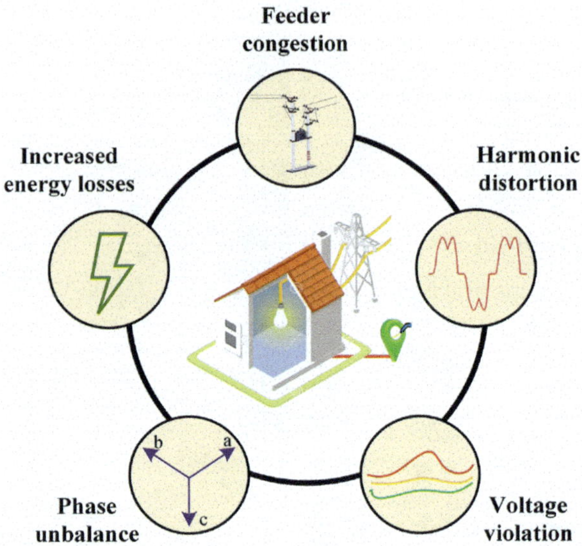

FIGURE 6.3 Power quality challenges.

distribution systems, as stated in the Energy Production and Consumption Revolution Strategies (2016–2030) of China [30].

According to statistics, these old residential communities without daily maintenance are more than 170,000, involving over 100 million residents in China [31]. Residents in old residential communities have the right to use electric vehicles, and thus meeting their charging needs is of great importance from the perspective of government policies orientation and EV promotion. In July 2020, the Chinese State Council has implemented policies for the renovation of old residential communities and the improvement of residents' quality of life, explicitly stating the need to take measures to meet the charging demands of EV users in old residential areas.

6.3 Shared fast charging scheme for old residential communities

6.3.1 Shared fast charging mechanism

In this chapter, a shared charging scheme based on DC fast charging points is proposed for old residential communities. Shared charging typically means one EV charger could service different EV owners when the charger is not in use or has completed the current charging task. The short time consumption of DC fast charging makes the shared destination charging service for residents feasible. Unlike the slow charging point, the fast EV charger could provide charging services for multiple EVs in the same period. We can imagine that if EVs could be scheduled for fast charging in order and leave the charging point for others when fully charged, the needed number of charging points could be much decreased compared with the conventional slow charging scheme. Accordingly, the needed occupied area of the built charging stations would decrease. Obviously, these merits are critical for old residential communities.

In order to implement the shared fast charging service, an interactive system used for information collection and EV schedule should be established. Fig. 6.4 presents an interactive interface for EV users embedded in a mobile APP or applet. Before the EV charging service, the EV users could upload related charging information through the interactive interface, for example, battery capacity, current state of charge (SoC), and desired SoC value. According to the rated charging power and service price, the estimated charging time and cost could be provided for EV users. Moreover, the interactive system could also order the registered EVs according to the corresponding registration time. At a certain time point, the EV with an earlier registration time would be arranged to charge preferably. The interface would display the reserved charging point, estimated waiting time, and charging starting time. When it comes close to the starting time, the EV user would receive related notifications to remind them to start charging in the reserved charging spot. Before the end of the charging period, the interactive interface would remind EV owners to leave the parking spot timely, thus facilitating the service provided for the following registered EV. Since the overnight charging demand of residents is not urgent, it can be imagined that EV owners have enough time to participate in the shared charging service and park their cars in other parking spots of the

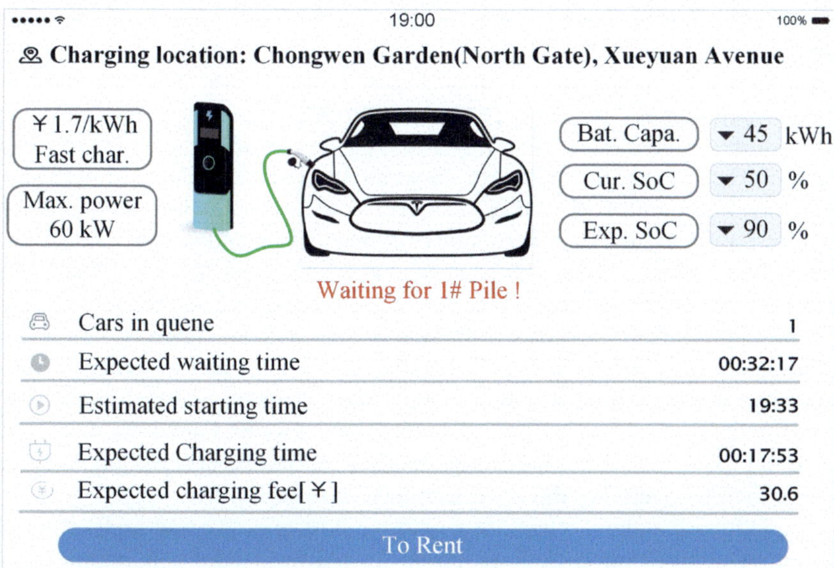

FIGURE 6.4 Interactive interface used for shared fast charging service.

community when the service finishes. Certainly, to increase the utilization efficiency of shared charging points, a specific economic penalty should be applied for EV users in the prolonged occupation of charging spots when the charging is completed.

6.3.2 SIU-based hybrid microgrid reconstruction

The conventional charging station architecture is illustrated in Fig. 6.5a. It can be seen that the typical low-voltage distribution grid of old residential communities usually comprises two or more radial feeders and operates in an open-loop state. Specifically, feeder-1 supplies 1#building and 2#building and nearby existing slow charging points, and feeder-2 supplies other load nodes. As depicted in Fig. 6.5b, the power distribution room comprises two individual 10 kV/400 V transformers (T1 and T2) to supply the two feeders. Concerning the conventional charging station architecture, a new-built AC feeder is induced from the power distribution room to power the terminal charging points, including the AC-powered slow or fast charging points. Regarding the AC-powered fast charging points, the used off-board charger usually comprises the AC/DC and DC/DC modules. The AC/DC module converts the AC voltage to the intermediate DC voltage, shaping the current flow for power factor correction and harmonic mitigation. The DC/DC module aims to generate the regulated voltage for EV battery charging and realizes the reliable isolation between the utility grid and EV battery oriented to the safety concern [32]. If the slow charging points are to be deployed, they are used for the AC line connection to EV charge inlet and communication with the on-board charger (OBC) [33]. Notably, the essential AC/DC conversion and charging control is realized by OBC.

Although it is a common solution to satisfy more EVs' charging demand through building AC-powered charging stations, some deficiencies in community application indeed remain. Initially, the conventional architecture lacks the power coordination capability between individual feeders, resulting in relatively low control flexibility and grid resiliency. A series of auxiliary services for girds, such as reactive power compensation and voltage regulation, can not be achieved without advanced power electronic devices. Then, the dedicated AC charging feeder usually adopts the three-phase, four-wire type, leading to prominent line costs if the transfer distance reaches a certain level. Additionally, the charging load generally shares a large proportion of the residential electrical load. The low-voltage, long-distance, and high-power AC power transfer would cause a significant power loss of lines [34]. Last, the remaining power capacity in different feeders can not be utilized flexibly and timely, thereby declining the EV accommodation capability.

To address the above mentioned issues, a cost-effective and high-efficient charging station architecture is proposed for old residential communities. As illustrated in Fig. 6.5b, the SIU would be integrated in the power distribution

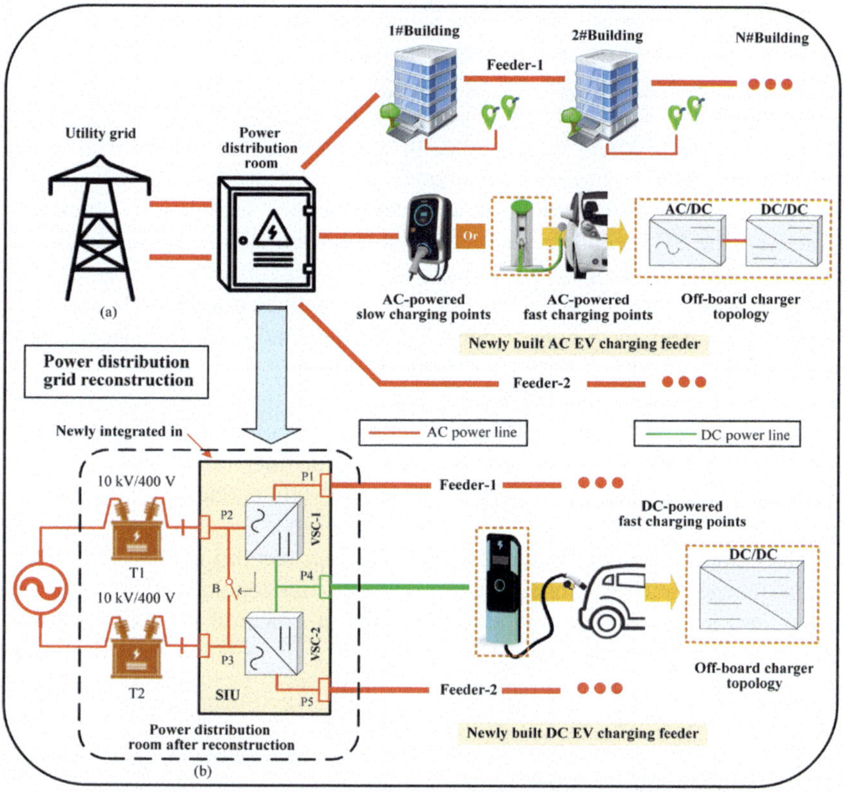

FIGURE 6.5 Grid reconstruction in old residential communities. (a) Conventional charging station architecture; (b) SIU-based charging station architecture.

room during the power distribution grid reconstruction. The SIU is an advanced power electronic device, which features multiple power conversion stages, various AC/DC ports, and flexible control capability. The contained voltage source converter 1 (VSC-1) and voltage source converter 2 (VSC-2) form a back-to-back structure comprising an intermediate DC link. The origin feeder-1/feeder-2 is accessed to AC port 1/port 5 (P1/P5), respectively. Similarly, the transformer T1/T2 is linked to AC port 2/port 3, respectively. Port 4 is a DC interface and provides a power supply for DC/DC charging modules in the terminal fast charging feeder. Notably, all configured ports have the function of supervision and protection. Thus, the SIU could acquire essential information for system operation management in real time, such as voltage magnitude, instantaneous power, and etc. Unlike the AC-powered fast charging points, the DC-powered fast charging points could only comprise the DC/DC conversion stage. Thus, the occupied space and cost of EV chargers could decline. In addition, the used two-wire DC line is regarded to be more efficient and economical compared with the AC counterpart when undertaking the same power transfer [34]. By the SIU integration in the power distribution room, the flexible power interaction between feeders could be realized, thus enhancing system flexibility and reliability. For instance, the emerging power deficit in a feeder could be mitigated by the surplus capacity of other feeders through the SIU's power dispatching. In this architecture, the existing transformers, AC lines, and power resources can be reused in the SIU reconstruction, thereby decreasing the reconstruction cost and complexity. Moreover, the SIU regulates the controllable switch B, which could enhance the power supply reliability. Specifically, if one transformer or SIU encounters a failure or maintenance situation, this switch would be closed, and the normal transformer covers the whole or partial power demand of feeders according to load classification, thus avoiding the power outage.

Table 6.1 shows a detailed comparison between the conventional charging station architecture and SIU-based charging station architecture. It is concluded that the proposed architecture has distinct merits over the conventional architecture in terms of control functionality and system performance.

The detailed configuration of DC port is illustrated in Fig. 6.6. The VSC-2 adopts the general three-phase two-level converter topology (VSC-1 as the same), and a symmetrical configuration without a neutral line is utilized for DC power transfer. The voltage on the positive or negative pole to ground (V_{DC}) is controlled to be equal and is half of the total DC voltage. Compared with the conventional AC power transfer through three-phase four-wire type, the two-wire DC power transfer could decrease the number of cores, thus decreasing the used line costs.

TABLE 6.1 Conventional charging station architecture versus SIU-based charging station architecture.

Item	Conventional architecture	SIU-based architecture
Power flow control capacity	×	√
Auxiliary services	×	√
DC interface	×	√
Interface supervision	×	√
Network losses	H[a]	L[b]
Line costs	H	L

[a]*H-high.*
[b]*L-low.*

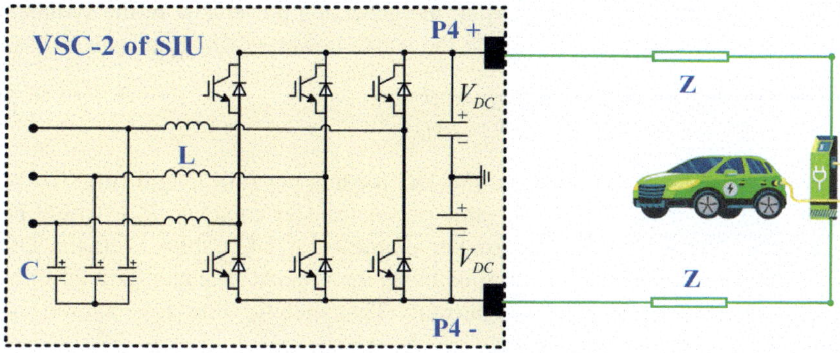

FIGURE 6.6 DC port configuration.

In China, the common rated voltage of low-voltage cables is 0.6/1 kV. And it has been demonstrated that the same cable insulation is able to withstand twice the RMS AC voltage in the condition of DC voltage applied [21]. Thus, the potential of withstand voltage capability of cables could be further exploited when executing DC power transfer. China Association of Building Energy Efficiency has released a design standard for emerging DC power distribution of civil buildings, which recommends the nominal voltage of DC feeders ($V_{DC\_BUS}^{nom}$) as 750 V. This recommended voltage could accommodate EV battery voltage and facilitate fast charging process.

For AC power transfer, the maximum allowed AC power is expressed as:

$$P_{AC} = \sqrt{3} U_{AC}^{nom} I_{AC} \cos\theta \tag{6.1}$$

where U_{AC}^{nom} denotes the rated line-line voltage of AC distribution grids (380 V); I_{AC} denotes the flowing AC current (A); $\cos\theta$ means the power factor. According to the EV integration standards of China (GB/T36278/GB/T29316), a unity power factor operation of EV chargers is preferred.

For DC power transfer through the symmetrical structure as shown in Fig. 6.6, the maximum allowed DC power is expressed as

$$P_{DC} = 2 V_{DC} I_{DC} \tag{6.2}$$

where I_{DC} indicates the flowing DC current (A).

Assumed the required maximum power transfer of AC and DC cable is equal, Equ. (1) and Equ. (2) could derive that the maximum I_{AC} is 1.18 times of I_{DC}, indicating that a smaller diameter of cables could be adopted in DC power transfer. The decreased core number and diameter of lines contribute to the significant decline in line costs.

In terms of the AC power transfer, the generated power loss is related to the active and reactive power flow *P*/*Q*. The corresponding equation is expressed as:

$$P_{\text{loss}}^{AC} = \frac{P^2 + Q^2}{U_{AC}^2} \tag{6.3}$$

where R_{AC} indicates the resistance of AC feeders per unit length (m).

In terms of the DC power transfer, the generated power loss in the symmetrical structure is only associated with the active power P, which is achieved as:

$$P_{\text{loss}}^{DC} = \frac{P^2}{4U_{DC}^2} R_{DC} \tag{6.4}$$

where R_{DC} indicates the resistance of DC feeders per unit length (m). As for the AC and DC feeders with the same power transfer capacity, R_{DC} would be higher than R_{AC} owing to the smaller diameter of DC cables. Yet, the DC power transfer may exhibit fewer line losses than the AC transfer derived from the significant decline of flow current. The specific line loss should be analyzed based on real scenarios.

6.3.3 Reconstructed grid features

As shown in Fig. 6.7a, the typical low-voltage distribution grid architecture of old residential communities often comprises two (or more) radial feeders and

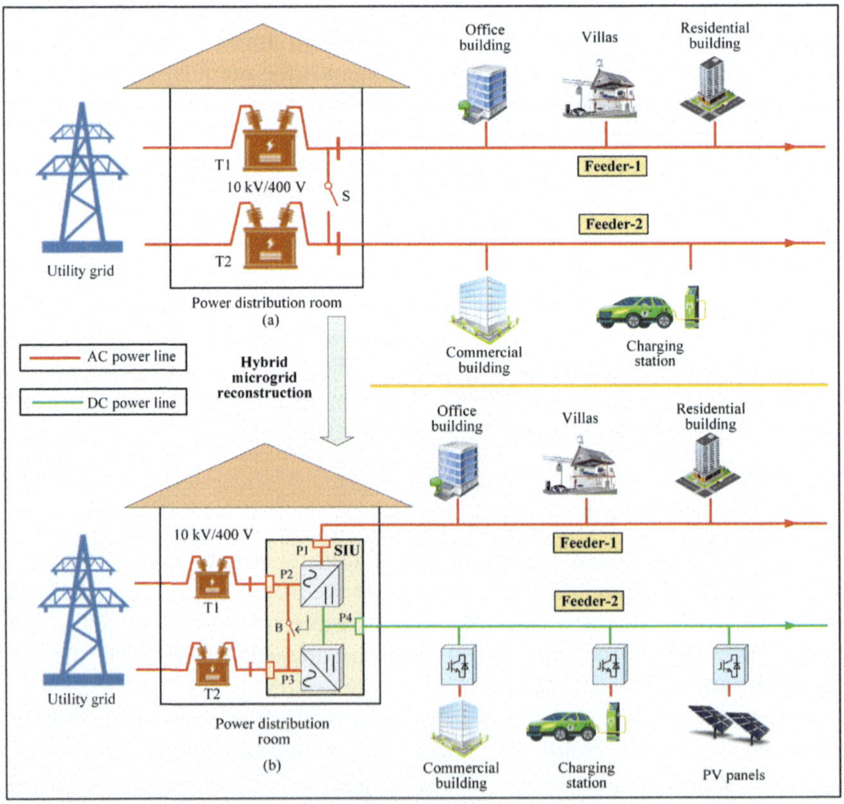

FIGURE 6.7 Distribution grid architecture of old residential communities. (a) Typical AC architecture; (b) SIU-based reconstructed architecture.

operates in an open-loop state. In feeder-1, the office building, villas, and residential building are clustered, and the commercial building, as well as a charging station are interfaced to the grid by the feeder-2. A power distribution room including two 10 kV/400 V transformers is used for powering the two feeders. In normal condition, the transformers T1 and T2 work individually, powering the loads of feeder-1 and feeder-2, respectively. Yet, in case of specific conditions, such as a transformer in fault or maintenance state, or happening power source failure, the normally open switch S would be closed, and the normal working transformer is responsible for supplying all or partial load demand of feeders depending on load classification. This dual-source redundant design conforms to the specifications of distribution grids (GB50053 published by the National Standardization Committee), and is widely adopted in residential grids, improving the reliability of power supply significantly.

Although it is a natural way to construct AC distribution grids through redundant configuration of power lines and transformers, there exist several deficiencies in reality when encountering EV penetration increment. First, there is no power coordination capability between different feeders, thus decreasing the control flexibility and grid resiliency. Second, as EV penetration rises, a series of power quality issues mentioned before would appear. Third, to accommodate more EVs, the update of AC distribution grids is regarded as a consensus. However, this scheme needs considerable investment and may not be feasible in the condition of space scarcity, especially for old residential communities. Last, considering the intermittent feature of EV charging behavior, the load rate of feeders would decrease furtherly after grid reinforcement.

To mitigate the defects of conventional AC grid architectures, a cost-effective and easy-to-implement hybrid microgrid reconstructed architecture is proposed for old community distribution grids. As shown in Fig. 6.7b, the previous AC feeder-2, which needs to comprise more DC loads and sources, is favorable for the DC reconstruction and operation. The existing two transformers and power lines are reused in this reconstruction scheme. And the major system reconstruction only needs to be executed in the power distribution room. The core component of this architecture is the SIU, which is of multiple AC/DC interfaces, power conversion stages, and flexible control functionalities. The voltage source converter 1 (VSC-1) and voltage source converter 2 (VSC-2) form a back-to-back (B2B) topology that encompasses a common DC bus. Without replacing the power lines of feeder-2, the DC feeder is formed by accessing the dedicated interfaces P4. The AC feeder-1 is remained, which links to the interface P1. The P2 and P3 are interfaced to the transformer T1 and T2, respectively. Notably, all of the interfaces possess supervision and protection functionalities, thus, a series of critical information such as frequency, voltage, and current magnitude is sampled by SIU in real time. Moreover, the controllable switch B is regulated by the SIU, and is to realize the redundancy design of the whole system. Specifically, when the failure of one transformer or utility source occurs, the normal-open switch B will be closed to guarantee the normal power supply from the other resource. Even if the SIU was in a fault state, this design also satisfies the redundant supply for AC feeders. The detailed circuit topology of SIU can be referred to Ref. [26]. Referred to the contents in the following sections, Table 6.2 shows a comprehensive comparison between the conventional grid update scheme and SIU-based reconstruction scheme. It can be concluded that the proposed reconstruction scheme shows prominent advantages over the conventional grid update scheme in terms of implementation difficulty and a series of technique indexes.

6.4 Hierarchical energy management with coordinated operation strategy

The enabled shared DC fast charging service and existing uncoordinated AC slow charging behavior would contribute to a higher peak-to-valley difference of load curves, thereby enhancing the risk of feeder congestion and lowering the operation efficiency. To realize the peak-shaving and valley-filling, as well as the coordinated system operation, a hierarchical management structure is established for the SIU-based architecture as depicted in Fig. 6.8. The

TABLE 6.2 Conventional grid update scheme versus SIU-based reconstruction scheme.

Item	Conventional update scheme	SIU-based reconstruction scheme
Power flow control capacity	×	√
Redundant design	√	√
DC interface	×	√
Power transfer capacity	M[a]	H
Power transfer capacity	H[b]	L[c]
Voltage drop	H	L
Implementation difficulty	H	L

[a]*M-medium.*
[b]*H-high.*
[c]*L-low.*

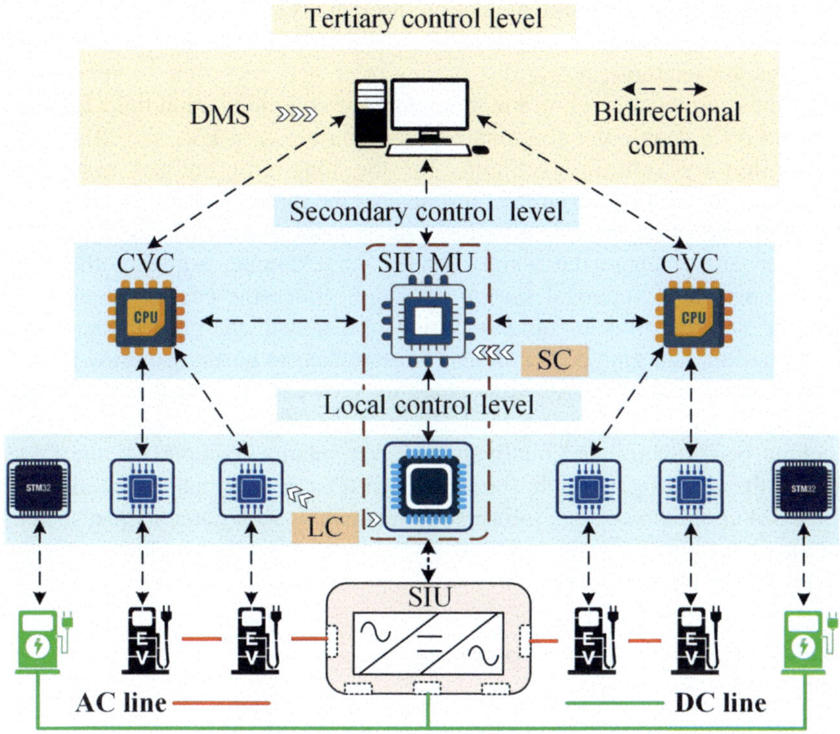

FIGURE 6.8 Hierarchical management structure.

hierarchical structure contains three control levels, viz., the local control level, secondary control level, and tertiary control level. The first two levels can be included in the lower level category, and the tertiary level belongs to the upper level.

Regarding the local control, it is performed locally based on measured signals and features a fast response (ms level) to ensure real-time system stability upon disturbance [35]. Specifically, the SIU management unit's (MU's) local controller (LC) handles the DC bus voltage stabilization and power flow control between feeders. As for the charging points, the embedded LCs are responsible for outputting needed voltage/current/power considering different charging demand. In this level, the controlled objects could remain in the desired operation state according to commands from upper control levels.

Regarding the secondary control, it features a relatively slower response (ms-s level). In this level, multiple clusters of vehicle controllers (CVCs) can be deployed in certain charging regions to conduct V2G schedules for slow-charged EVs. For this goal, the necessary EV information would be collected by CVCs and uploaded to the upper-level controller for dispatching reference. Furthermore, the generated schedule command is also transmitted to each charging point for implementation through CVCs' communication links. Additionally, the system controller (SC) of SIU MU would also make control decisions, for example, power dispatching reference, according to upper-level commands and local-level information. And the communication links between SC and CVCs enable the coordinated operation between EVs and SIU.

The tertiary control level aims for the long-term optimal systematic operation and exhibits the slowest response of the three levels (s-min level). In this level, the configured distribution management system (DMS) is used to supervise and manage the whole system, for example, realizing the V2G dispatching and economical system operation. Since the enabled shared fast charging service aims to quickly satisfy the added charging demand, the corresponding charging power should be stabilized in normal conditions. Yet, the remaining long-time AC slow charging EVs can be scheduled appropriately to flatten the emerging load climb. To compromise the desired dispatching performance and relatively low computation complexity, the DMS decides the charging schedule for supervised EVs in 15-minute intervals [36]. The DSM could also realize information interaction with outer agents, such as the distribution grid operator (DSO), to facilitate EVs' and buildings' participation in the electricity market [37].

6.4.1 Upper-level V2G dispatching

To mitigate the uncertainty of system operation in affecting the V2G dispatching results, the V2G schedule for slow-charged EVs adopts the rolling optimization pattern. The applied real-time V2G dispatching strategy is conducted by the DMS to realize the regulated charging profile. The following principles of the designed dispatching strategy should be abided:

- The charging demands of EV owners need to be satisfied before the set departure time.
- The load peak comprising the conventional load and added fast charging load should be flattened as low as possible.
- The surplus power capacity in the load valley period should be utilized properly for EV charging.
- The shifted slow charging load should not contribute to the new load peaks during the previous load valley period.

The principle of the proposed V2G dispatching strategy is illustrated in Fig. 6.9a. Just illustrated as the red line, the conventional load plus the added DC fast charging load, as well as the uncoordinated AC slow charging load, would elevate the whole load level significantly. The brown profile denotes the predicted total load comprising the conventional load and DC charging load, which of the uppermost/minimum value is expressed as $p_{\mathrm{upper}}/P_{\mathrm{lower}}$. Thus, the partial load profile that is over P_{upper} could be transferred to the valley period through the V2G dispatching, just marked as the red section. The core

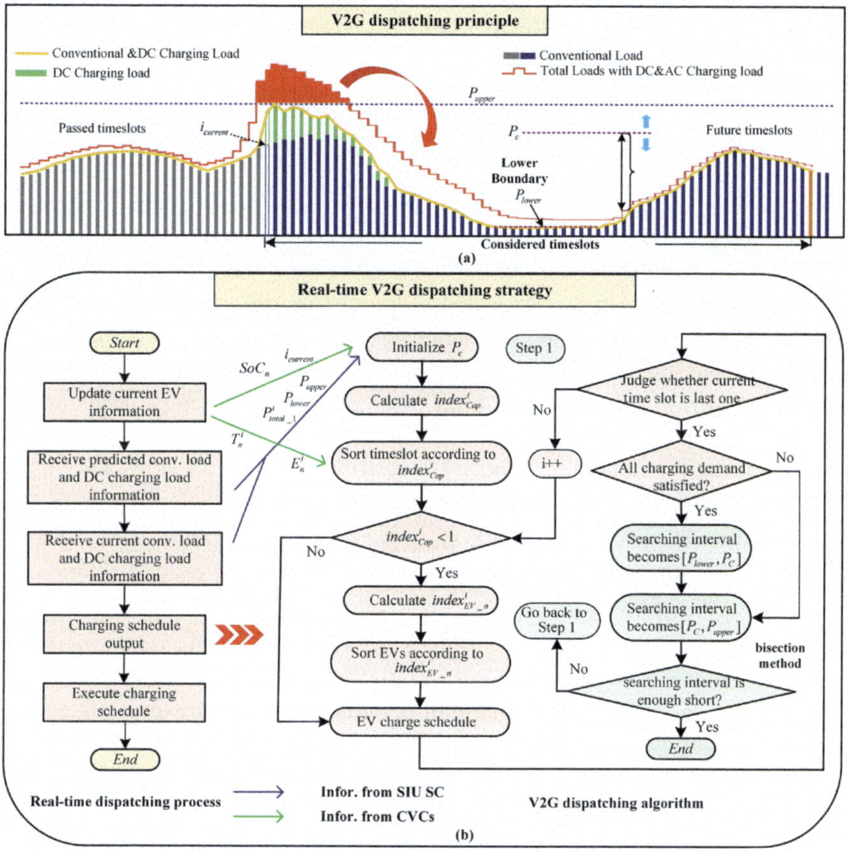

FIGURE 6.9 Real-time V2G dispatching strategy. (a) Principle of introduced V2G dispatching strategy; (b) V2G dispatching process and algorithm.

task of the proposed strategy is to realize the proper arrangement for the exceeding charging load.

After the EV charging schedule conducted, the elevated total load profile must be within the upper boundary P_{upper} and lower boundary P_{lower}. A key parameter P^i_{sup} is defined in Eq. (6.5) to represent the available power capacity for EV slow charging in each timeslot. Notably, the length of the timeslot is 15 min. Thus, there are 96 time slots in a 24-h period. And $i = 1$ indicates the first timeslot, viz., 12:00.

$$P^i_{\text{sup}} = P_c - P^i_{\text{total}}, P_c \in \left[P_{\text{lower}}, P_{\text{upper}}\right] \tag{6.5}$$

where P_c indicates the optimized criterion to satisfy all the EV charging demands and decrease the peak-to-valley difference (kW). indicates the total load comprising the conventional load and DC fast charging load at ith timeslot (kW).

Then, the total slow-charging load P^i_{dem} required by the plugged EV at timeslot i is expressed as:

$$P^i_{dem} = \sum_{n=1}^{N} P^{AC_{nom}}_{EV_n} \delta_N$$

$$\delta = 1, i \in \left[t^{\text{in}}_n, t^{\text{out}}_n\right] \tag{6.6}$$

where $P^{AC_{nom}}_{EV_n}$ denotes the nominal AC charging power for nth EV (kW); t^{in}_n/ t^{out}_n denotes the plug-in and expected plug-out timeslot of nth EV. When the considered timeslot i is within the range $\left[t^{\text{in}}_n, t^{\text{out}}_n\right]$, the coefficient δ_n is set to 1, indicating that the referred EV is presumed to be charged in the whole plugged-in period. Notably, this kind of charging state is only used for the dispatching calculation.

To indicate the remaining capacity of distribution grids for slow charging at each timeslot, a key index called capacity margin $\text{index}^i_{\text{cap}}$ is defined as Eq. (6.7). Obviously, the timeslot with the highest capacity margin should be arranged for EV charging first from the perspective of load valley filling.

$$\text{index}^i_{cap} = \frac{P^i_{\text{sup}}}{P^i_{\text{dem}}} \tag{6.7}$$

Another key index called charging priority $index^i_{EV_n}$ is expressed as Eq. (6.8).

$$index^i_{EV_n} = \frac{E^i_n}{T^i_n \Delta T P^{AC_{nom}}_{EV_n}}, i \in \left[t^{\text{in}}_n, t^{\text{out}}_n\right] \tag{6.8}$$

where E^i_n means the remaining charging demand of nth EV at ith timeslot (kWh); T^i_n denotes the remaining number of time slots to the end of nth EV charging period; means the length of a timeslot (hour).

This index indicates the priority of EVs to be charged in a timeslot. If the remaining capacity at a timeslot cannot meet the whole charging demand, the EV with higher needs to be charged preferential to the EVs with a lower charging priority index. Obviously, if $index^i_{EV_n}$ equals 1, it means that this EV needs to be charged in this timeslot to meet the expected charging demand.

The overall real-time V2G dispatching process is illustrated in the left diagram of Fig. 6.9b. Before the dispatching process begins at the ith timeslot, the DMS would update the plugged EV information according to the CVCs' return message, such as current timeslot i_{current} , nth EV soc, and designated t^{in}_n, t^{out}_n. Then, the DMS would receive the current and predicted conventional & DC charging load information sent from the SIU's SC. Thus, the key parameters, viz., P^i_{total}, P_{upper} , and P_{lower} could be determined, which would affect the V2G dispatching performance. After the data preparation, the V2G dispatching command is generated by the DMS according to the proposed algorithm. Last, the dispatching result is sent to and executed by local AC charging points. The above real-time dispatching process is triggered in every 15 min.

Specifically for the charging schedule output stage, the introduced V2G dispatching algorithm is illustrated in the right diagram of Fig. 6.9b. Initially, in step 1, the value of the optimized criterion P_c is determined, which is set as the midpoint of the interval $\left[P_{\text{lower}}, P_{\text{upper}}\right]$ in the first iteration. Then, the capacity margin index at each timeslot is acquired according to Equ. (7). Then, the timeslot with different value is ordered from most to least. The timeslot with higher would be scheduled with higher priority. After the targeted timeslot is decided, the corresponding charging load at this timeslot is to be scheduled. If $\text{index}_{\text{cap}}^{i}$ is less than 1, the EV charging priority is calculated and ordered from most to least. The scheduling principle is to guarantee the charging demand of EVs with higher charging priority. That is to say, a part of EVs with lower $\text{index}_{EV\_n}^{i}$ would not be allowed in charge when the accumulated charging load exceeds P_{sup}^{i}. Otherwise, if $\text{index}_{\text{cap}}^{i}$ is more than 1, it means the remaining power capacity in this timeslot is enough, and all EVs that need charging at this timeslot would be charged. After updating the charging schedule, we should judge whether the dispatching process for all considered time slots is executed. The considered timeslot range can be from the timeslot of the first EV plugged in to the 12 o'clock in the next day. If the whole dispatching process is not finished, the V2G dispatching for the next timeslot according to $\text{index}_{\text{cap}}^{i}$ order result is conducted. If yes, the next step aims to judge whether all EVs' charging demand is satisfied. If all EVs' charging demand can be satisfied, it indicates that the set available power capacity is enough to meet EV charging need, and the optimized criterion could try to decline to decrease the load peak at valley periods. If no, the can be promoted to meet all the charging demand. The bisection method is adopted to search the minimum solution of the criterion. When all the charging demand is satisfied, the search interval of P_c will be shorten to $\left[P_{\text{lower}}, P_c\right]$. Otherwise, the searching interval of P_c will be expanded to $\left[P_c, P_{\text{lower}}\right]$. Last, it should judge whether the length of the searching interval is short enough. If the searching interval is short enough, the V2G dispatching algorithm is to be ended. Otherwise, the process will go back to step 1 to start the new calculation turn. In this context, the algorithm sets a tolerance of 1 kW to increase the calculation accuracy. And the detailed code of the bisection method can be referred to. According to the above process, the optimum could be determined ultimately to optimize the peak-to-valley difference and satisfy the EV charging demand.

6.4.2 Lower-level coordinated operation

The lower-level coordinated operation is conducted in the local and secondary control levels to realize the transformer load balance and DC voltage restoration. Since the DC fast charging features a high-power demand, the feeder power coordination should be executed by SIU to mitigate the over-burden risk of transformers. Furthermore, due to the existing impedance of the DC line, the current flow would cause a voltage deviation from the nominal value at the terminal charging points. The enabled voltage restoration aims to avoid the terminal voltage overlimit, and aids in the transfer efficiency improvement owing to the decreased current flow after voltage improvement. The overall lower-level coordination strategy is illustrated in Fig. 6.10.

As illustrated in the left diagram of Fig. 6.10, when the power deviation of two feeders is less than the boundary value (P_{band} set as 5 kW), the power reference of VSC-2 is expressed as Eq. (6.9), which indicates an even power sharing of two transformers is to be achieved.

$$P_{\text{VSC}-2}^{\text{ref}} = \frac{1}{2} P_{DC} \tag{6.9}$$

If the SIU-measured power deviation of two feeders exceeds the allowed boundary, a proper power rescheduling for charging should be executed to mitigate the existing load imbalance between transformers. Corresponding $P_{\text{VSC}-2}^{\text{ref}}$ follows the below Eq. (6.10):

$$P' = \min\left(\frac{1}{2}\left(P_{DC} + P_{f1} - P_{f2}\right), P_{DC}\right)$$

$$P_{\text{VSC}-2}^{\text{ref}} = \max\left(P', 0\right) \tag{6.10}$$

where P_{f1}/P_{f2} means the power value of feeder-1/feeder-2. means an intermediate coefficient.

According to Eq. (6.10), a coordinated load balance of two transformers would be implemented by adjusting two VSCs' power share of DC charging. The transformer in a light load condition would undertake more EV charging

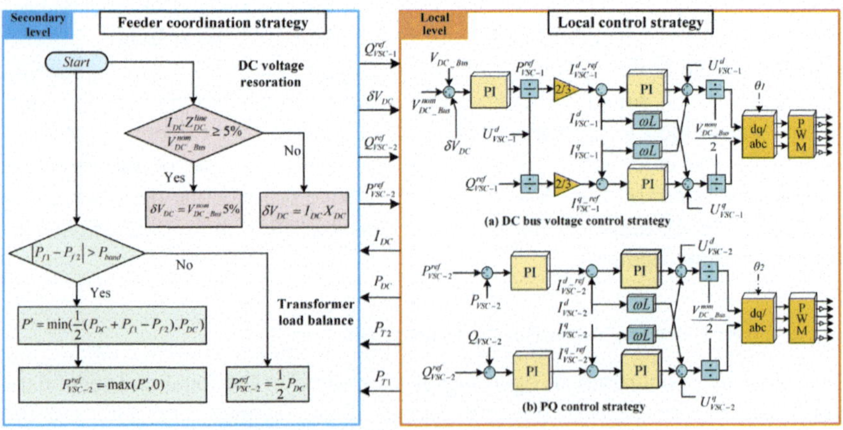

FIGURE 6.10 Lower-level coordination strategy.

power. The adjusted range of VSC-2 is restricted to $[0, P_{DC}]$, indicating that the transformer load balance will be activated when the DC charging power exists. And the upper bound of adjustment capability equals the DC charging load.

In the local control level, to achieve accurate power tracking for power command from the secondary level, the typical PQ control strategy is adopted by VSC-2. As depicted in Fig. 6.10b, the generated power reference from Equ. (9) or Equ. (10) would be processed by dual-closed loops to realize power control. The existing cross-coupling terms $\omega LI^{d}_{\mathrm{VSC}-x}/\omega LI^{q}_{\mathrm{VSC}-x}(x=1,2)$ in inner-current loops are eliminated by the feed-forward loop for better control performance. Notably, since the VSC-1 operates in the DC bus control mode, the remaining power imbalance is covered by VSC-1 autonomously.

As for the control flow of the terminal DC voltage restoration, initially, the SIU can easily measure the DC current value through the DC port. Referred by the prior measured line impedance, X_{DC} the line voltage drop ΔV_{DC} can be calculated as:

$$\Delta V_{DC} = I_{DC}X_{DC} \tag{6.11}$$

Obviously, to recover the terminal voltage to the nominal value $V^{nom}_{DC_{Bus}}$, the controlled DC bus voltage $V_{DC_{Bus}}$ should be improved accordingly. The DC bus voltage variation should be within the required range [-5%, 5%]. Thus, the control flow limits the voltage compensation value δV_{DC} under 5% of $V^{\mathrm{nom}}_{DC\_\mathrm{Bus}}$ value. When is within the range, the value of δV_{DC} expressed in Eq. (6.12) is assigned for the voltage compensation loop.

$$\delta V_{DC} = I_{DC}X_{DC} \tag{6.12}$$

In condition of ΔV_{DC} over 5% of $V^{nom}_{DC_{Bus}}$, to avoid the initial DC voltage of SIU exceeding the upper limit after compensation, δV_{DC} is calculated as Eq. (6.13). By this kind of voltage compensation, the DC voltage of the whole line could be limited to the acceptable range.

$$\delta V_{DC} = V^{nom}_{DC_{Bus}} 5\% \tag{6.13}$$

Regarding the DC bus voltage stabilization, Fig. 6.10a depicts the adopted DC bus voltage control strategy of VSC-1. The sum value of $V^{nom}_{DC_{Bus}}$ and forms the bus voltage reference. This strategy comprises the outer voltage loop and inner current loop. The outer voltage loop generates the d-axis current reference $I^{d_{\mathrm{ref}}}_{\mathrm{VSC}-1}$ to realize the closed-loop power control ultimately. In general, the reactive power reference $Q^{\mathrm{ref}}_{\mathrm{VSC}-1}/Q^{\mathrm{ref}}_{\mathrm{VSC}-2}$ is set to zero to realize the unity power factor operation of VSC-1/2.

6.5 Case study: Peak-shaving and valley-filling oriented strategy

The introduced shared fast charging scheme to accommodate more EVs is validated in a real old residential community in Nanshan District, Shenzhen,

China. As depicted in Fig. 6.11a, before the grid reconstruction, the residential community comprises two feeders, viz., feeder-1 and feeder-2, which are supplied by corresponding dedicated transformers located in a public distribution room. The feeder-1 holds five load nodes of residential buildings and four AC slow charging areas. Moreover, the feeder-2 includes four load nodes with three slow charging areas. The rated power capacity of each transformer is 800 kVA. And the specific conventional load parameter of each node is listed in Table 6.3. The specific data is calculated according to the configured power capacity per household multiplying the number of households in a building node. For example, the households in node 1 are designed with 4-kVA distribution power capacity, and the household number is 32. These node data would reveal the maximum load level of corresponding buildings.

Fig. 6.11b illustrates the grid topology after reconstruction. It can be seen that if the SIU is integrated in the power distribution room, the configured DC feeder could power the terminal DC fast charging points. The supply length is marked in this figure. To approximate the real power electronic system, the switching model of SIU-based architecture is built in the MATLAB/Simulink platform. The related model parameter is listed in Table 6.4.

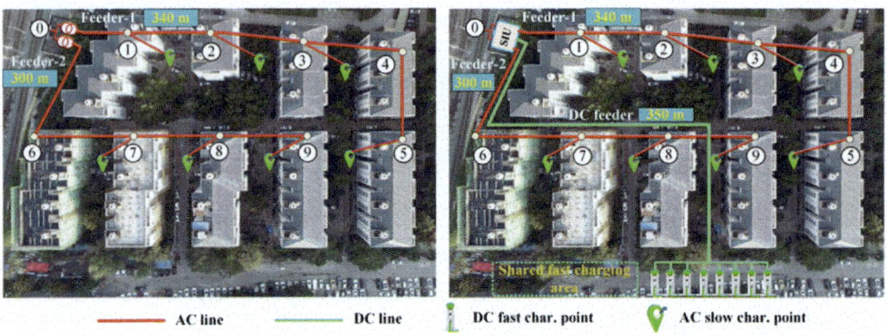

FIGURE 6.11 Grid reconstruction in a real old residential community. (a) Grid topology before reconstruction; (b) grid topology after reconstruction.

TABLE 6.3 Load levels of different nodes.

Feeder-1	Load (kVA)	Feeder-2	Load (kVA)
Node 1	128	Node 6	128
Node 2	72	Node 7	128
Node 3	96	Node 8	128
Node 4	96	Node 9	128
Node 5	96	/	/

TABLE 6.4 Parameters of built model.

Symbol	Value	Symbol	Value
$S_{VSC\text{-}1}/S_{VSC\text{-}2}$	300 kVA	$V_{DC_{Bus}}^{nom} / U_{AC}^{nom}$	750V/380 V
C	5e-4 F	L	10e-5 H
$P_{EV_n}^{AC_{nom}}$	7 kW	ΔT	0.25 h
P_{band}	5 kW	$Q_{VSC-1}^{ref}/Q_{VSC-2}^{ref}$	0
$L_{feeder\text{-}1}$	340 m	$L_{feeder\text{-}2}$	300 m
$L_{feeder\text{-}DC}$	350 m	R_{AC}	1.02e-4 Ω
S_{T1}/S_{T2}	800 kVA	R_{DC}	1.26e-4 Ω
X_{AC}	9.41e-5 Ω	X_{DC}	9.7e-5 Ω

We conducted an extensive survey on the charging behavior of different EVs. The surveyed data reveals that overnight charging is the primary type of charging behavior in residential community scenarios. Specifically, in the overnight charging mode, EV users usually start charging their cars after returning home in the evening and stop charging when they leave in the morning of the next day. According to real-world data, the charging feature of overnight charging is expressed as multiple statistical distributions, just outlined in Table 6.5. It can be seen that the EV plug-in time/plug-out time follows a normal distribution, and the plug-in SoC/plug-out SoC follows a uniform distribution. In real, the used AC slow charging points adopt the typical 7-kW rated power. And the designed DC fast charging station adopts the typical 60-kW fast charging point, which is widely applied in home or community scenarios. According to statistics, the car stock in the studied community is 280, of which 20% of the stock is EV. Presuming the existing AC charging points in each node could satisfy the current charging demand. According to the EV parameter in Table 6.5, the charging profile could be derived through the Monte Carlo simulation [35]. As shown in Fig. 6.12, the

TABLE 6.5 Charging feature of overnight charging.

Plug-in time	Plug-out time	Plug-in SoC	Plug-out SoC	Battery capacity (kWh)	AC slow char. Rate (kW)	DC fast char. Rate (kW)
N (19,1)	N (8,0.92)	U (0.2,0.5)	U (0.5,0.9)	45	7	60

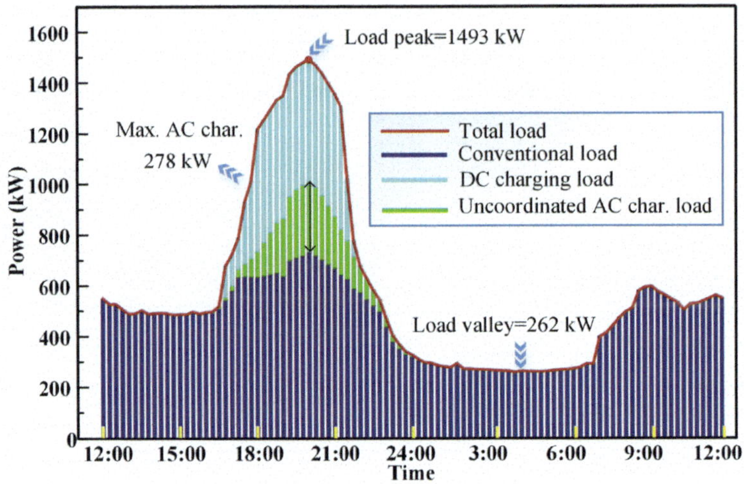

FIGURE 6.12 Overall load profile without V2G dispatching.

typical peak value of the conventional load plus the uncoordinated charging load is 1004 kW. Assuming a power factor of 0.9, the surplus power capacity of transformers is 484 kVA. Thus, eight 60-kW fast charging points could be deployed in the proposed architecture.

6.5.1 Shared fast charging scheme vs. conventional charging scheme

This case would comprehensively compare the proposed shared fast charging scheme and the conventional charging scheme in terms of the EV accommodation capability, occupied area, and system cost. As for the introduced shared fast charging scheme, the inherent registration-based queuing mechanism would affect the charging comfort if EV users should wait a long time for EV charging. Assuming that with the help of charging schedule systems, the shared fast charging point could service the next EV quickly once the previous charging task is completed. A comparison of the average charging end time and maximum charge waiting time as per different EV penetration is shown in Fig. 6.13. It can be observed that in the condition of added 20% EV penetration (56 EVs), the maximum waiting time for charging is 0.5 h, and the end time of total charging service is 21:00. Along with the EV penetration reaching 40%, the maximum waiting time/charging end time is 1 h/22:00, respectively. At last, if the added EV penetration is 80%, the maximum waiting time reaches as long as 4.05 h. Obviously, under the proposed shared charging scheme, a higher EV penetration would sacrifice charging comfort. In this context, 40% EV penetration is a reasonable upper boundary for EV accommodation in this community. This is because the maximum 1 h waiting time is acceptable for

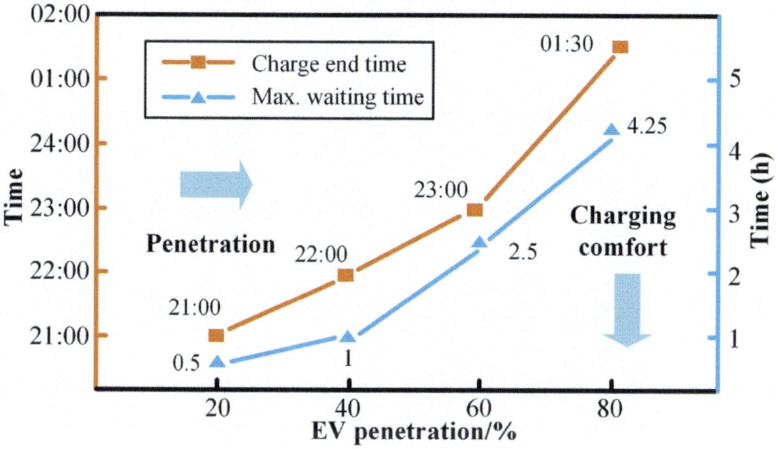

FIGURE 6.13 EV penetration comparison.

residents who generally have no urgent travel demand. Community residents can drive to the dedicated fast charging point when they are at home or nearby places receiving a notification for EV charging availability. According to the Chinese Sleep Study Report 2023 released by the Chinese Academy of Social Sciences, the charging end time of 22:00 is relatively matched with the regular rest schedule of Chinese (average sleep time 23:00–01:00). It can be imagined that after the registration charging is completed, the resident has time to park their EV in the temporary parking spot downstairs without affecting regular rest schedule. As for the slow charging scheme, considering the remaining 484-kVA power capacity mentioned before, most 69 7-kW slow charging points could be deployed. Thus, the improved EV penetration is 24%, which is significantly lower than that of the shared fast charging scheme.

Then, we will conduct a system cost analysis based on system configuration. For the slow charging scheme of 69 charging points, according to Equ. (1), the maximum current flow is calculated as 735 A. Therefore, two 185 mm^2-diameter 4-cores lines (ZC-YJV22–0.6/1 kV-4*185) could be adopted for configuration, of which single line capacity is 409 A, and the resistance/reactance per meter is 1.02e-4/9.41e-5 according to specifications as listed in Table 6.6. Concerning the introduced DC power transfer, the maximum current flow is 640 A according to Equ. (2). In light of this, two 150 mm^2-diameter 2-cores lines (ZC-YJV22–0.6/1 kV-2*150) can be used, of which a single line capacity is 359 A, and the resistance/reactance per meter is 1.26e-4/9.7e-5. The survey indicates that the price per meter of the line type ZC-YJV22–0.6/1 kV-4*185/ZC-YJV22–0.6/1 kV-2*150 is ¥612/¥270, respectively. Considering the 350 m power supply length in Fig. 6.11, the main line cost of the proposed architecture and slow charging station architecture equals ¥94,500/¥214,200, respectively. Moreover, thanks to the rapid development of the power electronic industry, the price of VSCs drops

TABLE 6.6 EV penetration, system investment and occupied area estimation for three charging schemes.

Charging scheme	Cost					Occupied area (m^2)	EV Penetration
	Char. Points	Lines	SIU	Total	Cost per 1% EV penetration		
Conventional slow charging scheme	¥241,500	¥214,200	/	¥455,700	¥18,987.5	862.5	24%
Conventional fast charging scheme	¥264,000	¥214,200	/	¥478,200	¥11,955	100	40%
Proposed shared fast charging scheme	¥144,000	¥94,500	¥120,000	¥358,500	¥8962.5	100	40%

to ¥0.2/W, and DC/DC for fast charging is about ¥0.3/W. Thus, the cost of 600-kW SIU is roughly estimated at ¥120,000. The total cost of 8 60-kW DC charging points is ¥144,000. Referred by the price of a single 7-kW slow charging point ¥3500, the total cost of 69 AC charging points is ¥241,500. Based on the above data, the comparison result regarding the system investment is shown in Table 6.6. It can be derived that the proposed shared fast charging scheme holds merits over the conventional slow charging scheme in terms of the total system cost and cost per 1% EV penetration. In addition, one charging point-equipped parking spot generally occupies an area of 12.5 m^2. Thus, the proposed scheme could realize higher EV penetration at the cost of less occupied areas.

If the AC-powered fast charging station architecture is adopted along with the introduced shared fast charging mechanism, 40% EV penetration could also be achieved when accommodating the charging end time of 22:00. Accordingly, the estimated occupied area would be the same as the proposed architecture. In addition, since both the topology and power capacity of SIU-based and conventional AC-powered fast charging station architecture have the same AC/DC + DC/DC structure, the equipment cost can be roughly regarded as the same. In terms of line deployment, due to the merit of DC line costs, the SIU-based architecture also exhibits the overall cost advantage as listed in Table 6.6. In conclusion, the cost and compactness merits of the proposed shared fast charging scheme facilitate its application in old residential communities compared with the conventional AC charging scheme.

6.5.2 Upper-level V2G dispatching process

This case shows the V2G dispatching process of AC slow-charged EVs in 15-minute intervals. As shown in Fig. 6.14, at 18:45 (viz., $i = 76$th timeslot), the

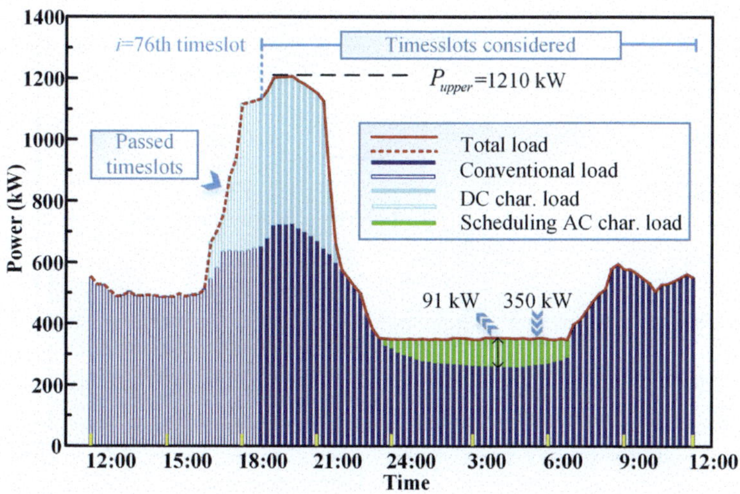

FIGURE 6.14 V2G dispatching results for slow-charged EVs at 76th timeslot.

V2G dispatching calculation begins. Referred to the predicted/current conventional load and DC fast charging load, as well as the plugged EVs information, the DMS would decide the proper charging time slots for supervised EVs. As shown in Fig. 6.14, the time slots considered for dispatching calculation are between the current timeslot and 12:00 in the next day. The dispatching results show that all the uncoordinated AC slow charging load is transferred to the valley period, viz., 23:15–7:00, accompanied with the maximum charging load equaling 91 kW. And the valley value of the total load is elevated from the previous 263–350 kW. When it comes to the next timeslot (viz., $i = 77$th timeslot), the plugged EV and load information would update. In Fig. 6.15, the dispatching results in the new interval illustrate that the scheduled AC charging load changes dynamically with the maximum charging load increasing to 105 kW. And the valley value of the total load also adjusts to 370 kW. Similarly, in Fig. 6.16, along with more EVs connected at 19:15 (viz., $i = 78$th timeslot), the maximum AC charging load climbs to 126 kW, and the valley value of the total load equals 383 kW. It can be seen that thanks to the AC charging load transfer, the predicted total load curve in three dispatching intervals is below the desired peak load (1210 kW) even with the DC fast charging load incorporated, indicating a prominent peak-shaving and valley-filling performance. And the enabled real-time dispatching process would reduce the impact of forecast uncertainty on scheduling results. These results demonstrate the effectiveness of the proposed upper-level strategy and rolling dispatching process.

6.5.3 Lower-level coordinated operation

In this case, a short-timescale real-time simulation is conducted to verify the effectiveness of the proposed architecture and lower-level coordination

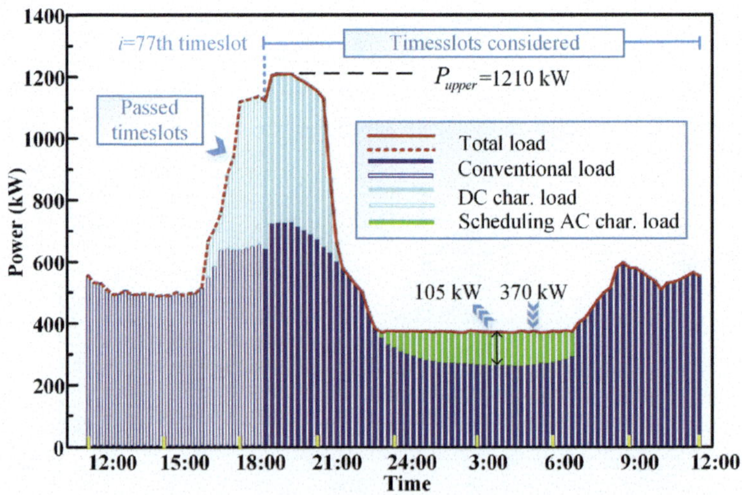

FIGURE 6.15 V2G dispatching results for slow-charged EVs at 77th timeslot.

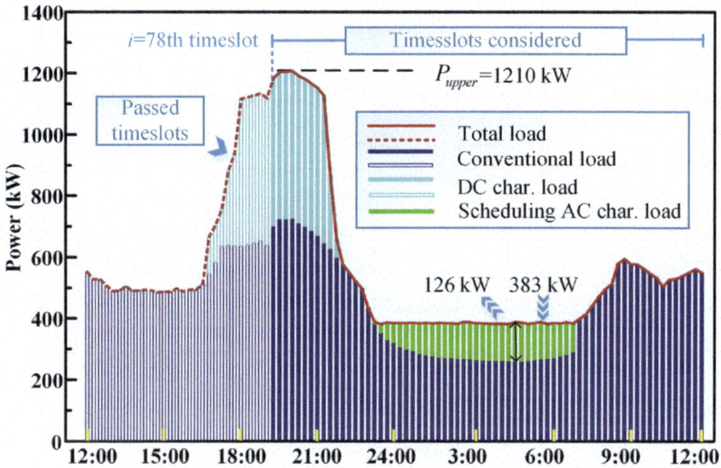

FIGURE 6.16 V2G dispatching results for slow-charged EVs at 78th timeslot.

strategy. Three time slots, viz., 76-78th interval as discussed in case 2, are included in this case study. Moreover, the initial and last stages are also designed to show the transient operation process. In the initial stage, the load levels of AC and DC feeders are depicted in Fig. 6.17(a). In Fig. 6.17b, with the proposed lower-level strategy, the existing AC load deviation between feeder-1 and feeder-2 (50 kW) is covered by the coordination of VSC-1 and VSC-2 with 55-kW/5-kW output for DC charging need. Owing to the enabled feeder power coordination, the two transformers (T1/T2) load remains the same 306-kW power output. The autonomous load balance of transformers is to be achieved. In Fig. 6.17d, under the proposed DC voltage restoration control, the initial voltage of the DC bus remains 753 V, which is slightly higher than the nominal value to keep the terminal voltage as 750 V. At 0.5 s, the simulation steps in the 76-78th timeslot, viz., 18:45–19:30. The DC charging power increases sharply to 480 kW. As depicted in Fig. 6.17d, after the emerging transient voltage drop, the terminal voltage resumes to 750 V within 300 ms with the proposed control strategy. The initial DC bus voltage is elevated to 778 V to compensate for the impedance-caused voltage drop. During 0.5–2.0 s, although the AC load in feeder-1 and feeder-2 fluctuates frequently, the power support for DC charging from VSC-1 and VSC-2 adjusts accordingly to compensate for the existing AC load deviation, as depicted in Fig. 6.17c. In the last stage, the related load levels resume the initial value, and the corresponding system performance returns to the initial state of 0–0.5 s. Specifically, the emerging transient DC voltage peak at 2 s could be flattened in 300 ms. The stabilized DC voltage in a steady state indicates the operation stability of the whole system.

A DC voltage control performance is compared under the typical voltage restoration method. As discussed in Ref. [22], the typical voltage restoration

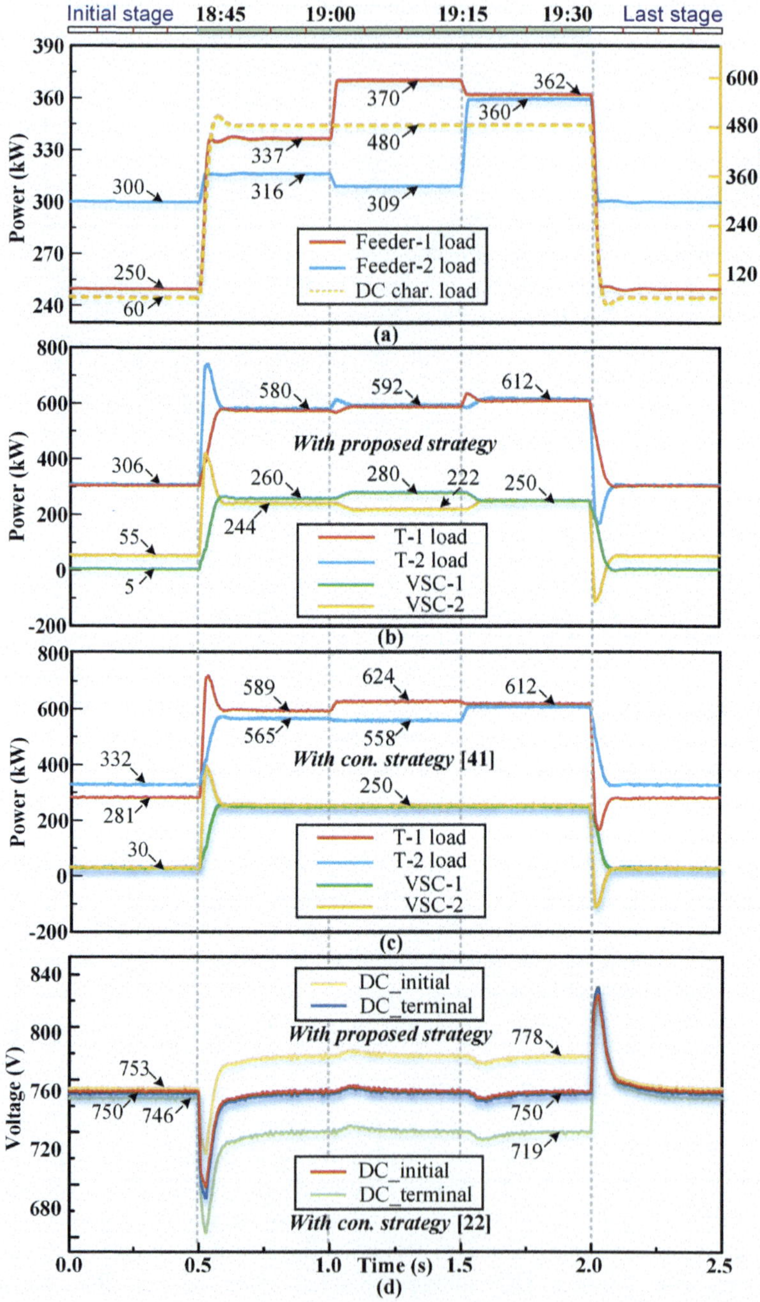

FIGURE 6.17 Lower-level coordinated operation performance during 76th-78th timeslot. (a) Load levels of AC and DC feeders; (b) power levels of VSC-1/2 and transformer T1/T2 with the conventional strategy; (c) power levels of VSC-1/2 and transformer T1/T2 with the proposed strategy (d) DC bus initial voltage and terminal voltage curve.

control would elevate the node voltage level within the required range when the voltage violation occurs. By adopting this method, the corresponding DC curve is illustrated in Fig. 6.17d. It can be seen that although the initial DC bus voltage could be stabilized at the nominal value (750 V), the terminal DC bus voltage would be close to the lower bound. Accordingly, the proposed lower-level strategy exhibits prominent voltage restoration performance.

Fig. 6.17c illustrates the power levels of VSCs and transformers when the conventional power coordination method is adopted. This method aims to realize proportional power sharing among individual VSCs based on the droop control mechanism. As seen in Fig. 6.17c, an even power sharing between VSC-1 and VSC-2 is to be achieved. Nevertheless, a 66-kW power unbalance between transformer T1 and T2 occurs during 1.0–1.5 s. In comparison with Fig. 6.17b, the proposed strategy is verified to cover the existing load unbalance between transformers, thereby fully utilizing the remaining power capacity of transformers and decreasing the potential overburden risk.

This case demonstrates that a stabilized and coordinated system operation could be implemented with the lower-level coordination strategy in the proposed architecture. The desired transformer load balance and prominent terminal voltage restoration could be realized, simultaneously.

6.5.4 Overall operation performance

This case shows the overall operation performance after all dispatching turns. In Fig. 6.18, in the condition of no V2G dispatching conducted, the uncoordinated AC charging load, plus the newly added fast charging load, would significantly elevate the total load level. In Fig. 6.18, the emerging load peak

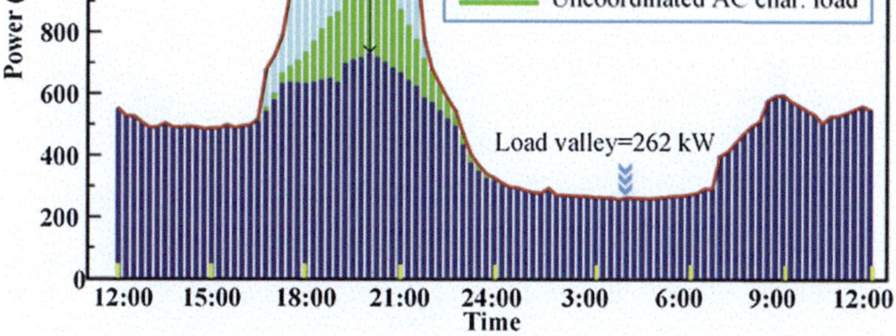

FIGURE 6.18 Overall load profile without V2G dispatching.

would reach 1493 kW, which is close to the capacity boundary of transformers. Specifically, the maximum AC charging load reaches 278 kW. Yet, the load valley is only 262 kW. Fig. 6.19 illustrates the overall load profile with the proposed real-time V2G dispatching strategy executed. It can be seen that all of the AC charging load is transferred to the valley period (23:30–7:00), accompanied with the load valley reaching 412 kW. And the maximum AC charging load is limited to 154 kW. The peak-to-valley difference is decreased from the previous 1231–803 kW, viz., dropping by 37%.

In order to manifest the effectiveness of the proposed V2G strategy, we also made a comparison with the conventional load variance optimization strategy discussed in Ref. [38]. Since the built formulation belongs to the convex quadratic program, the optimization result could be obtained [39]. It can be seen in Fig. 6.19, with the conventional strategy adopted, the obtained curve closely aligns with the curve obtained under the proposed strategy. Accordingly, the proposed strategy is verified to achieve an optimized peak-shaving and valley-filling performance.

Then, the computational performance of the proposed upper-level strategy and conventional load variance optimization strategy is discussed. To be fair, the two built models are executed on the same platform (CPU 2.2 GHz, RAM 16 GB), and the average solving time is depicted in Fig. 6.20 when different EV penetrations are involved. It can be seen that the fitting curve based on the data from the conventional strategy belongs to the exponential function and is significantly higher than the curve generated by the proposed strategy. The average solving time of the proposed strategy keeps a small value (0.1 s)

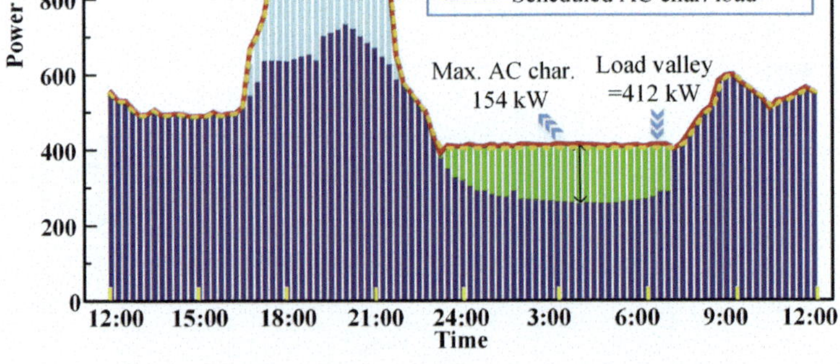

FIGURE 6.19 Overall load profile with V2G dispatching.

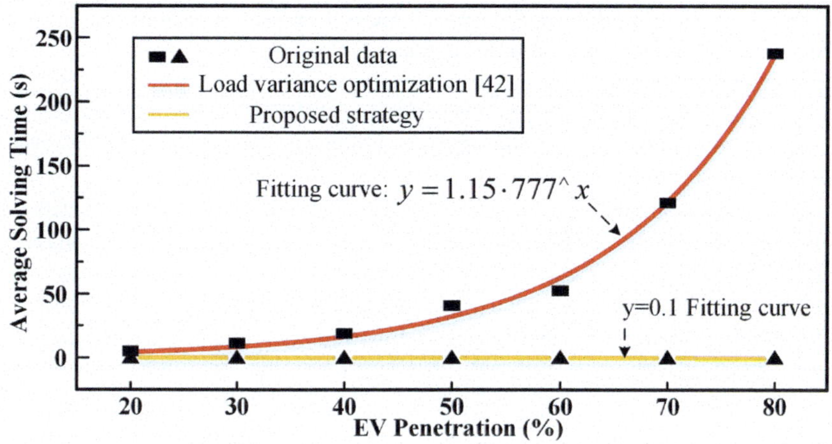

FIGURE 6.20 Average solving time versus EV penetration when different methods employed.

whatever the EV penetration changes. These results demonstrate the proposed strategy exhibits a highly efficient computation capability compared with conventional strategies.

Finally, we make an efficiency comparison between the AC-powered fast charging architecture and the SIU-based charging station architecture. According to the fast charging product data from many manufacturers [40], the typical efficiency value of fast charging converters is about 95%. In addition, for the comparison process, the two architectures is presumed to have the same DC power spectrum just drawn in Fig. 6.19. Thus, in this context, the converter efficiency of the two architectures is estimated at 95%. In light of this, the overall architecture efficiency curve considering the line and converter losses is depicted in Fig. 6.21.

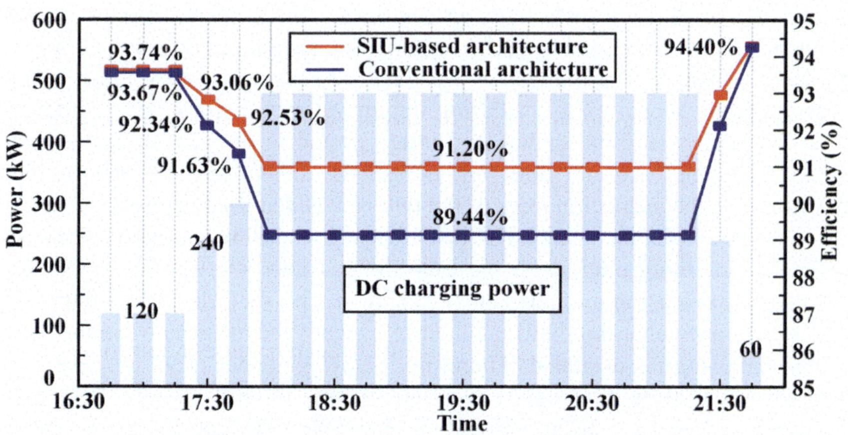

FIGURE 6.21 Overall fast charging efficiency comparison.

It can be seen that the supply efficiency curve of the SIU-based architecture is over that of the conventional architecture. Along with the power increment, the efficiency deviation increases, for example, in the condition of 480-kW power output, the efficiency deviation reaches 1.76%. The overall mean efficiency of the SIU-based architecture/conventional architecture is 91.95%/ 90.65%, respectively.

This case demonstrates the effectiveness and low solution complexity of the proposed real-time V2G dispatching strategy. An optimized peak-shaving and valley-filling performance will be achieved with all uncoordinated AC charging loads transferred to the valley period. Moreover, for the overall DC fast charging period, the SIU-based architecture exhibits 1.3% average efficiency merit over the conventional fast charging architecture derived from the line losses decline.

6.6 Case study: Price incentives driven strategy

Minimizing the operator's electricity procurement cost yields the target charging load for EVs. Using the target charging load as the guiding objective, dynamic electricity prices can be calculated under different penetration rates. This calculation considers the operational hours constraint of the established shared DC fast charging station. The resulting dynamic electricity prices only consider the period from 12:00 to 24:00 due to the working hours limitation. For penetration rates of 10% and 15% of EVs, as shown in Figs. 6.22a and 6.23a, the charging prices before and after guidance are illustrated Table 6.7. During the period from 12:00 to 24:00, the original industrial electricity prices from Table 6.4 only have peak and shoulder periods. Based on the dynamic electricity prices, the number of vehicles requiring charging during different time periods before and after guidance can be determined, thereby deriving the utilization rates of charging stations before and after guidance, as depicted in Figs. 6.22b and c and 6.23b and c. Observationally, due to the optimized scheduling of dynamic electricity prices, there is a notable decrease in the number of EVs charging during the origin usage period compared with before guidance. However, the number of EVs charging during the new added usage period slightly increases. Considering the arrival rate of EVs, the number of EVs charging during the new added usage period in the afternoon does not dramatically increase and does not exceed the arrival rate for that period. Additionally, the utilization rates of the two new added usage periods increase.

In the example, the shared DC fast charging station is designed with only seven fast charging piles, with the time divided into intervals of 15 min for calculation. From Figs. 6.22b and 6.23b, it can be seen that the number of people requiring charging before guidance exceeds the maximum number = 7 of charging stations designed. Therefore, before guidance, EVs may need to wait for a long time to charge or may not be able to charge at all.

In Fig. 6.24a, the charging prices before and after guidance are depicted for a maximum EV penetration rate of 22.3%, as per the designed fast charging

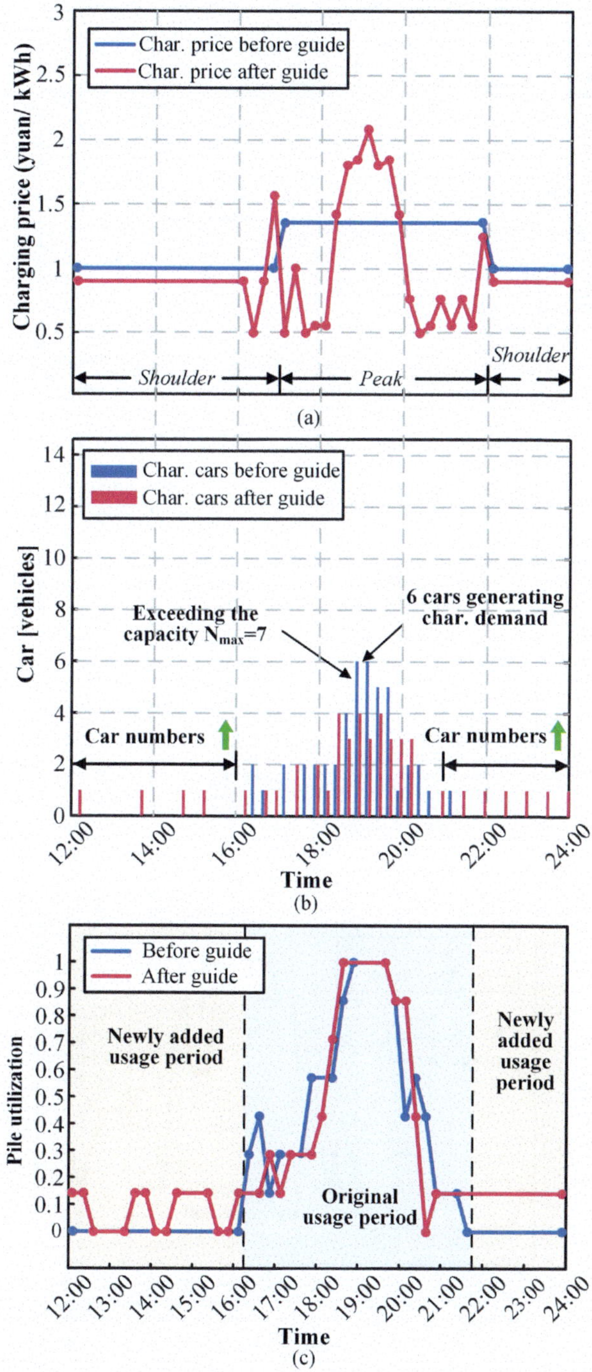

FIGURE 6.22 Optimization dispatching results at 10% penetration: (a) Char. prices; (b) the number of EVs generating char. demand; (c) utilization of char. piles.

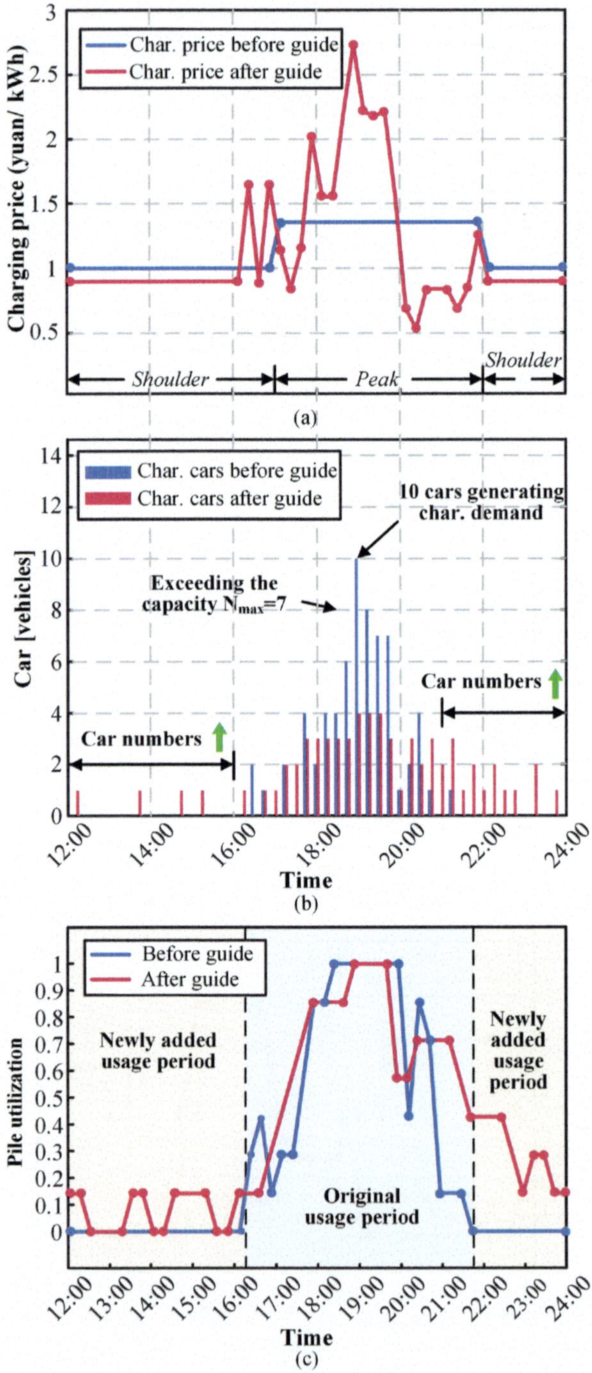

FIGURE 6.23 Optimization dispatching results at 15% penetration: (a) Char. prices; (b) the number of EVs generating char. demand; (c) utilization of char. piles.

TABLE 6.7 **Shenzhen's general industrial TOU electricity price.**

Period attribute	Period division	Price (yuan/kWh)
Peak hours	10:00–12:00, 17:00–22:00	1.3553
Shoulder hours	8:00–10:00, 12:00–17:00, 22:00–24:00	1.0000
Valley hours	0:00–8:00	0.2894

station capacity. Correspondingly, the number of vehicles requiring charging during different time periods and the utilization rates of charging stations before and after guidance are obtained, as shown in Fig. 6.24b and c. Compared with penetration rates of 10% and 15%, the number of EVs charging during the origin usage period sharply decreases after guidance, while the number of EVs charging during the new added usage period significantly increases. Considering the arrival rate of EVs, although the penetration rate has greatly increased, the number of EVs charging during the new added usage period in the afternoon does not show a significant increase. Most EVs are scheduled to charge during the evening new added usage period. Additionally, the utilization rates of charging stations during these two new added usage periods notably increase, leading to a significant decrease in the idle rate of charging stations.

6.7 Concluding remarks

This chapter proposes a cost-effective and high-efficient shared fast charging scheme for old residential communities to address the existing difficult-to-charge issue. This scheme comprises two aspects: charging mechanism and charging station architecture. The following conclusions have been arrived:

1. The registration-based queuing charging mechanism is designed for the fast charging schedule. Moreover, an integrated and multistage power electronic device called smart interlinking unit (SIU) is introduced for the grid reconstruction. The enabled DC power transfer replaces the typical AC type to supply the terminal fast charging points.
2. A hierarchical management structure with multilevel coordination strategies is introduced for the SIU-based charging station architecture. In the upper control level, this chapter introduces a real-time vehicle-to-grid (V2G) dispatching strategy considering the conventional and fast charging load incorporated. A coordination strategy in the lower control level is proposed to enhance the grid operation performance while realizing a stabilized system operation.
3. For the targeted real old residential community, the introduced shared fast charging scheme could increase the EV penetration to 40% when the

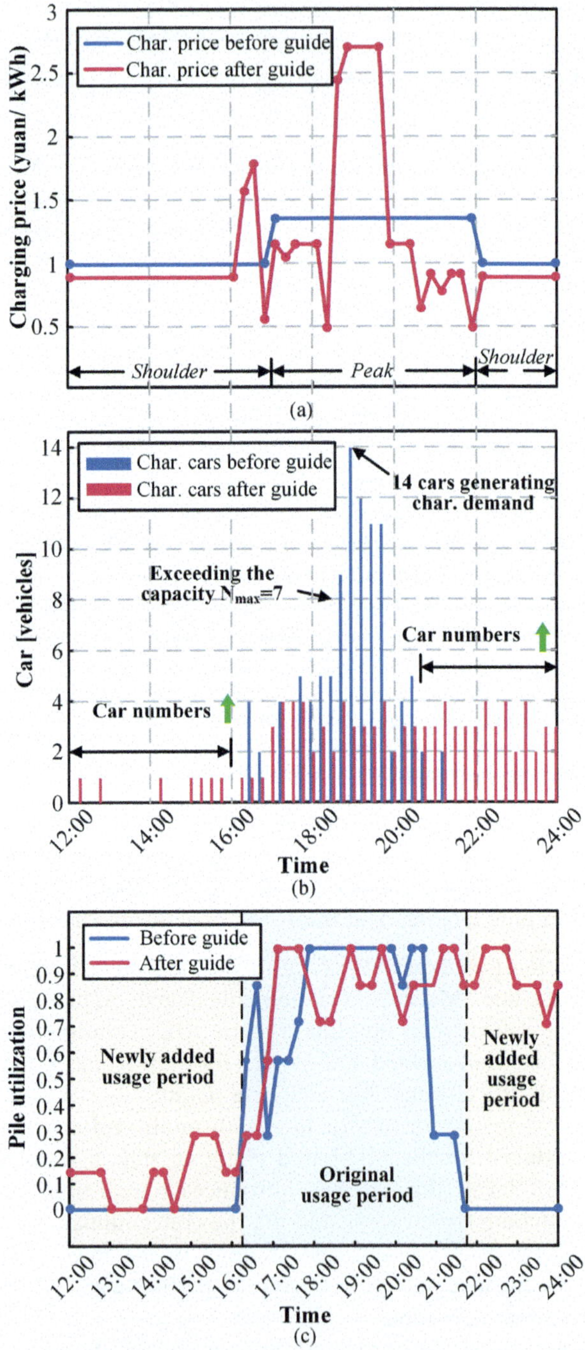

FIGURE 6.24 Optimization dispatching results at 22.3% penetration: (a) Char.prices; (b) the number of EVs generating char. demand; (c) utilization of char. piles.

charging end time and maximum waiting time are limited to 22:00/1 h, respectively. This charging comfort is believed to be feasible in residential scenarios. Additionally, the needed system investment per 1% EV penetration and total occupied area decreases by 37% and 88% compared with the conventional AC slow charging scheme. For the whole fast charging period, the introduced SIU-based charging station architecture could realize 1.3% average transfer efficiency improvement compared with the conventional AC-powered charging architecture. In conclusion, this scheme favors the application in old residential communities.

4. The upper-level real-time V2G dispatching strategy has been verified to have low solution complexity. It can achieve an optimized peak-shaving and valley-filling performance, reducing the peak-to-valley difference by 37% compared with the case with no V2G dispatching. The desired transformer load balance and terminal voltage restoration could be realized with the lower-level coordination strategy, enhancing the power quality and effectively decreasing the overburden risk of transformers.

However, there remain limitations in this study. The proposed shared fast charging mechanism is currently in its preliminary stage and does not account for complex scenarios such as vehicle owners delaying their charging. Additionally, the power rating and quantity of charging points are provisionally configured. In future research, we will conduct a comprehensive assessment of the efficiency of implementing shared fast charging in old residential communities while considering more real-world factors. Furthermore, we will also explore the optimal configuration of shared charging stations, taking into account factors such as solar power and energy storage [41].

References

[1] He Y, Chen Y, Yang Z, He H, Liu L. A review on the influence of intelligent power consumption technologies on the utilization rate of distribution network equipment. Protect Contr Modern Power Syst 2018;3:1–11.

[2] Patil P, Kazemzadeh K, Bansal P. Integration of charging behavior into infrastructure planning and management of electric vehicles: a systematic review and framework. Sustain Cities Soc 2022:104265.

[3] Zheng Y, Niu S, Shang Y, Shao Z, Jian L. Integrating plug-in electric vehicles into power grids: a comprehensive review on power interaction mode, scheduling methodology and mathematical foundation. Renew Sustain Energy Rev 2019;112:424–39.

[4] Das HS, Rahman MM, Li S, Tan C. Electric vehicles standards, charging infrastructure, and impact on grid integration: a technological review. Renew Sustain Energy Rev 2020;120:109618.

[5] İnci M, Büyük M, Özbek NS. Sliding mode control for fuel cell supported battery charger in vehicle-to-vehicle interaction. Fuel Cell 2022;22:212–26.

[6] Liu Y, Zhu Y, Cui Y. Challenges and opportunities towards fast-charging battery materials. Nat Energy 2019;4:540–50.

[7] Khalid MR, Alam MS, Sarwar A, Asghar MJ. A Comprehensive review on electric vehicles charging infrastructures and their impacts on power-quality of the utility grid. eTransportation 2019;1:100006.

[8] Ji Z, Huang X. Plug-in electric vehicle charging infrastructure deployment of China towards 2020: policies, methodologies, and challenges. Renew Sustain Energy Rev 2018;90:710–27.

[9] Li Z, Miao L. Automated stereo-garage with multiple cache parking spaces—structure, system and scheduling performance. Autom ConStruct 2020;119:103377.

[10] Belmeziti A, Cherqui F, Kaufmann B. Improving the multi-functionality of urban green spaces: relations between components of green spaces and urban services. Sustain Cities Soc 2018;43:1–10.

[11] Zheng Y, Shao Z, Shang Y, Jian L. Modeling the temporal and economic feasibility of electric vehicles providing vehicle-to-grid services in the electricity market under different charging scenarios. J Energy Storage 2023;68:107579.

[12] Sachan S, Deb S, Singh SN. Different charging infrastructures along with smart charging strategies for electric vehicles. Sustain Cities Soc 2020;60:102238.

[13] Rezaeian-Marjani S, Talavat V, Galvani S. Impact of soft open point (SOP) on distribution network predictability. Int J Electr Power Energy Syst 2022;136:107676.

[14] Yang Y, Pei W, Kong L, Pu T, Fan S, Huang R. A comprehensive optimization of load transfer strategy for distribution network with flexible interconnect device. In: 2018 IEEE/PES transmission and distribution conference and exposition (T&D). IEEE; 2018. p. 1–9.

[15] Yang Y, Pei W, Huo Q, Sun J, Xu F. Coordinated planning method of multiple micro-grids and distribution network with flexible interconnection. Appl Energy 2018;228:2361–74.

[16] He Y, Wu H, Bi R, Qiu R, Ding M, Sun M, et al. Coordinated planning of distributed generation and soft open points in active distribution network based on complete information dynamic game. Int J Electr Power Energy Syst 2022;138:107953.

[17] Samet H, Azhdari E, Ghanbari T. Comprehensive study on different possible operations of multiple grid connected microgrids. IEEE Trans Smart Grid 2016;9:1434–41.

[18] Strasser T, Andren F, Kathan J, Cecati C, Buccella C, Siano P, et al. A review of architectures and concepts for intelligence in future electric energy systems. IEEE Trans Ind Electron 2014;62:2424–38.

[19] Ronanki D, Kelkar A, Williamson SS. Extreme fast charging technology—prospects to enhance sustainable electric transportation. Energies 2019;12:3721.

[20] Chen Q, Liu N, Hu C, Wang L, Zhang J. Autonomous energy management strategy for solid-state transformer to integrate PV-assisted EV charging station participating in ancillary service. IEEE Trans Ind Inf 2016;13:258–69.

[21] Zhang L, Chen Y, Shen C, Tang W, Liang J. Releasing more capacity for EV integration by DC medium voltage distribution lines. IET Power Electron 2017;10:2116–23.

[22] Yu H, Lei X, Niu S, Shao Z, Jian L. Enhancing electric vehicle penetration and grid operation performance in old residential communities through hybrid AC/DC microgrid reconstruction. Appl Energy 2023;347:121459.

[23] Inci M. Technoeconomic analysis of fuel cell vehicle-to-grid (FCV2G) system supported by photovoltaic energy. Energy Technol 2023;11:2201162.

[24] İnci M, Büyük M, Savrun MM, Demir MH. Design and analysis of fuel cell vehicle-to-grid (FCV2G) system with high voltage conversion interface for sustainable energy production. Sustain Cities Soc 2021;67:102753.

[25] Yu H, Shang Y, Niu S, Cheng C, Shao Z, Jian L. Towards energy-efficient and cost-effective DC nanaogrid: a novel pseudo hierarchical architecture incorporating V2G technology for

both autonomous coordination and regulated power dispatching. Appl Energy 2022;313:118838.

[26] Shang Y, Liu M, Shao Z, Jian L. A centralized vehicle-to-grid scheme with distributed computing capacity engaging internet of smart charging points: case study. Int J Energy Res 2021;45:841–63.

[27] Ioakimidis CS, Thomas D, Rycerski P, Genikomsakis KN. Peak shaving and valley filling of power consumption profile in non-residential buildings using an electric vehicle parking lot. Energy 2018;148:148–58.

[28] Xu Y, Juan Y-K. Optimal decision-making model for outdoor environment renovation of old residential communities based on WELL Community Standards in China. Architect Eng Des Manag 2022;18:571–92.

[29] Yu H, Niu S, Shang Y, Shao Z, Jia Y, Jian L. Electric vehicles integration and vehicle-to-grid operation in active distribution grids: a comprehensive review on power architectures, grid connection standards and typical applications. Renew Sustain Energy Rev 2022;168:112812.

[30] Ji Z, Huang X. Plug-in electric vehicle charging infrastructure deployment of China towards 2020: policies, methodologies, and challenges. Renew Sustain Energy Rev 2018;90:710–27.

[31] Xu Y, Juan Y-K. Optimal decision-making model for outdoor environment renovation of old residential communities based on WELL Community Standards in China. Architect Eng Des Manag 2022;18:571–92.

[32] Gupta J, Maurya R, Arya SR. Designing an on-board charger to efficiently charge multiple electric vehicles. Chin J Electr Eng 2023;9:38–56.

[33] Zheng Y, Shang Y, Shao Z, Jian L. A novel real-time scheduling strategy with near-linear complexity for integrating large-scale electric vehicles into smart grid. Appl Energy 2018;217:1–13.

[34] Savrun MM, İnci M, Büyük M. Design and analysis of a high energy efficient multi-port dc-dc converter interface for fuel cell/battery electric vehicle-to-home (V2H) system. J Energy Storage 2022;45:103755.

[35] Yu H, Niu S, Zhang Y, Jian L. An integrated and reconfigurable hybrid AC/DC microgrid architecture with autonomous power flow control for nearly/net zero energy buildings. Appl Energy 2020;263:114610.

[36] Jain C, Singh B. A three-phase grid tied SPV system with adaptive DC link voltage for CPI voltage variations. IEEE Trans Sustain Energy 2015;7:337–44.

[37] Li Y, Wang Z, Xu W, Gao W, Xu Y, Xiao F. Modeling and energy dynamic control for a ZEH via hybrid model-based deep reinforcement learning. Energy 2023;277:127627.

[38] Yang S, Wang X, Ning W, Jia X. An optimization model for charging and discharging battery-exchange buses: consider carbon emission quota and peak-shaving auxiliary service market. Sustain Cities Soc 2021;68:102780.

[39] Sortomme E, Hindi MM, MacPherson SJ, Venkata S. Coordinated charging of plug-in hybrid electric vehicles to minimize distribution system losses. IEEE Trans Smart Grid 2010;2:198–205.

[40] Aretxabaleta I, De Alegria IM, Andreu J, Kortabarria I, Robles E. High-voltage stations for electric vehicle fast-charging: trends, standards, charging modes and comparison of unity power-factor rectifiers. IEEE Access 2021;9:102177–94.

[41] Zhang L, Liang J, Tang W, Li G, Cai Y, Sheng W. Converting AC distribution lines to DC to increase transfer capacities and DG penetration. IEEE Trans Smart Grid 2018;10:1477–87.

Chapter 7

Internet of smart charging points for vehicle-to-grid operation

7.1 Introduction

Over the past few years, the primary environmental issue in current China and other developing countries is atmospheric pollution. To address the above challenges, China has done a great deal in air pollution regulations, like prohibiting straw burning [1] and controlling vehicle exhaust [2]. Moreover, PEV concentrating on environment-friendly and energy efficiency can also be an alternative way to mitigate the stress on air pollution and energy crisis [3]. It is predicted that the stock of EVs is projected to exceed 250 million units worldwide by 2030 [4]. Ref. [5] shows that the predicted number of EVs can reach as high as 880,129 at 2025 in Shenzhen, China's innovation hub, and the peak load of the power grid will be increased by 11.08%. This growing electricity demand raises significant concerns about grid stability and energy sourcing, particularly when EVs are charged using electricity generated from fossil fuels such as coal. In such cases, the environmental benefits of EVs are largely undermined, thereby highlighting the need for low-carbon charging pathways. In this context, the integration of renewable energy source (RES) into PEV charging systems presents a viable solution. However, the intermittent and variable nature of RES poses additional operational challenges for the grid. Therefore, the future development of transportation electrification cannot be discussed independently from the evolution of power-system operation. The wider the deployment of EVs becomes, the more urgent it is to establish an effective framework in which vehicle charging demand, renewable-energy utilization, and grid security can be coordinated in a unified manner.

The V2G technology involving PEVs and RES is a vital theme to reduce the fuel-consuming and alleviate the air pollution [6]. The strategy of V2G scheduling determines the charging and/or discharging decisions of PEV to fulfill the energy requirement of PEV, mitigate the intermittency of RES, and

Smart Penetration of Electric Vehicles in Distribution Grids. https://doi.org/10.1016/B978-0-443-27706-1.00001-5

improve the stability of the power grid [7]. Besides, the V2G technology can provide several services like voltage regulation [8], frequency regulation [9] and spinning reserve [10]. As PEV penetration continues to rise, these services will become increasingly essential for maintaining grid performance under high load and high volatility scenarios. There are lots of schemes to implement the V2G technology, wherein the architecture and the algorithm are two main components. It is worth noting that the architecture and the algorithm are not two separate individuals. In other words, the scheduling algorithm is designed based on the V2G architecture. Without a proper PEV charging and discharging architecture, the effectiveness of scheduling algorithm will be greatly discounted.

From the V2G architectures perspective, there are many researches focused on the scalability, reliability, and information security. First of all, the V2G architecture should be scalable so that a large number of elements, including the CPs and the PEVs for internal access as well as the RESs for external access, can conveniently participate in the V2G scheduling. In Ref. [11], a scalable publish/subscribe communication architecture for information exchange between PEVs and the power grid is proposed, in accordance with a charging management algorithm to coordinate PEV users' charging plans. Considering the small amount of data in the scheme, a scalable demand response architecture is proposed in Ref. [12]. The proposed architecture can be extended to 1000 households and the solving time is less than 3 minutes. In order to simulate the uncertain situation in power system, this architecture also integrates plenty of RESs with intermittent behaviors. In Ref. [13], a charging station involving 100% RESs supplied is designed to avoid the emission of pollutants. Moreover, the construction cost of charging station is relatively low and the payback period is merely 16 months. These studies underline the feasibility of scaling up V2G participation in both user-side and energy-side dimensions. Nevertheless, scalability in the architectural sense should not be interpreted merely as the ability to connect more devices. It should also include the ability to maintain acceptable communication efficiency, computational tractability, and operational reliability as the number of participating charging points and vehicles increases. Otherwise, a scheme that works well at a small scale may encounter severe performance degradation when deployed in large practical systems.

In addition, the reliability of V2G architecture should be guaranteed when considering real-time scheduling and voltage regulation. The authors from Ref. [14] present a real-time PEV charging architecture to motivate PEV users to utilize more RESs. In Ref. [15], a real-time architecture for integrating plenty of PEVs into power system is proposed. And based on the proposed architecture, a highly efficient algorithm is designed for scheduling to realize load flattening. In Ref. [16], a novel architecture is proposed for the charging coordination of PEVs. It has been shown that this architecture could conduct the voltage regulation by using only the current and historical voltage magnitudes at the CP, without communication with other CPs or agency. Such decentralized reliability is especially important in settings where communication infrastructure is limited or vulnerable to failure. In fact, the requirement for reliability becomes even more stringent when V2G operation is expected to support grid-side ancillary services. Under such circumstances, delayed or unstable control responses may not only reduce operational benefits, but may also jeopardize the secure operation of the distribution network. Therefore, reliability should be regarded as an intrinsic design requirement of V2G architecture rather than a supplementary performance target.

In recent years, information security and privacy protection have also become critical considerations in V2G systems. Thus, the process of message exchange between the PEV and the energy provider should be secured. A distributed online architecture is proposed in Ref. [17], which not only satisfies each PEV user but also preserves the privacy of the users to avoid security pitfalls. Similarly, in Ref. [18], a privacy-preserving online architecture is proposed to allow PEVs to bid and charge. Based on this architecture, the optimal decision is obtained without information of PEV entering and leaving, and the privacy information of PEV location is differentially ensured by utilizing the Laplace mechanism. With the increasing digitalization of charging infrastructures, privacy protection is no longer limited to safeguarding user identity alone. Charging time, charging location, energy demand, and even travel behavior may all be inferred from operational data if the information flow is not properly designed.

From the scheduling algorithms of V2G perspective, decentralized method and centralized method are the two major types, which are compared in Ref. [19]. The decentralized scheduling method is the basic dispatch program, which is broadly utilized in the V2G problems because of the directness for PEV users [20]. The rule is that let the PEV users make their own charging decisions via the electricity price mechanism [21] However, the decentralized method may not maintain the optimal result if a lot of PEVs are concentrated to charge at low electricity price periods. Hence, a game theory method analyzing the benefits of all stakeholders is utilized to help make interactive decision-makings [22]. Ref. [23] designed a bargaining game by investigating EV charging stations and incorporating energy systems with renewable energies, so the decision-making of V2G scheduling is not merely about the charging cost of PEV users. Nonetheless, electricity price-based decentralized methods face implementation barriers in China, where grid operations are dominated by two state-owned utility companies that maintain centralized control of dispatching [24]. Furthermore, although pilot reforms in electricity retail markets have been initiated in select regions, electricity retailers are still largely absent across most of the country [25]. This practical background implies that the applicability of a scheduling strategy is strongly influenced by the institutional and market environment in which it is deployed. A method that is attractive in a liberalized electricity market may not be directly transferable to systems characterized by centralized dispatching and limited retail competition. Therefore, in addition to theoretical effectiveness, the design of V2G scheduling should also consider the regulatory and operational context of the target power system.

The centralized scheduling method, such as NLP and MIP, are generally applied to the global optimization scheduling [26]. However, MIP achieves the optimal outcomes, depending on the complicated model with the objective function and different constraint conditions. Besides, NLP, as well as MIP, suffer from the "curse of dimension". In addition to the above methods, numerous intelligent algorithms, such as GA, PSO, EA, SDP, NN, and RL are utilized to accommodate the real-world implementation [27]. Mozafar etc. [28] utilized hybrid GA and PSO to address the scheduling problem considering uncertainties of solar radiation, the wind speed and the EV load demand. A novel optimized charging model is proposed to conduct the EA, in which the efficiency of the charging station and the damage of the PEV battery are selected as the joint objective function [29], and the results show a considerably shorter computational time. Jian developed an SDP model with event-triggered, which conducts the high uncertainty of the PEV; the algorithm can significantly reduce the load variance of the power grid [19]. The NN is employed to forecast the controllable charging load, then the predicted charging load is incorporated to V2G charging to reduce the power losses [30]. Ref. [31] developed an online reinforcement learning method to adaptively handle the frequency fluctuation caused by the deployment of irregular PV and wind park, and the results provide a satisfying learning ability. However, PSO and GA are enclosed in sub-optimization frequently, and artificial intelligent methods require sufficient PEV and RES information for the training and validation samples [32]. Furthermore, although the computation time of the intelligent algorithms has an improvement over that of NLP and MIP, the calculation speed of such methods is still long in comparison to the requirement for the real-time V2G scheduling. This reveals a persistent trade-off in existing V2G studies. On the one hand, more sophisticated algorithms are often able to capture uncertainty, multi-objective operation, and nonlinear system characteristics more accurately. On the other hand, the resulting increase in computational complexity may weaken their practical suitability for real-time implementation. Accordingly, improving the computational efficiency of V2G scheduling is not merely an algorithmic preference, but a key prerequisite for translating theoretical coordination schemes into deployable engineering solutions.

For enhancing the computational performance and adapting the national conditions, here, this chapter formulates a V2G scheme involving a scalable, reliable and secure architecture as well as an associated effective and efficient scheduling algorithm. Among them, the CPs under the V2G architecture can be expanded to ensure that the algorithm dispatch different number of PEVs according to the various actual situation. The PVs can be accessed externally in the V2G architecture, which can provide the physical foundation of scheduling algorithm concerning on the PV self-consumption. The reliability of V2G architecture is considered so that real-time control and voltage regulation can be realized in the scheduling algorithm. Most importantly, the V2G architecture involves edge computing, which provides the rapid analysis and privacy-preserving for the scheduling algorithm [33]. In this edge computing scheme, we consider that the computing and storage units are equipped in the CP, that is, the smart charging point (SCP), which merely schedules the connected PEV charging problem. Besides, we consider "burn after scheduling" for the PEV charging information, so as to avoid privacy leakage caused by uploading to the cloud server. Although this privacy feature is not directly quantifiable in simulation results, its architectural superiority is self-evident in terms of security and scalability. Moreover, the V2G scheme aims to follow a target load profile; in this chapter, objectives such as maximizing PV self-consumption and performing peak-shaving and valley-filling are considered to improve grid operating efficiency. It is worth noting that the proposed architecture and algorithm are extensible and can be generalized to accommodate various other grid-oriented objectives, such as economic dispatch, carbon footprint minimization, or local grid resilience enhancement. Such extensibility is particularly meaningful for book-chapter discussion, because the value of the proposed scheme lies not only in solving a single scheduling problem, but also in illustrating a broader design logic for future smart charging infrastructures. By combining scalable architecture, edge-based computing capability, privacy-preserving information handling, and grid-oriented operational objectives, the proposed scheme offers a practical pathway toward coordinated EV-grid interaction under increasingly complex distribution-level conditions.

Current publications actually only provide a limited amount of ways for V2G scheduling which refers to the edge computing. For example, Zhou proposed an efficient V2G energy trading framework based on edge computing, in which the scheduling problem is figured out by the game theory [34]. However, as explained earlier, the transaction under the electricity retail market are not easy to adapt to the current situation of China. Ref. [35] analyzed the information leakage of the computing offloaded tasks; defined the computing offloaded tasks with privacy protecting as a classic optimization problem; and adopted the heuristic algorithm to realize optimal result. But in fact, both [34] and [35] have neglected the RES to integrate V2G scheduling. This indicates that, although edge computing has begun to attract attention in the V2G field, the existing literature still leaves considerable room for improvement in terms of jointly addressing renewable-energy integration, real-time control, privacy preservation, and practical grid-support functions within a single framework. Table 7.1 shows the comparison of the proposed V2G

TABLE 7.1 Comparisons of the proposed V2G scheme with other works.

Considered issues	RES Access	Voltage regulation	Real-time control	Privacy Preserving	Load flattening	High Efficiency	Edge computing
Ref.[11]			√	√		√	√
Ref.[12]	√		√	√	√	√	
Ref.[13]	√	√	√			√	
Ref.[14]	√		√	√	√		
Ref.[15]			√		√	√	
Ref.[16]		√	√		√	√	√
Ref.[17]	√		√	√	√	√	
Ref.[18]			√	√		√	
Ref.[20]		√	√			√	
Ref.[21]				√	√	√	
Ref.[23]	√		√	√	√	√	
Ref.[28]	√	√	√		√	√	
Ref.[29]			√		√	√	
Ref.[19]			√		√	√	
Ref.[30]		√	√		√		

Ref.[31]	√		√				
Ref.[32]		√	√		√	√	
Ref.[34]			√	√		√	√
Ref.[35]			√	√		√	√
This chapter	√	√	√	√	√	√	√

scheme with other literature in terms of RES access, voltage regulation, real-time control, privacy preserving, load flattening, high efficiency and edge computing. To the best of our knowledge, it is the first time to propose an V2G scheme concerning on all the above characteristics. The main contributions of this chapter are summarized as follows.

A V2G scheme concerning on edge computing, that is, the internet of smart charging points with photovoltaic integration (ISCP-PV) is built to dramatically decrease the computational complexity and preserve the privacy of PEV users. Within the architecture of ISCP-PV, the PEV sensitive information is merely uploaded to the connected SCP, which conducts the decision making of PEV charging and/or discharging locally, and carries out "burn after scheduling" for such sensitive information.

The cyber-physical co-modeling are developed to fulfill the fundamental insights into ISCP. And the load peak-shaving and valley-filling, the self-consumption of PV output, the voltage regulation, the up/down regulation, and the communication delay are all considered, comprehensively.

The optimal energy dispatching of ISCP is implemented in an extremely efficient way. First, the distributed computing is adopted by decomposing the complex optimization problem into several simple problems. Second, the parallel computing is applied. Finally, the power flow calculation, which is a heavy task, is handled uniformly by the high-performance computer.

The network performance of ISCP is fully discussed. The small-world network is employed for the cyber topology to improve the efficiency of communications and decrease the wiring costs. Besides, the privacy-preserving of both the energy coordinator and the PEV users are guaranteed by processing and storing the sensitive information of the two participants nearby.

The developed cyber-physical co-modeling are simulated in MATLAB® and NS2. The total time consuming of the proposed work for internal V2G scheduling is 0.864,726 seconds, which is much shorter than the case without considering the parallel computing (1.748,102 seconds) and the case proposed in in our previous work (3.001,348 seconds).

The remainder of this chapter is organized as follows. This chapter begins with the system architecture and key characteristics of ISCP, including the physical scheme, communication network, distributed computing mechanism, and user privacy-preserving features. It then develops the cyber-physical co-modeling framework, with emphasis on the physical level, the cyber level, and the integration of energy and information. The subsequent part presents the optimal energy dispatching of V2G operation, covering internal V2G scheduling, external V2G regulation, and the role of parallel computing in improving dispatching efficiency. Case studies are then provided to evaluate the framework in terms of energy dispatching and network simulation, followed by concluding remarks.

7.2 System architecture and characteristics

7.2.1 Physical scheme

ISCP-PV is a cyber-physical system which has a remarkable device: SCP. The key difference of SCPs from the traditional CPs lies in that they equip computing and storage units to conduct edge computing for V2G scheduling. It is noteworthy that the current computing and storage units installed in the CP, such as STM32F103 and TMS320F28335, has a computing frequency of up to megahertz, and the storage unit has reached a very high level. Thus, the current CP can easily deal with a single V2G problem without additional cost for purchasing new hardware. The objective of edge computing is to diminish processing latency and provide privacy-preserving, since the data needn't be transformed from the SCP to a central EC, then back to the SCP. The main functions of ISCP-PV are implemented in a function area, which is regarded as the basic unit of V2G operation.

ISCP-PV consists of several entities: PEVs, on-site solar panels, various sensors, ECs of different levels, PEV users, SCPs, and so on, as shown in Fig. 7.1. The PEVs and solar panels are the protagonists of energy management. The various sensors provide information perception. Although no charging information is sent to ECs, the ECs can still facilitate the V2G scheduling via define how edge computing should work, publish the key materials used for system initialization, and supervise ISCP-PV for security reasons. The PEV users connect the PEV to the SCPs and upload their charging requirement to the SCP through the smart mobile terminal. The SCPs perform the data processing and decision making via edge computing. Particularly, each SCP in a functional area is equally privileged in the V2G operation.

7.2.2 Communication network

The communication network is one of the majorities of crucial components for ISCP-PV implementation. Classified by coverage range, the communication network of ISCP-PV consists of three levels: 1 LAN that is, the communication network in a functional area, 2 UAN, and 3 WAN.

Fig. 7.2 illustrates the comparison of various communication technologies in terms of coverage and rate of data. At present, the commonly used wired communication technologies that could be arranged in ISCP-PV include fiber optic, coaxial cable, Ethernet, DSL, serial (RS-232/422/485), PLC, CAN and other wired communication technologies. Wired communication has the advantages of high speed, modest delay, low attenuation. Besides, the wired communication is hard to jam and eavesdrop from a distant location. However, there are some drawbacks such as complex wiring and lack of mobility.

Compared with the wired communication, wireless communication can be placed at locations where wiring would not be possible. Correspondingly, the wireless communication technology is well suited for mobility such as fast

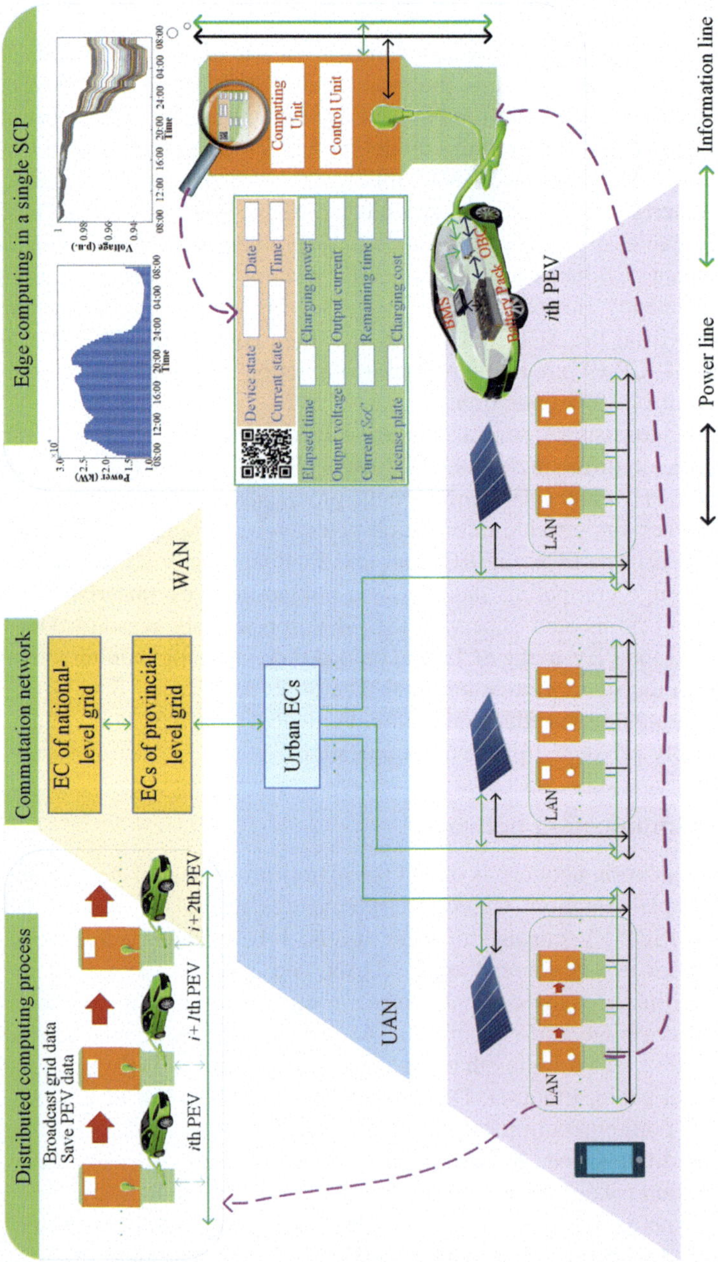

FIGURE 7.1 Architecture of ISCP-PV.

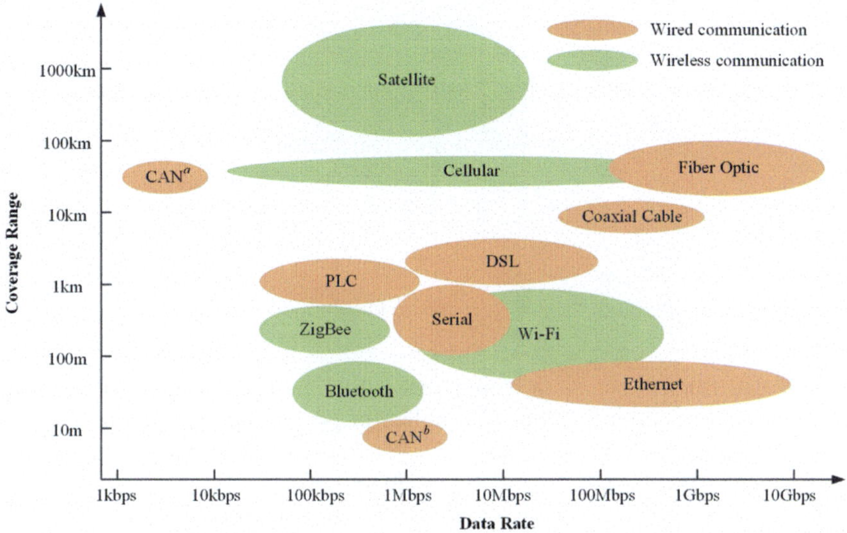

FIGURE 7.2 Comparison of various communication technologies concerning data rate and coverage distance.

reconfiguration, fast expansion and strong destruction resistance. However, wireless communication is vulnerable to a cyber-attack, for example, eavesdropping, jamming and triangulation. Furthermore, another limiting factor of wireless communication is that the wireless devices need sufficient power supply. The commonly used wireless communication technologies include Wi-Fi, Cellular, ZigBee, Bluetooth, Satellite, etc.

Table 7.2 summarizes the network demand for various V2G implementations in terms of participants, information, latency, coverage range, and security. For charge request application, communication technologies that bring smooth and safe experience for customer use are necessary to enable charging requests such as PEV ID, GPS, departure time and energy demand from the smart mobile terminal of PEV user to the connected SCP. Accordingly, mature Cellular, Wi-Fi and Bluetooth technologies can be utilized to satisfy the above needs.

For battery status detection applications, communication technology with stable, safe and professional are required. CAN as a robust vehicle bus standard could allow SCP and BMS to communicate with each other. Hence, it's an appropriate choice for this type of application.

For release V2G strategy application, the data size of a typical control commands is about 50 bytes, and the demand of latency is under 1 seconds. Hence, the data rate is preferably over 400 bps (i.e., 50 bytes * 8 bits/bytes/1 sec $\approx$ 400 bps).

Concerning the distributed computing application, the data rate is influenced by the number of SCPs. Assuming that there are 3000 SCPs for V2G

TABLE 7.2 Specific network requirements for V2G applications.

Network	LAN				UAN		WAN
Application	**Charge request**	**Battery detection**	**Release strategy**	**Distributed compute**	**Data preparation**	**System monitoring**	
Participants	Mobile, SCP	BMS, SCP	SCP, PEV	SCPs	UAN-EC, SCP	UAN-EC, WAN-EC	
Information	ID, energy demand, GPS	Battery *SoC*, capacity	Control commands	Updated grid information	Load and voltage profiles	Power consumption data	
Latency	<1 s	<1 s	<1 s	< Few secs	<< 1 s	<50 ms	
Range	About 10 m	Up to 10 m	Up to 10 m	10 m-3 km	About 10 km	About 10 km	>100 km
Security	Very high	Very high	High	High	High	Extremely high	
Technologies	Cellular, Wi-Fi	CAN	RS 485	Fiber optic			

scheduling, 96-time intervals in a one-day cycle, and the load as well as PV output information is assumed to adopt the format of single-precision floating-point (32 bytes) in one-time interval. If the total message of 32*96 bytes requires preferred several second latency, the data rate is preferably greater than 25 Mbps (i.e., 32*96 bytes * 8 bits/bytes * 2999/3 sec $\approx$ 24.568 Mbps). Considering that the coverage range in a function area is at most 3 km, fiber optic and coaxial cable can serve the above purpose.

For data preparation application, a communication technology is necessary to allow the information of load and voltage profiles from an urban EC to a function area to initiate the V2G scheduling. Similar to the distributed computing application, the grid information employee single-precision floating-point and there are 96-time intervals in a one-day cycle. Therefore, the data rate is preferably greater than 250 kbps (i.e., 32*96 bytes * 8 bits/bytes/0.1 sec $\approx$ 245.76 kbps) if the latency requirement is about 0.1 seconds. Considering that the coverage range in a function area is about 10 km and the wireless technologies have the drawback of unstable, fiber optic and coaxial cable can meet the requirements.

For system monitoring applications, the communication technology requires higher data rate, lower latency, extremely high security and long coverage distance (100 km or more). Fiber optic can provide such services. Also, satellite communications can be used as a backup communication mean to provide redundant communications at critical conditions.

7.2.3 Distributed computing

With the large PEV fleet involving in the V2G scheduling, the traditional centralized method shows very low computational efficiency due to the curse of dimensionality, and create delay issues for the PEV far from the EC in practical application. Different from the traditional computing method, the edge computing is a distributed computing paradigm which brings computation and data storage closer to the PEV. Based on this idea, the edge computing can offer high computational performance by decomposing the high dimensionality V2G scheduling problem into many sub-problems with low complexity (one-PEV one-problem). Besides, the edge computing can improve the speed of data transportation because of the edge-to-edge communication. At last, the edge computing exhibits outstanding merit in privacy-preserving because the data need not to be uploaded to the upper-level ECs. The processes of edge computing in one-time interval are described as follows:

- Jumping-off point for edge computing, urban-EC predicts the information of PV output, load profile and delivers to an SCP in the LAN.
- The SCP collects PEV battery status, charging request of PEV user and grid information as illustrated in the upper-right part of Fig. 7.1. This PEV could be denoted as the *i*th PEV as presented in the upper-left part of Fig. 7.1.

- This SCP computes the V2G scheduling strategy for the *i*th PEV and updates the grid information according to the V2G decision.
- As depicted in the upper-left part of Fig. 7.1, this SCP broadcasts the updated grid information to the neighboring SCP, which connects to the $i+1$st PEV. At the same time, the PEV charging information of *i*th PEVis saved locally.
- Repeat Step 2-4 until the V2G scheduling for all PEVs is finished in a one-time interval.

7.2.4 User privacy preserving

As public awareness of personal privacy continues to grow, it is increasingly necessary to develop schemes that safeguard sensitive information of electric vehicle EV users—such as arrival time, departure time, charging location, and energy demand. Without adequate privacy protection, user participation in V2G energy interactions may decline significantly. This chapter addresses the issue from the perspective of scheduling strategy: by equipping each SCP with local storage and computational units, the proposed system enables decentralized scheduling. In contrast to centralized approaches—which require aggregating user data at a central coordinator and thereby increase the risk of privacy breaches—the decentralized architecture allows each SCP to independently perform scheduling operations based on local information, as detailed in Section 7.2.3. This localized decision-making process significantly enhances the protection of user privacy.

Furthermore, the decentralized scheduling proposed in this chapter fundamentally differs from conventional decentralized strategies. Traditional decentralized scheduling often lacks a unified system-wide objective; individual users typically make charging and discharging decisions solely based on price signals issued by the grid operator, with the aim of maximizing their own benefits. Such self-interested behavior, while computationally efficient, often leads to suboptimal outcomes at the system level due to the absence of coordination. In contrast, the decentralized mechanism developed in this chapter operates under a globally coordinated objective, wherein each SCP performs local optimization that contributes to an overarching scheduling objective defined for the entire network.

7.3 Cyber-physical co-modeling

7.3.1 Physical level: From distribution grid to smart charging points

The architecture of cyber-physical co-modeling within ISCP is depicted in Fig. 7.3. The proposed co-modeling is composed of several energy sources, 27 conventional electrical centers, 5 charging stations, and related communication

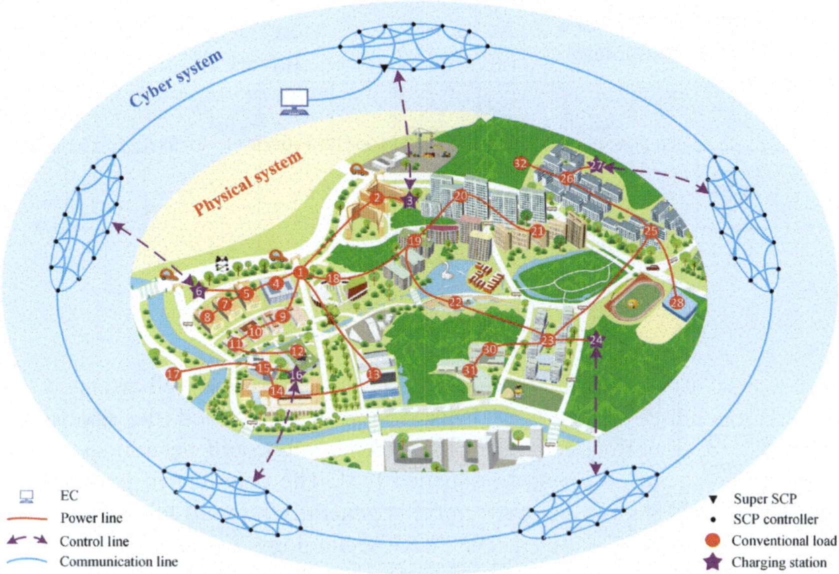

FIGURE 7.3 Architecture of cyber-physical integration within ISCP.

tools. The configuration and the data source of the distribution grid is derived from the Office of Facility Maintenance of Southern University of Science and Technology. And the detailed data of the distribution grid are presented in Appendix A. The energy sources, on the one hand, have a dual 110 kV power supply system and are connected to bus 1 and bus 32 through two 110 kV/10 kV transformers. On the other hand, it is assumed that the photovoltaic panels are installed on the roofs of all buildings within the distribution network to serve as the RES. Notably, the forecasting methodology of the conventional power load is not conducted since it is not the primary target of this chapter. The five charging stations, which are deployed at node 3, 6, 16, 24, 27, can not only absorb the PV energy, but also provide services to the distribution network. Moreover, the smart charging point (SCP) is installed in the charging station.

The SCP can not only be regarded as an energy supply device, but also be regarded intelligent computing system (von Neumann architecture) as depicted in Fig. 7.4. From the perspective of the circuit topology, it consists of a three-phase full-bridge AC/DC converter, DC-link capacitor, and a bidirectional buck-boost DC/DC converter. The SCP is connected to the utility grid via a passive filter at the front-end and interacted with the PEV battery at the back-end. Due to the space limitation, please refer to Ref. [36] for more detailed description of the circuit topology. From the perspective of the control hardware, first of all, the input devices of SCP are used for signal collection, such

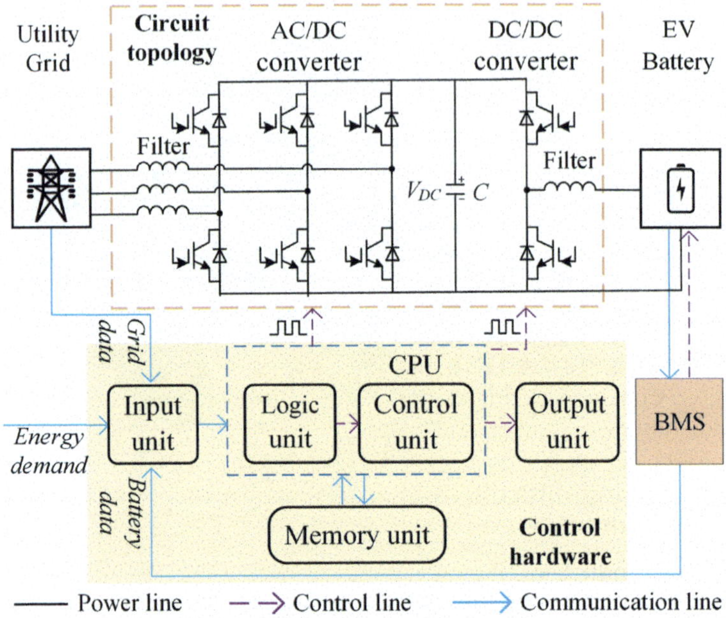

FIGURE 7.4 Overall structure of SCP.

as the virtual keyboard to gather the energy requirement and departure time of PEV user, the video camera to identify the license plate number and recognize the position of PEVs, and the sensor to monitor the voltage and current values during charging and/or discharging period. Second, the output devices of the SCP are used to convert signals into human-readable version, such as the screen to show the charging status of PEVs, the printer to print the payment information, and the sound card to send the alarm when out of order. Then, the processing unit of the SCP is used to calculate the PEV charging-discharging profile based on the data collected by the input devices and the target of the optimal energy dispatching. Next, the control unit of the SCP is used to drive the circuit to implement the operation of the charging or discharging. It is worth noting that above logic unit and processing unit can be collectively referred to as the central processing unit (CPU). Finally, the memory unit of the SCP is used to save data to serve the CPU. The circuit topology and the control hardware are interacted to achieve the desired functions and features: on the one hand, in control hardware, the input devices collect various kinds of information and send it to the processing unit. According to the optimization objective, the processing unit conducts energy dispatching to obtain the optimal decision making of the SCP. On the other hand, the control unit generates the PWM signal according to the decision making, and drives the power switching devices of the AD/DC converter and the DC/DC converter, so as to output the power target constantly within a certain period. The SCP can be connected with each other as a whole, that is, ISCP to coordinate the optimal energy dispatching. ISCP has several excellent features, such as ①Decentralization, there is no central management node responsible for unified computing in ISCP; ②Equality of status, the privilege of each charging pile is not lower than that of other charging points.

7.3.2 Cyber level: Topology design and latency analysis

For the purpose of excellent performance of the ISCP, the mesh network plays an important role in the communication sector. It is a kind of universal network for decentralization and peer-to-peer communication, wherein the nodes link directly and non-hierarchically to other nodes and cooperate with one another to efficiently route data.

There are three classic mesh networks as depicted in Table 7.3. Among them, the fully connected mesh network is a network topology that every node has a connection to each of the nodes within the network. The mean path length of the fully connected mesh network is fairly small (merely one); however, the total number of edges should reach $N(N\text{-}1)/2$. Hence, such a topology is fairly complex, and building such a topology is quite expensive. The lattice/regular network is one of the partially-connected mesh topologies, in which each node connects to all of its nearest neighbors. However, even if this topology is straightforward and the wiring cost is low, the mean path length is quite large, which increases the communication latency.

The small-world network is another partially-connected mesh topology, in which the mean path length increases proportionately to the logarithm of the total number of the nodes N. Hence, both the communication latency and the wiring cost get a good result. Three parameters of the small-world network need to be considered, which are N, K, and β, respectively. N is the total number of nodes. Assuming two nodes connected by an edge are called neighbors, and the number of neighbors for a node j is defined as the degree k_j. Therefore, the average of the degrees of all nodes is defined as the mean degree K,

$$K = \frac{1}{N} \sum_{1}^{K} K_j \tag{7.1}$$

β $(0 \le \beta \le 1)$ is a special parameter that can transmit the lattice network to the small-world network by randomly rewiring each edge of the lattice with probability β. In this chapter, the Watts–Strogatz model, which is one of the technologies for constructing the small-world network, is adopted. Due to the space limitation, please refer to Ref. [37] for understanding the details of network construction if you are interested.

TABLE 7.3 Description for different networks.

Name	Fully meshed	Lattice/regular	Small world
Typology	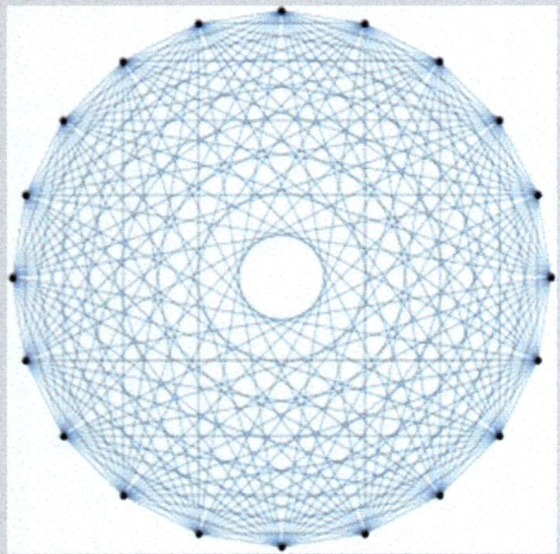	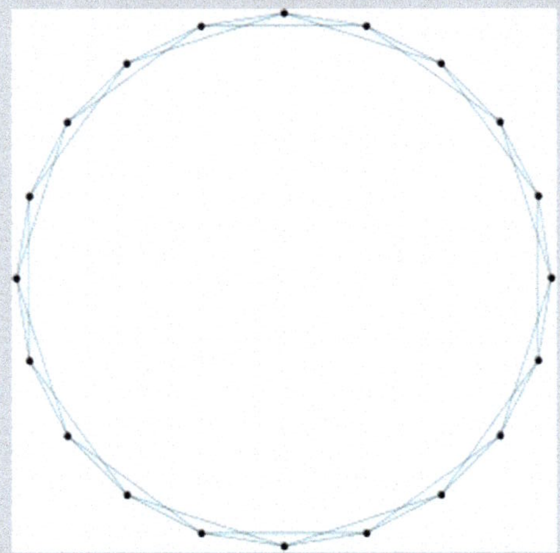	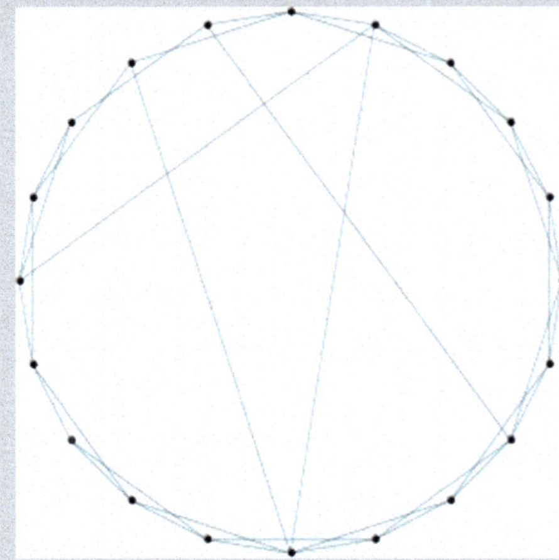
Edge	*N*(*N*-1)/2	*NK*/2	*NK*/2
Cost[a]	Large	Small	Small
MPL[b]	Small	Large	Small
Latency	Small	Large	Small

For reliable communication between the nodes, the wired link with duplex capacity is considered. The delay of the communication process from one node to another node can be divided into three parts and estimated as follows:

$$D^{total} = D^{propa} + D^{trans} + D^{queue} \tag{7.2}$$

$$D^{paopa} = L/c \tag{7.3}$$

$$D^{trans} = P \times 8/D \tag{7.4}$$

where D^{total} is the total delay of the communication process, in second; D^{propa} is the propagation delay which is calculated by the propagation speed of light via wired link and the link distance, in second; L is the distance of the link, in km; c is the propagation speed of light via wired link technologies, such as the fiber optic ($2*10^8$ m/s), the twisted pair ($1.77*10^8$ m/s), the coaxial ($2.31*10^8$ m/s); D^{trans} is the transmission delay which is calculated by the data size and the data rate, in second; P is the packet size which varies according to the different situation, in byte; D is the data rate, in bit/s; D^{queue} is the data queuing delay, which is different for different situation, in second.

7.3.3 Integration of energy and information

There are eight steps of the integration of energy and information in each time slot, wherein six steps for cyber level and two steps for physical level (optimal energy dispatching in green font). It should be noted that the status of the super SCP described below is not higher than other SCPs, and it is merely a medium to communicate with the EC on behalf of the ISCP:

Step1(information): at t-1st time slot, the EC sends the grid dispatching instructions of the tth time slot to the super SCP.
Step2(information): distributed communication between SCPs.
Step3(Energy): the distributed computing between SCPs for optimal energy dispatching corresponding to step 2.
Step4(information): each SCP sends the V2G decision makings of the optimal energy dispatching to the super SCP.
Step5(information): the super SCP sends all of the decision makings of ISCP to the EC.
Step6(Energy): the EC carries out a security check for the decision makings of ISCP.
Step7(information): the EC sends the verified decision makings to the super SCP.
Step8(information): the super SCP sends the verified decision makings to each of SCP.

According to the above discussions, the scope of execution can be divided into three categories as illustrated in Fig. 7.5. Among them, one step is

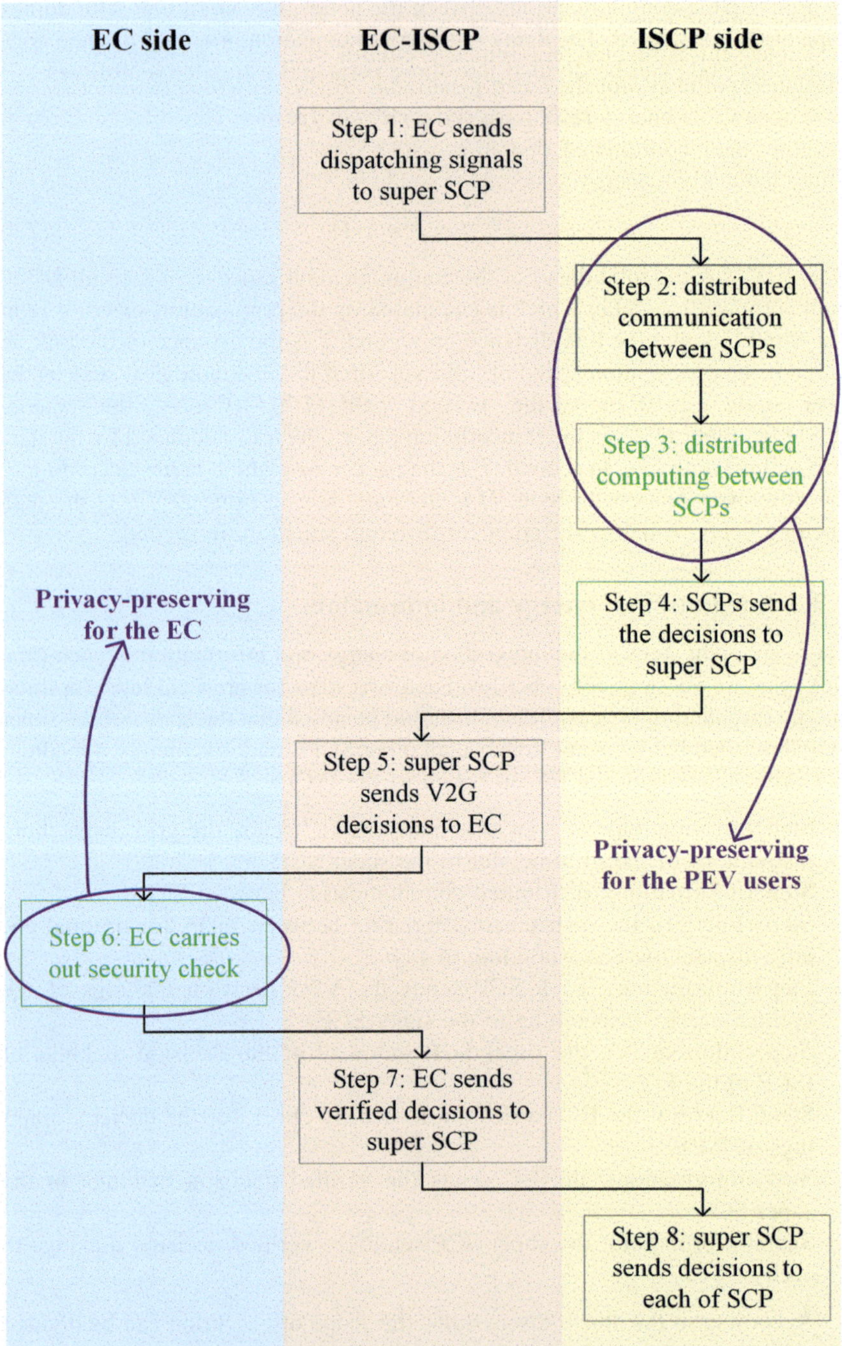

FIGURE 7.5 Flowchart of the integration of energy and information.

executed in the EC side (Step 6), four steps are executed in the ISCP side (Step 2, 3, 4, 8), and three steps are executed between the EC and the ISCP (Step 1, 5, 7). Steps 2 and 3 described above are the key steps for implementing the optimal energy dispatching with the egalitarian, distributed, and edge computing characteristics. First, the SCPs are sorted according to the principle of "first come, first dispatch" for PEVs. Second, the super SCP carries out the edge computing locally to achieve the V2G decision making based on the grid information provided by EC and the charging information provided by PEV users. Next, the super SCP updates the grid information based on the V2G decision making. And then, the updated grid information is transmitted from the super SCP to the next SCP, and the PEV charging information is saved locally. Finally, the Steps 2 and 3 are finished when the optimal energy dispatching for all SCPs are conducted.

Particularly, both sensitive information of the PEV users and the EC are protected. On the one hand, in Step 2 and Step 3, the sensitive information of the PEV users, such as the arriving time slot, the departure time slot, the energy requirement, and the location, are uploaded to the connected the SCP. And the SCP conducts the "burn after dispatching" for such information. Hence, the privacy-preserving for the PEV users is greatly enhanced. On the other hand, there are few literature focused on the privacy-preserving for the EC. Compared with our previous work [38], the voltage regulation is conducted near the EC rather than the SCP in Step 6. In this situation, several sensitive information of the EC, such as the power flow information, does not need to be transferred to the SCPs. Hence, the privacy-preserving for the EC is considerably improved.

In this chapter, two super SCP are set according to the dual 110 kV power supply system. Hence, the integration of energy and information becomes more flexible. And the following functions can be provided:

(a) **Hot standby**, one super SCP is responsible for V2G scheduling, and one super SCP serves as a hot standby, which can be started quickly in the event of another super node failure.
(b) **Double-check**, one super SCP is responsible for V2G scheduling, and one super SCP carries out the double-check to avoid errors.
(c) **Parallel computing**, the ISCP can be divided into two regions, and the two super SCPs are responsible for the optimal energy dispatching in one region simultaneously to improve efficiency.

The details of the optimal energy dispatching for parallel computing are discussed and shown in Section 7.4.3.

7.4 Optimal energy dispatching of V2G operation

The optimal energy dispatching for V2G is divided into two layers, that is, the internal V2G scheduling and the external V2G regulation. Among them, the "internal" and the "external" are relative to the distribution grid. For the energy dispatching of the distribution grid itself, we call it "internal energy dispatching". In this chapter, the load peak-shaving and vallry-filling as well as the PV self-consumption of the distribution grid are discussed, which can improve the stability of the distribution grid and the efficiency of renewable energy. Besides, the internal V2G scheduling is scheduled every 15 minutes. For the instructions of the energy dispatching received from outside the distribution grid, that is, the EC, we call it "external V2G regulation". In this chapter, the up/down regulation, whose signals received from the EC, is discussed. Moreover, the response time of the external V2G regulation is minutes- or even second-level.

7.4.1 Internal V2G scheduling

According to whether there is PV output in the dispatching period of PEVs, there are three optimization strategies: maximum self-consumption of PV output, minimum load variance, and maximum self-consumption of PV output & minimum load variance of distribution campus grid. If this chapter does not take the PV output into account, the optimization model of the internal V2G scheduling for the ith PEV is the load variance minimization, and the objective function is

$$\min \sum_{t=t_i^{arr}}^{t_i^{dep}} \left(p_i^{con} + p_{i,t}^{EV} - p^{ave}\right)^2 \tag{7.5}$$

$$p^{ave} = \frac{1}{T} \sum_{t=1}^{T} p_t^{con} \tag{7.6}$$

where t_i^{arr} and t_i^{dep} are the start and the end time slots for the ith PEV charging; p_t^{con} is the conventional load at the tth time slot; $p_{i,t}^{EV}$ is the V2G decision making need to be optimized of the ith PEV at the tth time slot; p^{ave} is the average conventional load in one day cycle; T is the total time slots in one day cycle and $T = 96$ since the time slot is set as 15 minutes. Besides, define

$$SoC_{i,t} = \frac{E_{i,t}}{B_i} \tag{7.7}$$

where $SoC_{i,t}$ is the battery SoC at the tth time slot; $E_{i,t}$ is the battery energy at the tth time slot; B_i is the energy capacity of the PEV battery. For the energy requirement of the PEV, the constraint function is

$$\eta_i^{ch} \Delta t \sum_{t=t_i^{arr}}^{t_i^{dep}} p_{i,t}^{EV} = SoC_i^{dep} - SoC_i^{arr} B_i = E_i \tag{7.8}$$

where η_i^{ch} is the charging efficiency when the connected SCP supply energy for the ith PEV; Δt is the time duration of a time slot, that is, 15 minutes;

SoC_i^{arr} is the battery SoC at the start time slot; SoC_i^{dep} is the battery SoC at the end time slot; E_i^{re} represents the energy requirement for the ith PEV. Furthermore, for any time slotfrom t_i^{arr} to t_i^{dep}, the constraint of PEV charging as well as discharging power and the constraint of battery state of charging are

$$p_i^{dis,max} \le p_{i,t}^{EV} \le p_i^{ch,max} \tag{7.9}$$

$$SoC_i^{min} \le SoC_{i,t} \le SoC_i^{max} \tag{7.10}$$

where $p_i^{dis,max}$ and $p_i^{ch,max}$ are the maximum discharging and charging power for the ith PEV; SoC_i^{min} and SoC_i^{max} are the maximum and minimum battery SoC for the ith PEV.

For the purpose of obtaining the decision making quickly, the optimization problem of the internal V2G scheduling for the ith PEV is split in six scenarios based on the conventional load profile, the average conventional load, and the start as well as the end time slots for a PEV charging, that is, whole valley, whole peak, valley-peak, peak-valley, valley-peak-valley, and peak-valley-peak. Furthermore, the scenario which the ith PEV belongs is selected based on the charging information such as t_i^{arr}, t_i^{dep}, and E_i^{re}. And the charging capacity in valley period E_i^{vf}, the start and end time slots for ith PEV charging in valley period $t_i^{s,vf}$ and $t_i^{e,vf}$, the discharging capacity in peak period E_i^{ps}, the start and end time slots for PEV discharging in peak period $t_i^{s,ps}$ and $t_i^{e,ps}$ are calculated.

According to the above discussions about the constraint conditions (Formula 7.5–7.10), six scenarios, the data preparation of PEV discharging at the peak period ($t_i^{s,ps}$, $t_i^{e,ps}$and E_i^{ps}) and charging at the valley period ($t_i^{s,vf}$, $t_i^{e,vf}$ and E_i^{vf}), the water filling algorithm is utilized for load flatting as shown in Tables 7.4 and 7.5. In particular, $P[a_1\text{-}\ P_t^{con}]$ and $P[P_t^{con}–a_2]$ are defined as:

$$P\left[a_1 - P_t^{con}\right] = \begin{cases} 0, a_1 - P_t^{con} < 0 \\ a_1 - p_t^{con}, 0 \le a_1 - P_t^{con} < P_t^{ch,max} \\ P_t^{ch,max}, a_1 - P_t^{con} \ge P_t^{ch,max} \end{cases} \tag{7.11}$$

$$P\left[P_t^{con} - a_2\right] = \begin{cases} 0, P_t^{con} - a_2 < 0 \\ a_2 - p_t^{con}, 0 \le P_t^{con} - a_2 < P_t^{dis,max} \\ P_t^{dis,max}, a_2 - P_t^{con} \ge P_t^{dis,max} \end{cases} \tag{7.12}$$

One of the highlights of this chapter is the proof of optimality for the internal V2G scheduling of a single PEV. The optimality proof for the load variance minimization (load flatting) is shown as follow.

Theorem 1 For each PEV, the solutions given by Tables 7.6 and 7.4 are the optimal solution of load variance minimization.

TABLE 7.4 Load peak-shaving algorithm for one-PEV one-problem conducted by one SCP.

1: **Input:** $p_i^{dis,max}$, E_i^{ps}, $t_i^{s,ps}$, $t_i^{e,ps}$ **and** P_t^{con}, $t = t_i^{s,ps}$, $t_i^{s,ps}+1$, ..., $t_i^{e,ps}-1$, $t_i^{e,ps}$
2: **Output**: a_2 and $p_{i,t}$, $t = t_i^{s,ps}$, $t_i^{s,ps}+1$, ..., $t_i^{e,ps}-1$, $t_i^{e,ps}$
3: Initialize $a^{min} = \min(P_t^{con} - p_i^{dis,max})$ and $a^{max} = \max(P_t^{con})$
4: **while** $a^{max} - a^{min} > 1.0e\text{-}6$ **do**
5: Choose $a_2 = (a^{max} + a^{min}) \div 2$
6: **for** $t = t_i^{s,ps}$: $t_i^{e,ps}$
7: $p_{i,t} = \mathrm{P}[P_t^{con} - a_2]$
8: **end**
9: **if** $\eta_i^{dis}\Delta t > E_i^{ps}$
10: set $a^{min} = a_2$
11: **elseif** $\eta_i^{dis}\Delta t < E_i^{ps}$
12: set $a^{max} = a_2$
13: **end**
14: **end**

To prove Theorem 1, we first examine three Lemmas and then give the conclusion.

Lemma 1 Peak shaving at the time slots of load peak periods ($p_t^{EV} \le 0$, p_t^{con}-$p_t^{EV} \ge p^{ave}$) as well as valley filling at the time slots of load valley periods ($p_t^{EV} \ge 0$, $p_t^{con} + p_t^{EV} \le p^{ave}$) are the premise to get the optimal solution. Where p_t^{EV} is the is the V2G decision making for the PEV at the tth time slot.

Proof 1 The original optimal value of the load variance is

$$A = \sum_{t=t_i^{arr}}^{t_i^{dep-1}} \left(p_t^{con} + p_i^{EV} - p^{ave}\right)^2 = \sum_{t=t_i^{arr}}^{t_i^{dep-1,t\neq m}} \left(p_t^{con} + p_i^{EV} - p^{ave}\right)^2 + \left(p_m^{con} + p_i^{EV} - p^{ave}\right)^2$$

First, it is assumed that the peak shaving is carried out at the mth time slot of the load valley periods, and the V2G decision making at the mth time slot is defined as. The load variance is

TABLE 7.5 Self-consumption of PV output by water-filling algorithm for one-PEV one-problem conducted by one SCP.

1: **Input:** $p_i^{ch,max}$, E_i^{re}, t_i^{arr}, t_i^{dep} **and** P_t^{con}, $t = t_i^{arr}, t_i^{arr}+1, \ldots, t_i^{dep}\text{-}1, t_i^{dep}$
2: **Output**: a_3 and $p_{i,t}$, $t = t_i^{arr}, t_i^{arr}+1, \ldots, t_i^{dep}\text{-}1, t_i^{dep}$
3: Initialize $a^{min} = \min(P_t^{con} - p_i^{ch,max})$ and $a^{max} = \max(P_t^{con})$
4: **while** $a^{max} - a^{min} > 1.0e\text{-}6$ **do**
5: Choose $a_3 = (a^{max} + a^{min}) \div 2$
6: **for** $t = t_i^{arr}: t_i^{dep}$
7: $p_{i,t} = \mathrm{P}[P_t^{con} - a_3]$
8: **end**
9: **if** $\eta_i^{ch} \Delta t \sum_{t=t_i^{arr}}^{t_i^{dep}} p_{i,t} > E_i^{re}$
10: set $a^{min} = a_3$
11: **elseif** $\eta_i^{ch} \Delta t \sum_{t=t_i^{arr}}^{t_i^{dep}} p_{i,t} < E_i^{re}$
12: set $a^{max} = a_3$
13: **end**
14: **end**

$$B = \sum_{t=t^{arr}}^{t^{dep}-1, t\neq m} \left(p_t^{con} + p_t^{EV} - p^{are}\right)^2 + \left(p_m^{con} + \underline{p}_m^{EV} - p^{ave}\right)^2$$

$$\because p_m^{con} + p_i^{EV} - p^{are} < p_m^{con} + \underline{p}_m^{EV} - p^{ave}$$

$$\therefore \left(p_m^{con} + p_i^{EV} - p^{ave}\right)^2 < \left(p_m^{con} + \underline{p}_m^{EV} - p^{ave}\right)^2$$

$\therefore A < B$.

$\therefore$ The assumption is not true.

Second, it is assumed that the valley filling is carried out at the mth time slot of the load peak periods and the V2G decision making at the mth time slot is defined as $\overline{p_m^{EV}}$. The load variance is

$$C = \sum_{t=t^{arr}}^{t^{dep}-1, t\neq m} \left(p_t^{con} + p_t^{EV} - p^{ave}\right)^2 + \left(p_m^{con} + \overline{p_m^{EV}} - p^{ave}\right)^2$$

TABLE 7.6 Load valley-filling algorithm for one-PEV one-problem conducted by one SCP.

1: **Input:** $p_i^{ch,max}$, E_i^{vf}, $t_i^{s,vf}$, $t_i^{e,vf}$ and P_t^{con}, $t = t_i^{s,vf}, t_i^{s,vf}+1, \ldots, t_i^{e,vf}-1, t_i^{e,vf}$
2: **Output**: a_1 and $p_{i,t}$, $t = t_i^{s,vf}, t_i^{s,vf}+1, \ldots, t_i^{e,vf}-1, t_i^{e,vf}$
3: Initialize $a^{min} = \min(P_t^{con})$ and $a^{max} = \max(P_t^{con} + p_i^{ch,max})$, $t = t_i^{s,vf}+1, \ldots, t_i^{e,vf}-1, t_i^{e,vf}$
4: **while** $a^{max} - a^{min} > 1.0e\text{-}6$ **do**
5: Choose $a_1 = (a^{max} + a^{min}) \div 2$
6: **for** $t = t_i^{s,vf}: t_i^{e,vf}$
7: $p_{i,t} = \mathrm{P}[a_1 - P_t^{con}]$
8: **end**
9: **if** $\eta_i^{ch}\Delta t > E_i^{vf}$
10: set $a^{max} = a_1$
11: **elseif** $\eta_i^{ch}\Delta t < E_i^{vf}$
12: set $a^{min} = a_1$
13: **end**
14: **end**

$$\because \left|p_m^{con} + p_m^{EV} - p^{ave}\right| < \left|p_m^{con} + \overline{p_m^{EV}} - p^{ave}\right|$$

$$\therefore \left(p_m^{con} + p_m^{EV} - p^{ave}\right)^2 < \left(p_m^{con} + \overline{p_m^{EV}} - p^{ave}\right)^2$$

$\therefore A < C$

$\therefore$ The assumption is not true.

Hence, Lemma 1 is proofed.

Lemma 2 *T*he charging capacity of the valley period as well as the discharging capacity of the peak period in Formulas (7.6–7.15) in previous work [33] is optimal for load variance minimization.

Proof 2 It is assumed that the charging capacity within a valley period is not optimal. This case can be discussed in two categories. On the one hand, the charging capacity within the valley period can be decreased to reach an optimal solution. Correspondingly, the discharging capacity within the peak period should be decreased based on Formula 7.8 (energy requirement). The time slot of changed charging power due to the decrease of charging capacity in the valley period is defined as $\bar{t}$, and the corresponding changed charging power is defined as $\underline{p}t$. The time slot of changed discharging power due to the decrease of discharging capacity in peak period is defined as, and the corresponding changed discharging power are defined as $\overline{p_{\bar{t}}}$.

The original optimal value of the load variance is A.

The load variance due to the charging capacity within a valley period can be decreased to D.

$$D = \sum_{t=t_i^{arr}}^{t_i^{dep-1,t\neq t,t}} \left(p_t^{con} + p_t^{EV} - p^{ave}\right)^2 + \sum_{t}^{t=t} \left(p_t^{con} + \underline{p}_t^{EV} - p^{ave}\right)^2 + \sum_{t}^{t=t} \left(p_t^{con} + \overline{p_t^{EV}} - p^{ave}\right)^2$$

$$A = \sum_{t=t_i^{arr}}^{t_i^{dep-1,t\neq t,\bar{t}}} \left(p_t^{con} + p_t^{EV} - p^{ave}\right)^2 + \sum_{t}^{t=t} \left(p_{\underline{t}}^{con} + p_{\underline{t}}^{EV} - p^{ave}\right)^2 + \sum_{t}^{t=t} \left(p_t^{con} + p_t^{EV} - p^{ave}\right)^2$$

$\because t \in \underline{t},\ \underline{p}t < p\underline{t},\ \therefore \left|p_{\underline{t}}^{con} + \underline{p}_{\underline{t}}^{EV} - p^{ave}\right| > \left|p_{\underline{t}}^{con} + p_{\underline{t}}^{EV} - p^{ave}\right|.$

$\because t \in t,\ \overline{p_{\bar{t}}} > p_{\bar{t}},\ \therefore \left(p_{\bar{t}}^{con} + \overline{p_{\bar{t}}} - p^{are}\right) > \left(p_{\bar{t}}^{con} + p_{\bar{t}} - p^{are}\right).$

$\therefore A < D.$

$\therefore$ The assumption is not true.

On the other hand, the charging capacity within the valley period can be increased to reach an optimal solution. The charging capacity of the valley period is given by Formulas (7.8–7.10). If the charging capacity within the valley period is increased, the constraint cannot be satisfied, so this assumption is also not true.

It is assumed that the discharging capacity within a peak period is not optimal. The proof of optimality is similar to the above, so that it is no longer proved due to space limitation.

Lemma 3 For each peak or valley period, the algorithms given in Tables 7.3 and 7.4 are optimal according to the optimal charge or discharge capacity proved by **Lemma 2**.

Proof 3 For each peak or valley period, the energy dispatching problem can be regarded as a convex quadratic programming according to the given parameters. Take a valley period as an example, given $t^{s,vf}$ and $t^{e,vf}$, and E^{vf}. The programming of valley-filling can be formulated as follows:

$$\min \sum_{t=t_i^{s,vf}}^{t_i^{e,vf}-1} \left(p_t^{con} + p_t^{EV} - p^{ave}\right)^2 \tag{7.13}$$

s.t.

$$\eta^{ch} \Delta t \sum_{t=t_i^{s,vf}}^{t_i^{e,vf}-1} p_t^{EV} = E^{vf} \tag{7.14}$$

$$0 \le p_t^{EV} \le p^{ch,max} \tag{7.15}$$

It can be found that the constraints of PEV battery SoC have been disappeared due to the derivation of Formulas (7.6–7.15) in previous work [33]. Using the Lagrange multiplier method, the Lagrangian function for formulas (7.13 and 7.15) is given by

$$L\left(p_t^{EV}, \lambda_1\right) = \frac{1}{2} \sum_{t=t^{s,vf}}^{t^{e,vf}-1} \left(p_t^{con} + p_t^{EV} - p^{ave}\right)^2 + \lambda_1 \left(\eta^{ch} \Delta t \sum_{t=t^{s,vf}}^{t^{e,vf}-1} p_t^{EV} - E^{vf}\right) \tag{7.16}$$

whereis the Lagrangian multiplier for the constraint (14). The factor of 1/2 is for convenience. The KKT conditions of formula (7.16):

$$p_t^{EV} + p_t^{con} - p^{ave} + \lambda \eta \Delta t = 0 \tag{7.17}$$

$$\eta^{ch} \Delta t \sum_{t=t^{s,vf}}^{t^{e,vf}-1} p_t = E^{vf} \tag{7.18}$$

Define $\lambda_2 = \lambda_1 \eta \Delta t - p^{ave}$ is the modified Lagrangian multiplier. Hence,

$$p_t^{EV} = -\lambda_2 - p_t^{con} \tag{7.19}$$

However, the above optimal conditions are not considered the constraint (15). When constraint (15) is considered, the optimal solution is either (19) or

$$p_t^{EV} = 0 \wedge -\lambda_2 - p_t^{con} < 0 \tag{7.20}$$

or

$$p_t^{EV} = p^{ch,max} \wedge -\lambda_2 - p_t^{con} > p^{ch,max} \tag{7.21}$$

By comparing the algorithms given in Tables 7.4 and 7.6 and the above discussions, we can get that the final water surface a_l is the same as the negative modified Lagrangian multiplier $-\lambda_2$. The error of the final water surface a^{max} - a^{min} is extremely small (1.0e-6). Hence, the solutions given by Tables 7.4 and 7.6 are optimal. The proof of optimality for the valley period is similar to the above, and it is no longer proved due to space limitation.

If we merely takes into account the PV output periods, that is, the PEV staying time is always in the PV output periods, the optimization model of the internal V2G scheduling for the *i*th PEV is the self-consumption of PV output. The objective function is:

$$\min \sum_{t=t_i^{arr}}^{t_i^{dep}} \left(p_t^{pv} - p_{i,t}^{EV}\right)^2 \tag{7.22}$$

where p_t^{PV} is the PV output energy at the *t*th time slot; For the energy requirement of the PEV, the constraint is the same as Formula (7.8). For any time slotfrom t^{arr} to t^{dep}, the constraint of PEV charging power is the same as Formula (7.15). It can be found that the constraint of PEV discharging power and the constraint of PEV battery *SoC* have been disappeared because the PEV only needs to absorb the PV energy as much as possible during the PV output periods. The decision makings and the optimality proof of the self-consumption of PV energy are shown as follow:

Using the Lagrange multiplier method, the Lagrangian function for Formula (7.8), (7.15), and (7.22) is given by:

$$L\left(p_t^{EV}, \lambda_2\right) = \frac{1}{2}\sum_{t=t^{arr}}^{t^{dep}} \left(p_t^{pv} - p_t^{EV}\right)^2 + \lambda_3\left(\eta^{ch}\Delta t \sum_{t=t^{arr}}^{t^{dep}} p_t^{EV} - E^{re}\right) \tag{7.23}$$

where are the Lagrangian multipliers for the constraint (8). The factor of is for convenience. Next, the KKT conditions of formula (7.23) is

$$p_t - p_t^{pv} + \lambda_3 \eta \Delta t = 0 \tag{7.24}$$

$$\eta^{ch}\Delta t \sum_{t=t_i^{arr}}^{t_i^{dep}} p_t - E_i = 0 \tag{7.25}$$

Define, and is the modified Lagrangian multipliers. Therefore,

$$p_t = -\lambda_4 + p_t^{pv} \tag{7.26}$$

However, the above optimal conditions are not considered the constraint (15). When constraint (15) is considered, the optimal solution is either (26) or

$$p_t^{EV} = 0 \wedge -\lambda_4 + p_t^{pv} < 0 \tag{7.27}$$

or

$$p_t^{EV} = p^{ch,max} \wedge -\lambda_4 + p_t^{pv} > p^{ch,max} \tag{7.28}$$

By comparing the algorithm given in Table 7.5 and the above discussions, we can get that the final water surface a_3 is the same as the negative modified Lagrangian multiplier $-\lambda_4$. The error of the final water surface a^{max} - a^{min} is extremely small (1.0e-6). Hence, the algorithm given in Table 7.5 is optimal.

According to the above discussions, if the PEV staying periods are not in the PV output periods, the algorithm of the load variance minimization is utilized. If the PEV staying periods are merely in the PV output periods, the algorithm of the self-consumption of PV output is utilized. However, if PEV staying periods are not only in the PV output periods but also in the other periods, it is essential to calculate the charging capacity of the PEV during the PV output periods firstly; then identify the Scenario which the PEV belongs to, and conduct the data preparation of the remaining peak and valley periods. Finally, the decision making for the internal V2G scheduling is calculated by the algorithms of the load variance minimization as well as the self-consumption of PV energy, separately. In this chapter, because PEVs in the campus generally arrive in the morning, leave in the afternoon or in the evening, only the case of first PV output periods and then load peak periods is considered.

In this situation, if the energy requirement of the PEV cannot be guaranteed by charging during PV output periods, the charging capacity in PV output period is $(t_i^{pv,end}\text{-}t_i^{arr}+1)*p_i^{ch,max}$ for self-consumption of PV output. And the discharging capacity in other periods is $E_i^{re}\text{-}(t_i^{pv,end}\text{-}t_i^{arr}+1)*p_i^{ch,max}$, where $t_i^{pv,e}nd$ is the end of photovoltaic output during the staying periods of the ith PEV. On the contrary, if the energy requirement of the PEV can be guaranteed by charging during the PV output period, the charging capacity in PV output period $E_i^{ch,pv}$ need be evaluated. Such charging capacity is related to the PEV battery SoC, energy requirement, staying period, which is

$$z = \min\left(t_i^{pv,end} - t_i^{pv,start} - t_i^{need} + 1, t_i^{dep} - t_i^{pv,end}\right) \tag{7.29}$$

$$z = p^{ch,\max} \times \min\left(z, \left(SoC_i^{max} - SoC_{i,t}\right) \times B_i - t_i^{pv,end} - E_i^{\Re}\right) \tag{7.30}$$

$$E_i^{ch,pv} = E_i^{\Re} + z \tag{7.31}$$

where $t_i^{pv,\ start}$ is start end of photovoltaic output during the staying periods of the ith PEV; and t_i^{need} is the number of time slot required for the ith PEV to charge at $p_i^{ch,max}$ to meet the energy requirement. In addition, the discharging capacity E_i^{ps} is z during the subsequent load peak period for the load peak-shaving.

7.4.2 External V2G regulation

The external V2G regulation is utilized to continuously balance the supply and demand of the grid at a short time scale. In particular, maintaining the grid frequency at a constant value is one of the approaches for the regulation. Therefore, ISCP, as the utilization side of the gird, can provide the fast up/down regulation by absorbing/releasing power to the outside EC.

Firstly, the EC calculates the up/down regulation signal based on the frequency errors as well as the stability limits of the distribution grid; and sends the signal to the ISCP. Then, ISCP evaluates whether the regulation is possible. Finally, the decision makings of the external V2G regulation are sent to the EC. Remarkably, the down regulation is not appeared in the PV output period, since increasing energy consumption of the distribution grid will further decrease the self-consumption of PV output. The algorithm for external V2G regulation is shown in Table 7.7.

Instead of implementing the optimal energy dispatching of PEVs, the stability of the distribution grid needs to be considered. Hence, the node voltage, as an important indicator of the grid stability, is selected, and should be limited by:

$$V_n^{min} \leq V_{n,t} \leq V_n^{max} \tag{7.32}$$

where V_n^{min} and V_n^{max} is the security restrictions of voltage for the nth node, respectively; V_n^t is the voltage value for the nth node at the tth time slot. In this chapter, the fluctuation of voltage is restricted to $-7\%\sim+7\%$ of the rated value during 96% period in a day based on the GB/T 12325–2008.

The flow chart of the optimal energy dispatching for V2G operation is depicted in Fig. 7.6. In previous work, the power flow calculation of the voltage regulation was applied within each SCP. Such practice only protects the privacy of the PEV users. However, the sensitive data of the EC, that is, the power flow information, was leaked to the ISCP. In this chapter, the power flow calculation is implemented outside the ISCP as depicted in the blue circle of Fig. 7.6, that is, the steps 5-7 in Section 7.3.3.

If the optimal energy dispatching conducts the internal V2G scheduling, the EC receives the decision makings of the PEVs, such as the charging or discharging power and the node location of distribution grid; then conducts the voltage regulation in turn according to the ordinal number of V2G scheduling based on "first come, first dispatch". The algorithm for the voltage regulation of the internal V2G scheduling is shown in Table 7.8. If the decision making of the ith PEV at the tth time slot $p_{i,t}$ is zero, jump power flow process of this PEV. Otherwise, the node voltages of the distribution power grid for the tth time slot are computed by the backward-forward sweep method. If all node voltages are within the security restrictions, the decision making $p_{i,t}$ will not change and the grid information will be updated. If there is a voltage value exceeds the security restrictions, it is necessary to decide whether the energy demand of the ith PEV can be supported when the PEV is charged at $p_i^{ch,max}$ in all remaining stay periods. The decision making $p_{i,t}$ will not change and the grid information will be updated during the tth time slot if the condition cannot be satisfied. Otherwise, the decision making $p_{i,t}$ will be zero. After finishing the voltage regulation of the ith PEV during the tth time slot, the voltage regulation for the next PEV during will be conducted. The whole voltage regulation process will be finished until all PEVs are accomplished for the tth time slot. Since the up/down regulation signal is calculated according to the

TABLE 7.7 Algorithm for the external V2G regulation.

1: **Input:** t, R_t, P_t^{PV}
2: **Output:** $p_{i,t}$, $i = 1,2, \ldots, N\text{-}1, N$
3: **if** $P_t^{PV} > 0$
4: $R_t > 0$, only the up regulation is appeared in the PV output period
5: Calculate the number of PEVs that can participate in up regulation N_{re_up}
6: **if** $-R_t > P_t^{PV} + N_{re\_up}{}^* \, p^{dis,max}$
7: Break
8: **end**
9: **if** $-R_t < P_t^{PV}$
10: $P_t^{PV} = P_t^{PV} + R_t$
11: $p_{i,t} = P_t^{PV}/N_{re\_up}$, $i = 1,2, \ldots, N\text{-}1, N$,
12: **else**
13: $R_t = R_t + P_t^{PV}$
14: $p_{i,t} = R_t/N_{re\_up}$, $i = 1,2, \ldots, N\text{-}1, N$,
15: **end**
16: **else**
17: if $R_t > 0$
18: Calculate the number of PEVs that can participate in up regulation N_{re_up}
19: **if** $- R_t > N_{re\_up}{}^* \, p^{dis,max}$
20: Break
21: **end**
22: $p_{i,t} = R_t/N_{re\_up}$, $i = 1,2, \ldots, N\text{-}1, N$,
23: **else**
24: Calculate the number of PEVs that can participate in down regulation N_{re_dw}
25: **if** $R_t > N_{re\_dw}{}^* \, p^{dis,max}$
26: Break
27: **end**
28: $p_{i,t} = R_t/N_{re\_dw}$, $i = 1,2, \ldots, N\text{-}1, N$,
29: **end**
30: **end**
31: Update the charging information of PEVs

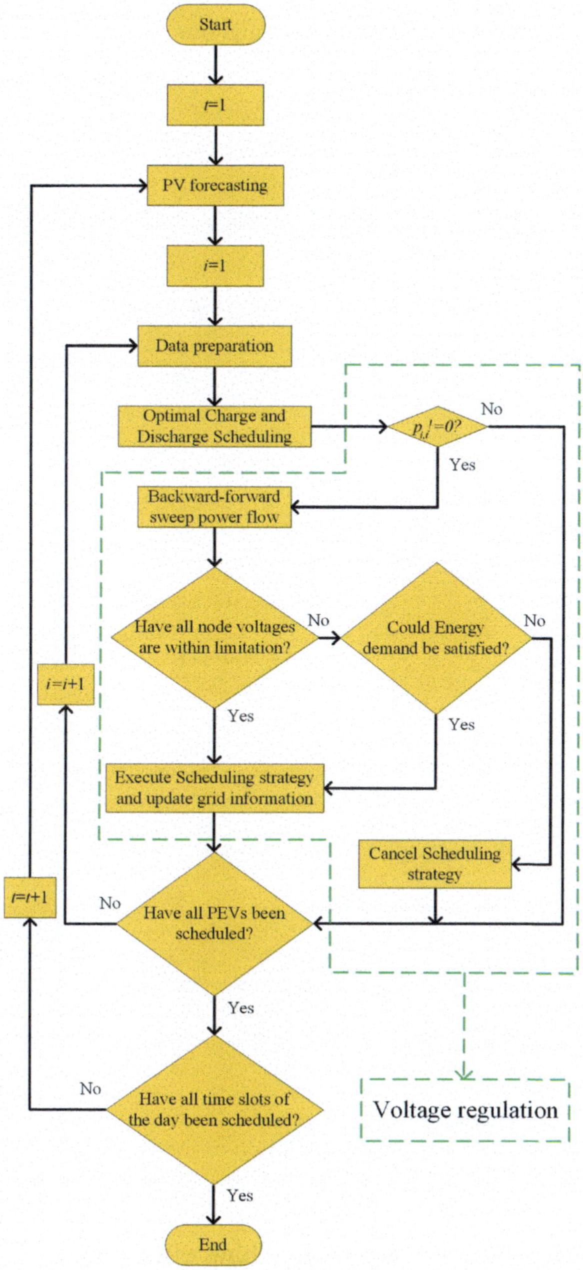

FIGURE 7.6 Flow chart of optimal energy dispatching for V2G operation.

TABLE 7.8 Algorithm for the voltage regulation.

1: **Input:** $p_{i,t}$, L_i, $i = 1,2, \ldots, N\text{-}1, N$
2: **Output:** $p_{i,t}$, $i = 1,2, \ldots, N\text{-}1, N$
3: **if** *performup/down regulation*
4: Voltage regulation is not considered since up/down regulation is calculated according to
5: The stability limits by EC.
6: **else**
7: **for** $i = 1: N$
8: **if** $p_{i,t} = 0$
9: **else**
10: Power plow calculation for $p_{i,t}$, L_i
11: **if** *no voltage value exceeds the limitation*
12: $p_{i,t} = p_{i,t}$
13: **else**
14: **if** E^{re} *can be satisfied within the remaining time slots*
15: $p_{i,t} = 0$
16: **else**
17: $p_{i,t} = p_{i,t}$
18: **end**
19: **end**
20: **end**
21: **end**
22: **end**

stability limits of the distribution grid, hence, the voltage value will not exceed the safe limitation. So, the voltage regulation of the external V2G regulation is not considered.

7.4.3 Parallel computing and dispatching efficiency

According to the two super SCP discussed in Section 7.3, this chapter divides the ISCP into two regions equally based on the scheduling algorithm in Section 7.4.1 and 7.4.2. After the optimal energy dispatching of each region is completed, the two super SCPs send the decision makings of PEVs to the EC. And then the EC conducts the voltage regulation according to the principle of "first come, first dispatch" for all PEVs. Finally, the EC sends the decision makings of PEVs after voltage regulation to the ISCP. The utilization of hierarchical computing, parallel computing and distributed computing in this chapter are described as follows.

First of all, hierarchical computing classifies scheduling tasks into two categories: load peak shaving and valley filling & self-consumption of PV output and up/down regulation. Moreover, only one scheduling task is implemented in each time period based on the instructions of EC. In this way, hierarchical computing is the subdivision of a multi-task problem on the temporal scale, and arrange single-task problems in corresponding time slots. Therefore, the hierarchical computing improves the computational efficiency on the temporal scale.

Secondly, in each time slot, parallel computing divides the dispatching task into two regions in the spatial scale. And the V2G schemes are implemented in the two regions synchronously. Besides, the scheduling results of two regions are coordinated into a whole. Therefore, parallel computing improves the computational efficiency on a spatial scale.

Finally, in the specific time slot the specific region, distributed computing carries out the distributed V2G scheduling among SCPs, and each SCP merely needs to calculate a single V2G scheduling problem of the connected PEV. Therefore, the huge, complex problem is decomposed into many subproblems, which improves the efficiency of computational. From the above discussions, it can be seen that each SCP only needs to solve a single V2G scheduling problem in each time slot. This is due to the fact that the huge, complex problem are continually refined based on three computing methods.

In addition, the current computing unit installed in the CP has the megahertz computing frequency, and the megabyte storage capacity, such as the STM32F103, which has the maximum frequency of 72 MHz and the maximum storage capacity of 1 MB; and the TMS320F28335, which has the maximum frequency of 150 MHz and the maximum storage capacity of 512 KB. Under this circumstance, the current computing and storage units can conduct a single V2G problem easily. Therefore, the existing hardware can support three kinds of computing applications. The results of the dispatching efficiency will be verified in the next Section.

7.5 Case study: Internet of smart charging points with photovoltaic integration

7.5.1 Case setup

In this chapter, a 32-bus distribution grid with five charging station bus as depicted in Fig. 7.3 is utilized to test the proposed cyber-physical co-modeling.

For the optimal energy dispatching, first of all, with the consideration of the practical circumstance and the parallel computing, two SCPs close to the substation are designated the super SCPs. And the node voltages of the distribution network is computed by the backward-forward sweep method.

Second, the PV energy is concentrated near charging stations to be consumed locally by PEVs. The real data of solar irradiance of the campus at 15-minutes intervals were derive from Solcast [39]. For the space limitation, only the typical sunny day in summer is selected in this chapter for optimal energy dispatching. It is assumed that the area of the photovoltaic panels of the campus is 40,000 m^2, the conversion efficiency of solar photovoltaic panels is 16%.

At last, the ownerships of PEVs are divided into three parts: 1 about 40% for the administrative staff part, 2 about 40% for the research staff part, and 3 about 20% for the family members of teachers who live on campus drive away. The parameters of PEV are supposed as follows: $B_i = 35.2$ kWh, -3.52 kW $\leq p_{i,t} \leq -3.52$ kW, $\eta_i^{ch} = \eta_i^{dis} = 0.98$, $SoC_i^{min} = 0.2$, $SoC_i^{max} = 0.8$. The probability distributions of the PEV charging behaviors are fitted based on the collected data and depicted in Appendix C.

For the network communication, Fig. 7.7 depicts the optimization result of the smart-world network for a charging station with 600 nodes. It can be found from Fig. 7.7a that the average path length has reached a very small value when the mean degree K is 6 and the rewiring probability β is greater than 0.2. Therefore, the mean degree $K = 6$ is selected for decreasing the wiring costs. Fig. 7.7b depicts the degree distributions of $K = 6$ for different rewiring probabilities, and $\beta = 0.5$ is selected in this chapter.

The network typology of ISCP utilizing smart-world network is illustrated in Fig. 7.8. It can be found at the right side of Fig. 7.8, the typology of the 1 CS with the 600 SCP nodes, 6 mean degrees, and 0.5 rewriting probabilities is fairly complex. For the conveniences to the readers, in the left side of Fig. 7.8, the scaled-down typology of 5 CSs is depicted, wherein each CS with the 20 SCP nodes, 6 mean degrees, and 0.5 rewriting probabilities, and five CSs connect with each other utilizing the fully meshed network. The CS 1, 2, 3, 4, and 5 correspond to nodes 3, 6, 16, 24, and 27 in Fig. 7.3, and the two SCP nodes are depicted by triangle.

Finally, the co-simulation is carried out on a desktop PC with an Intel Core i5, 2.80 GHz processor. The simulation of optimal energy dispatching is conducted in MATLAB®, and the network simulation is conducted in NS2. The packet exchanging among the EC, super SCP, and SCPs utilize the UDP protocol, and the bandwidth of the fiber optic is 100 Mbps.

7.5.2 Results of energy dispatching

Recall from Section 7.4 that the optimal energy dispatching involves the internal V2G scheduling and the external V2G regulation. The internal V2G

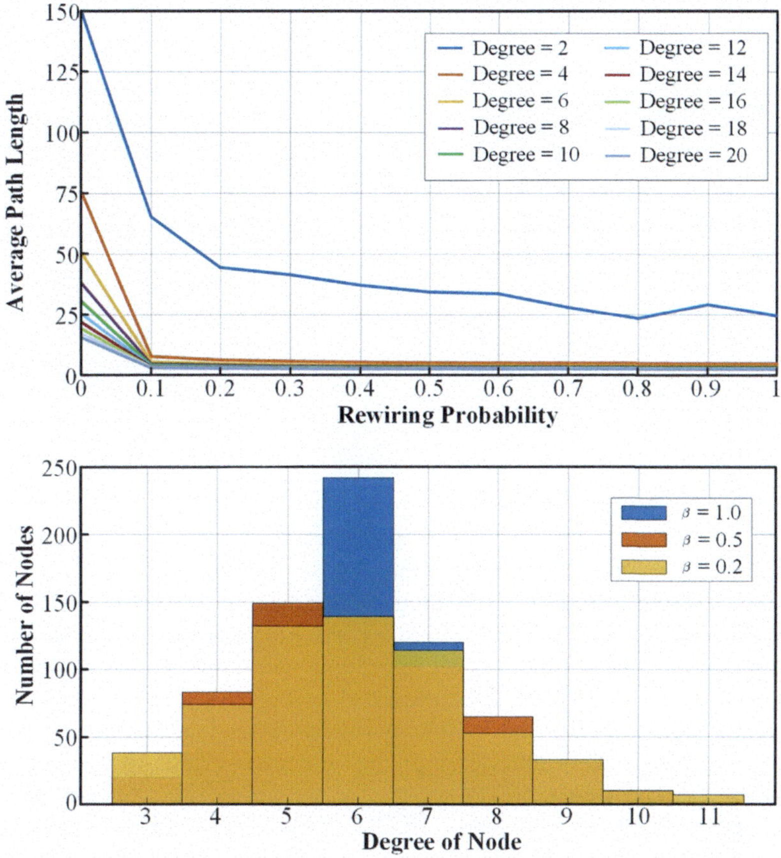

FIGURE 7.7 Optimization of smart-world network, (a) Average edges of each nodes (degree), (b) Rewiring probability.

scheduling is executed for each time slot. To be exact, it is calculated in t-1thtime slot, and carried out in tth time slot. Fig. 7.9 depicts the energy profiles for distribution grid in the internal V2G scheduling. It can be found in Fig. 7.9a that, when the uncoordinated charging is conducted, two new load peaks are created. Besides, the difference between the maximum and minimum load value becomes larger and a load valley period occurs between 12:00 to 15:00. Moreover, the performance of self-consumption of PV output by PEV charging is poor. This is because the PEV charging is conducted instantly when the PEVs plug in the SCP, which is missed the period when the PV output is high. On the contrary, it can be found in Fig. 7.9b that, when the internal V2G scheduling is conducted, the difference between the maximum load and minimum load becomes smaller, and the performance of self-consumption of PV output by PEV charging is very well. This is because

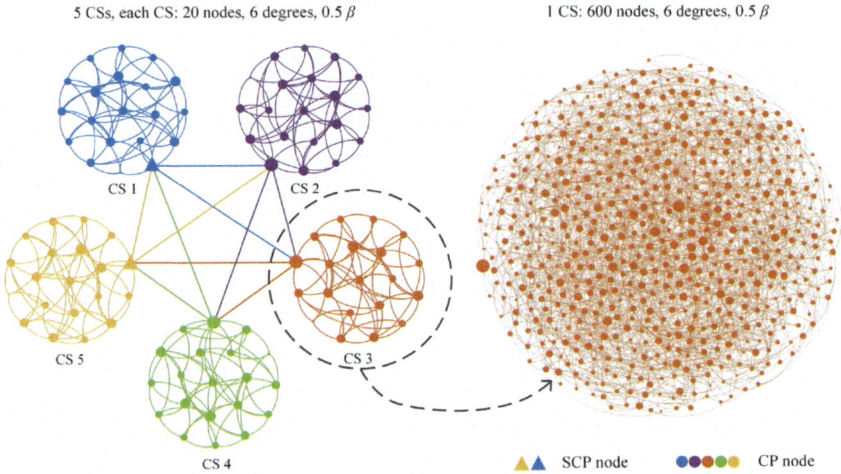

FIGURE 7.8 Network typology of ISCP utilizing smart-world network.

the ISCP utilizes the algorithm in Table 7.5, so that the profile of the PEVs charging and the profile of PV output are matched. In the other period, the ISCP utilizes the algorithm in Tables 7.4 and 7.6, and the load flatting are accomplished. In addition, the V2G scheduling and output curves reported in our previous work [38] are also illustrated in Fig. 7.9 as a benchmark. In particular, the maximum and minimum values of net load with the internal V2G scheduling algorithm during the daytime (7:00 to 19:00) are 27,066.95 kW and 19,036.46 kW, separately; the above two values of benchmark are 27,063.43 kW and 13,604.45 kW, separately; while are 31,810.52 kW and 12,618.18 kW under the uncoordinated charging. The maximum and minimum values of net load with the internal V2G scheduling algorithm at nighttime (19:00 to 7:00) are 26,573.85 kW and 12,827.6 kW, respectively; the above two values of benchmark are 26,825.15 kW and 12,770.87 kW, separately; while are 30,317.94 kW and 11,360.42 kW under the uncoordinated charging. In addition, the ratio of the self-consumption of PV output by PEV charging with the internal V2G scheduling algorithm are 83.97%; the value of benchmark is 82.72%; while 31.97% under the uncoordinated charging. It can be found that the results of this chapter and the previous work are exactly similar, or even slightly better. That is to say, the load flatting and the PV self-consumption of two works are equally perfect. Moreover, it's worth noting that, one of the highlights of this chapter is proposing a hierarchical distributed V2G coordination apt to parallel computing to achieve extremely high dispatching efficiency. The computational performance of this chapter is more than 200% higher than that of our previous work [38] (0.864,726 seconds for this chapter and 3.001,348 seconds for our previous work). For more detailed description, please refer to Section 7.5.4.

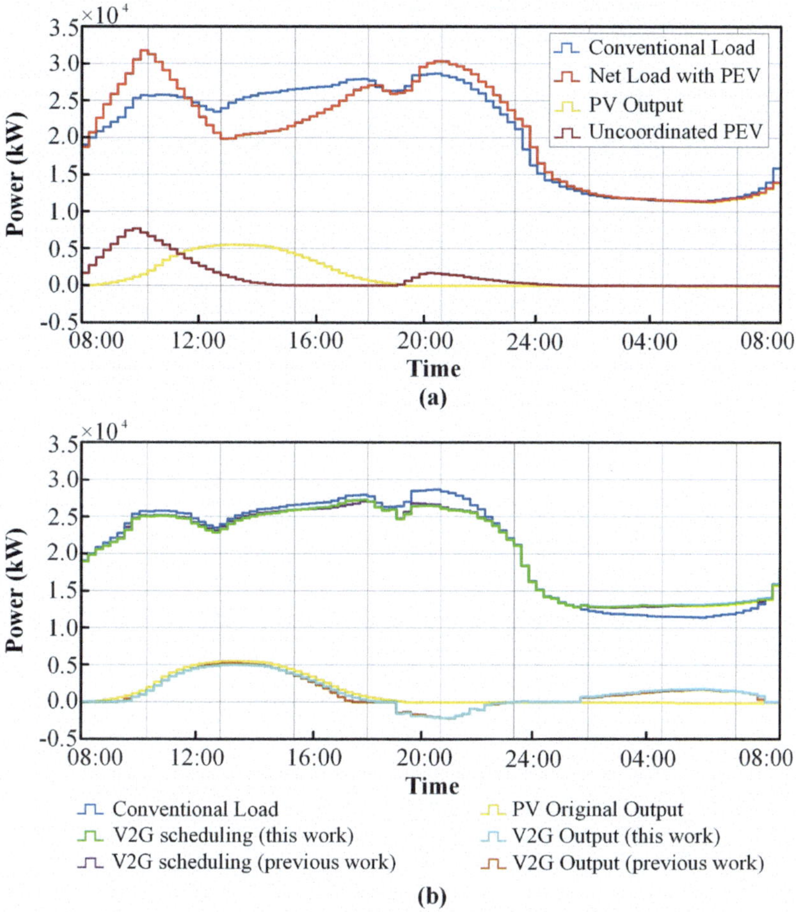

FIGURE 7.9 Energy profiles for distribution grid in the internal V2G scheduling, (a) Uncoordinated charging, (b) Internal V2G scheduling.

The voltage profiles of the internal V2G scheduling are illustrated in Fig. 7.10. It can be seen in Fig. 7.10a that, in the 48th time slot (12:00), the voltage profiles for both the uncoordinated charging and the internal V2G scheduling are within the safe range. The voltage profile of the internal V2G scheduling is lower than that of the uncoordinated charging. This is because the profile of the PEVs charging and the profile of PV output are matched. It can be found in Fig. 7.10b that, in the 80th time slot (20:00), the voltage profile of the uncoordinated charging is close to and even over the safe range. Nevertheless, the voltage profile of the internal V2G scheduling is above the safety range. This is because the 80th time slot is in the peak load period, and the PEV discharging is performed for load peak shaving, which contributes to

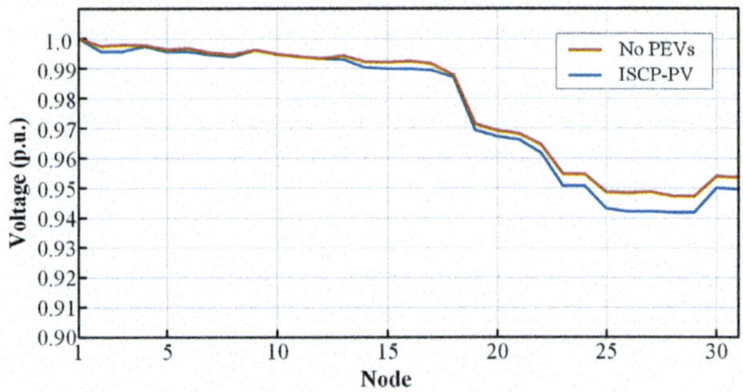

(a)

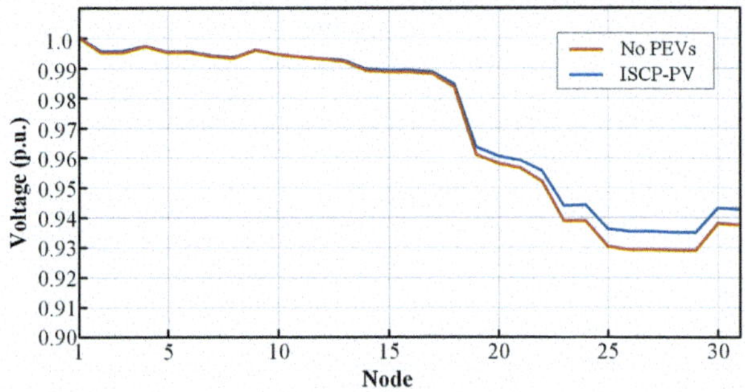

(b)

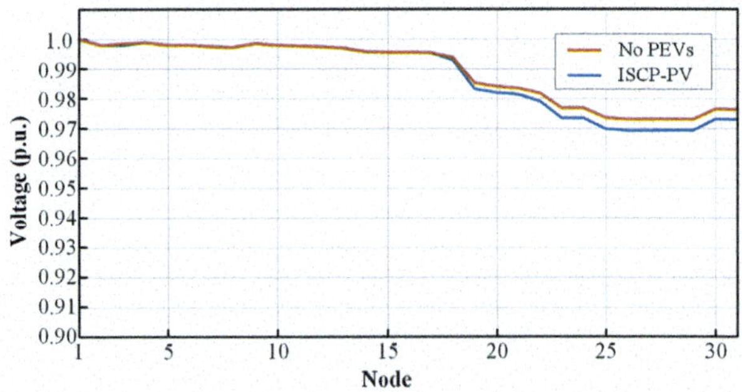

(c)

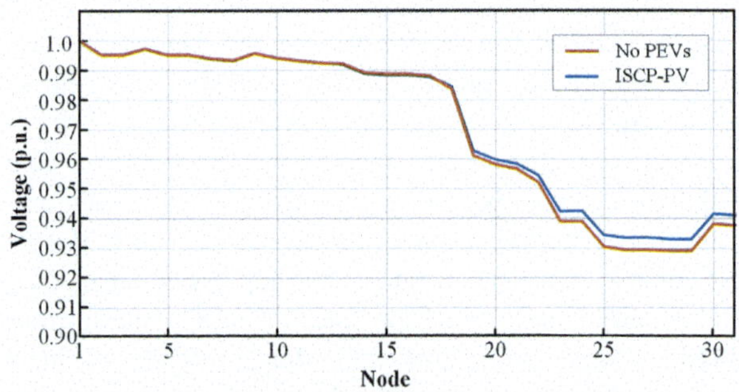

(d)

FIGURE 7.10 Voltage profiles for distribution network, (a) at 48th time slot, (b) at 80th time slot, (c) maximum voltage profile in 24 hours, (d) Minimum voltage profile in 24 hours increase the values of the voltage. It can be found in Fig. 7.10c and d that, the voltage profiles of the internal V2G scheduling are limited in the safe range, to be exact, the maximum voltage profile of the internal V2G scheduling is lower than that of the uncoordinated charging, and the minimum voltage profile of the internal V2G scheduling is higher than that of the uncoordinated charging.

The external V2G regulation is executed when the EC issues the up/down regulation signal to ISCP. To be exact, the ISCP receives the signal and calculate the decision makings in *t*-1st time slot, and carries it out in *t*th time slot. Fig. 7.11 depicts the energy profiles for distribution grid in the external V2G regulation. It can be found in Fig. 7.11a that, the up/down regulation signals are illustrated. The first regulation signal is the up regulation (−6000 kW), which starts at the 42nd time slot (PV generation period), and lasts for 15 minutes. The second regulation signal is the down regulation (2400 kW), which starts at the 85th time slot, and lasts for 15 minutes. Once

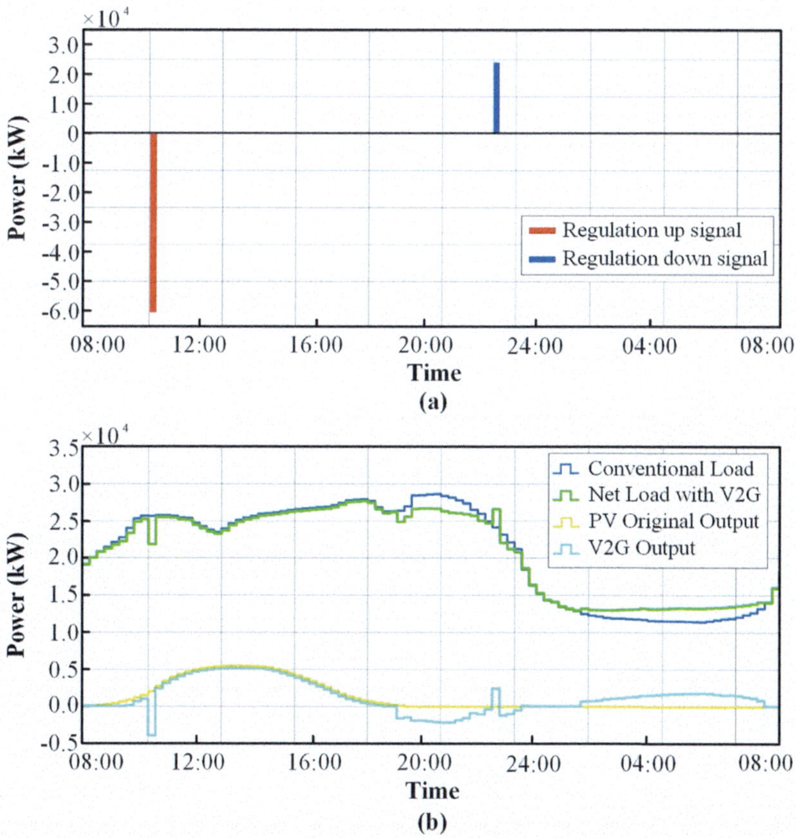

FIGURE 7.11 Energy profiles of distribution grid in the external V2G regulation, (a) Uncoordinated charging, (b) V2G scheduling.

receiving the signal, the ISCP utilizes the algorithm in Table 7.7 to satisfy the desired regulation. It can be found in Fig. 7.11b that both the desired regulations are achieved. In the first external V2G regulation, the number and discharging power of PEVs are 2342 and 1.70 kW, respectively, and the PV supports the discharging power of 2511.66 kW. In the second external V2G regulation, the number and charging power of PEVs are 710 and 3.42 kW, respectively, since there are 7 PEVs must discharging in the *t*th time slot based on the internal V2G scheduling.

7.5.3 Results of network simulation

Three cases are considered to demonstrate the effectiveness and efficiency of this chapter. Case 1 is the flow chart of Fig. 7.6 and parallel computing is applied; Case 2 is the flow chart of Fig. 7.6 but parallel computing is not applied; Case 3 is utilized from the previous work [38]. From the network simulation perspective, the specifications of the different steps of the three cases are depicted in Tables 7.9 and 7.10, respectively. For the internal V2G scheduling, in step 1, the packet size of Case 1 and Case 2 is 584 bytes (8 bytes for UDP header, 384 bytes for floating-point type of the conventional load prediction data in 96 time slots, and 192 bytes for the PV output prediction data in 48 time slots). However, the packet size of Case 3 is 1228 bytes since the grid information includes not only 8 bytes for UDP header, and 580 bytes for

TABLE 7.9 Specifications for the different steps of network simulation for the internal V2G scheduling.

			Packet size (bytes)		
Step	Participants	Information	Case 1	Case 2	Case 3
1	EC to super SCP	Grid information	584	584	1224
2	SCP to SCP	Updated grid information	584/1500 times	584/3000 times	1224/3000 times
4	SCPs to super SCP	V2G decisions	16/30 times	16/60 times	–
5	Super SCP to EC	V2G decisions	12,008	24,008	–
7	EC to super SCP	V2G decisions after check	12,008	24,008	–
8	Super SCP to SCPs	V2G decisions after check	12/30 times	12/60 times	–

TABLE 7.10 Specifications for the different steps of network simulation for the external V2G regulation.

Step	Participants	Information	Packet size (bytes)	
			Case 1	Case 2
1	EC to super SCP	Grid information	12	12
2	SCP to SCP	Updated grid information	12/1500 times	12/3000 times
4	SCPs to super SCP	V2G decisions	16/30 times	16/60 times
5	Super SCP to EC	V2G decisions	–	–
7	EC to super SCP	V2G decisions after check	–	–
8	Super SCP to SCPs	V2G decisions after check	12/30 times	12/60 times

the conventional load data the PV output data, but also 640 bytes for the power flow data. In step 2, the updated grid information is transmitted around SCPs so that the packet size for each Case is the same as step 1. It should be noted that the times of the data transmitting from SCP to SCP are 1500 for Case 1 due to the parallel computing applied. The packet size for Case 1 and Case 2 in step 4 is 16 bytes (8 bytes for UDP header, 4 bytes for the decision-making data, and 4 bytes for the node location). However, the link will be jammed and the packet loss will occur if all SCPs send data to super SCP at the same time. Therefore, the SCPs need to be classified in group equally, and the decision makings are delivered from the SCPs to super SCP by grouping. According to our simulation, the group within 50 SCPs sends information without jam. Correspondingly, the times of the data transmitting from the SCPs to the super SCP is 30 for Case 1. The times of the data transmitting from the SCPs to the super SCP is 60 for Case 3 since the parallel computing is not applied. The packet size for Case 1 in step 5 and 7 are 12,008 bytes (8 bytes for UDP header and 8*1500 bytes for the decision-making data for one super SCP) due to the parallel computing applied. However, the packet size for Case 2 in step 5 and 7 are 24,008 bytes (8 bytes for UDP header and 8*3000 bytes for the decision-making data for super SCP) due to no parallel computing applied. At last, the packet size for Case 1 and Case 2 in step 8 is 12 bytes (8 bytes for UDP header, 4 bytes for the decision-making data). The times of the data transmitting from the super SCP to the SCPs are the same as step 3. In particular, the packet size for Case 3 in steps 3, 4, 5, 7, and 8 is null since there is no external check in Case 3.

For the external V2G regulation, in step 1, the packet size of Case 1 and Case 2 is 12 bytes (8 bytes for UDP header, and 4 bytes for the signal of the capacity for up/down regulation). However, the packet size of Case 3 is 652 bytes since the grid information includes not only 8 bytes for UDP header, and 4 bytes for the signal of the capacity for up/down regulation, but also 640 bytes for the power flow data. In step 2, the updated grid information is transmitted around SCPs so that the packet size for each Case is the same as step 1. And the times of the data transmitting from SCP to SCP are 1500 for Case 1 due to the parallel computing applied. The specifications for the steps 4 and 8 are the same as those for the internal V2G scheduling. The steps 5 and 7 do not exist since the voltage regulation of the external V2G regulation is not considered. Therefore, the packet sizes of these steps are null.

The transmission media in this chapter is wired for network stability. And the fiber optic is selected for high transmission distance, data rate, and bandwidth. It should be noted that other transmission media such as the twisted pair and the coaxial cable can also be considered according to the different situations. Therefore, the specification for the different transmission media is depicted in Table 7.11 for convenience of readers.

We implemented the network simulation for the Step 1, 2, 4, 5, 7, 8 described in Section 7.3.3 to obtain the delays of the communication process. Fig. 7.12 depicts the delay and the packet loss with bandwidth = 100 Mbps, packet size = 100 bytes, link distance = 1 km for different data rates. It can be found in Fig. 7.12a that, when the data rate increases from 10 Mbps to 100 Mbps, the propagation and queueing delay is almost constant, and the transmission delay gradually decreases based on Formula (7.3). Hence, the total delay gradually decreases. When the data rate exceeds 100 Mbps, the transmission delay continues to decrease gradually. However, the propagation and queueing delay became very large suddenly. This is because the packet loss appears and the communication link is jammed. Fig. 7.12b depicts the fitted curves for the total delay and the packet loss. It can be found that although the data rate is small, the total delay can also be large. And it can be found that the

TABLE 7.11 Specification for the different transmission media.

Type	Distance	Data rate	Bandwidth	Propagation speed
Twisted pair	Up to 100m	Low	Low	59% c
Coaxial	Up to 500m	Moderately high	Moderately high	Up to 77% c
Fiber optic	Up to 80 km	Very high	Very high	67% c

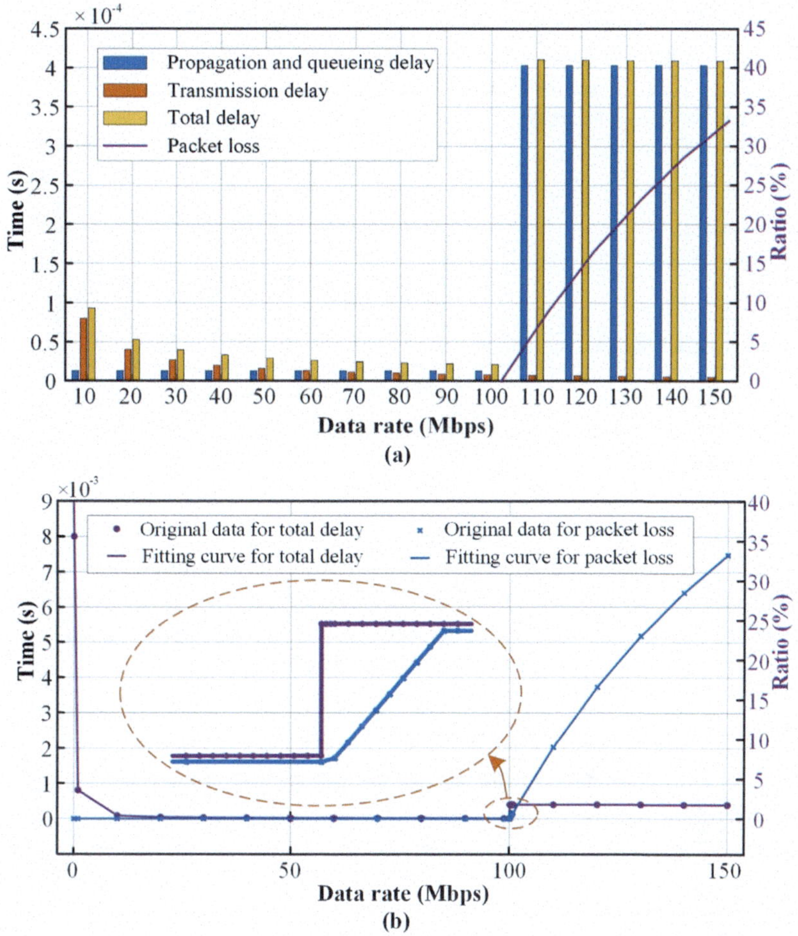

FIGURE 7.12 Network simulation for different data rates (bandwidth = 100 Mbps, packet size = 100 bytes).

details of the data rate around 100 Mbps, that is, when the data rate is just over 100, the packet loss and the link congestion will occur, and the total delay will appear a step. Therefore, in the following network simulation, the data rate is set as 40 Mbps to maintain a low delay and leave the margin of data rate for the other applications.

According to the bandwidth = 100 Mbps, link distance = 1 km, and the selected data rate = 40 Mbps, the communication delays of different packet sizes in Fig. 7.13. It can be found that the propagation and queueing delay, the transmission delay, and the total delay are all increased with the increase of packet size. In particular, the delays of Step 2, which is the key step, for three

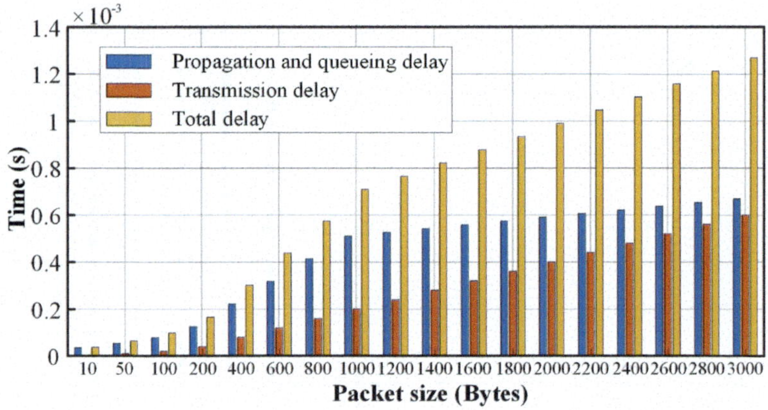

FIGURE 7.13 Network simulation for different packet size, (bandwidth = 100 Mbps, data rate = 40 Mbps).

cases are depicted in Fig. 7.14. It can be found that the delay of Case 1 is the best since the packet size is small and the parallel computing is utilized, the delay of Case 2 is the medium since the packet size is small but the parallel computing is not utilized, and the delay of Case 3 is the worst since the packet size is large and the parallel computing is not utilized.

7.5.4 Performance comparison

Combined with the delays of the energy optimization in Step 3 and Step 6, the time consuming of the different steps for the different cases are illustrated in Tables 7.12 and 7.13 For the internal V2G scheduling, it can be found in

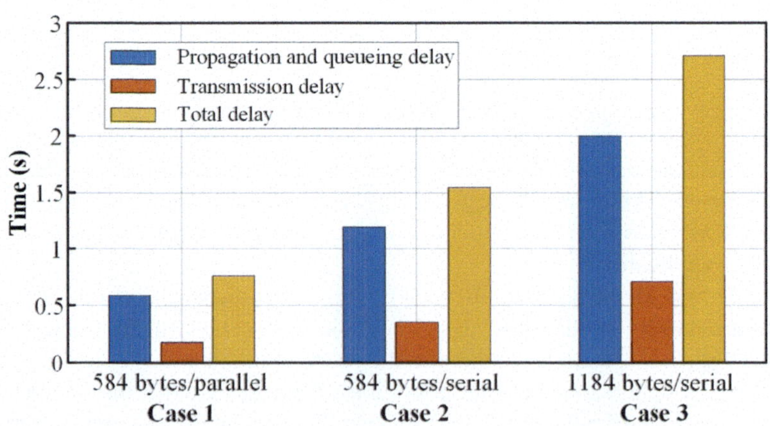

FIGURE 7.14 Network simulation of step 2 for different cases.

TABLE 7.12 Time-consuming of the different processes for the different cases within internal V2G scheduling.

Case		Step 1	Step 2	Step 3	Step 4	Step 5	Step 6	Step 7	Step 8	Sum
1	Trans & queue	0.000072	0.590652	–	0.001109	0.000986	–	0.000986	0.000931	0.594736
	Propagation	0.000117	0.175201	–	0.000096	0.002402	–	0.002402	0.003603	0.183821
	Total	0.000189	0.765853	0.041025	0.001205	0.003388	0.052060	0.003388	0.004534	**0.871642**
2	Trans & queue	0.000072	1.197065	–	0.002222	0.001946	–	0.001946	0.001862	1.205113
	Propagation	0.000117	0.350403	–	0.000192	0.004802	–	0.004802	0.007206	0.367522
	Total	0.000189	1.547468	0.089333	0.002414	0.006748	0.099969	0.006748	0.009068	**1.761937**
3	Trans & queue	0.000119	2.001321	–	–	–	–	–	–	2.00144
	Propagation	0.000237	0.710399	–	–	–	–	–	–	0.710636
	Total	0.000356	2.711720	0.289271	–	–	–	–	–	**3.001347**

TABLE 7.13 Time-consuming of the different processes for the different cases within external V2G regulation.

Case		Step 1	Step 2	Step 3	Step 4	Step 5	Step 6	Step 7	Step 8	Sum
1	Trans & queue	0.000026	0.027412	–	0.001109	–	–	–	0.000931	0.029478
	Propagation	0.000003	0.003601	–	0.000096	–	–	–	0.003603	0.007303
	Total	0.000029	0.031013	0.003929	0.001205	–	–	–	0.004534	**0.040711**
2	Trans & queue	0.000026	0.063003	–	0.002222	–	–	–	0.001862	0.067113
	Propagation	0.000003	0.007202	–	0.000192	–	–	–	0.007206	0.014603
	Total	0.000029	0.070205	0.007858	0.002414	–	–	–	0.009068	**0.089574**

TABLE 7.14 Comparison among different topologies concerning communication efficiency and wring cost.

Network topology	Lattice ($K = 2$)	Lattice ($K = 6$)	Small world	Fully meshed
Delay (s)	53.7149	18.6315	0.590653	0.109501
Wring cost	3×10^3	9×10^3	9×10^3	4.4985×10^6

Table 7.12 that the sum of time-consuming of Case 1 is the best, that is, 0.864,726 seconds, which shows the excellent performance of time consumption. The sum of time-consuming of Case 2 and Case 3 are 1.748102 and 3.001,348 seconds, respectively. For the external V2G regulation, it can be found in Table 7.13 that the sum of time-consuming of Case 1 is the best, that is, 0.040,711, which shows the excellent performance of time consumption. The sum of time-consuming of Case 2 is 0.089,574 seconds.

The communication efficiency and the wring cost are compared for different topologies as shown in Table 7.14. On the one hand, the delay of the Step 2 for the four topologies are simulated in NS2. It can be found that the delays of lattice network (both $K = 2$ and $K = 6$) are 53.7149 and 18.6315 seconds respectively. They are much larger than the delays of small world network (0.590,653 seconds) and full mesh network (0.109,501 seconds). The communication delay of small-world network is 90.94 times ($K = 2$) or 31.54 times ($K = 6$) faster than that of lattice network. The communication delay of small-world network is 490.54 times ($K = 2$) or 170.15 times ($K = 6$) faster than that of lattice network. This is because the number of data transfers is relatively small in the latter two networks. On the other hand, we use the number of edges as a measure of wiring cost. It can be found that the wring cost of the both lattice networks are relatively little, however, the long communication delay determines that the lattice network is unsuitable. In addition, although the delay of a full mesh network is smaller than that of the small world network, the wiring cost of the fully meshed network is extraordinarily high (4.4985×10^6), which is three orders of magnitude higher than the small world network (9×10^3). Therefore, the small network utilized in this chapter is cost-effective.

7.6 Concluding remarks

This chapter has examined the ISCP as a scalable and intelligent solution for V2G operation in modern distribution grids. Starting from the smart charging point as the basic functional unit, the chapter discussed how sensing, control, communication, storage, and local computation can be integrated at the charging interface, thereby transforming conventional charging infrastructure into a cyber-physical energy node. On this basis, the communication architecture, distributed computing process, and privacy-preserving mechanism of ISCP were introduced, providing the technical foundation for large-scale and real-time coordination of EVs.

A major theme of this chapter is that effective V2G implementation depends not only on optimization models in the physical energy system, but also on the organization of information flow in the cyber layer. For this reason, a cyber-physical co-modeling framework was further established to connect distribution-grid constraints, charging-point interaction, and communication-network behavior in a unified manner. Within this framework, internal V2G scheduling was used to support load peak shaving, valley filling, photovoltaic self-consumption, and voltage security, while external V2G regulation enabled the charging network to respond to grid-side up/down dispatch instructions. The combination of distributed coordination and parallel computing further improved real-time applicability by decomposing the large-scale scheduling task into manageable sub-problems and accelerating the execution process.

The case studies show that such an ISCP-based framework can deliver multiple benefits simultaneously. At the energy-management level, coordinated scheduling improves the operational compatibility between EVs and photovoltaic generation, while also enhancing load-shaping performance and supporting distribution-grid operation. At the communication level, appropriate cyber-topology design reduces latency and wiring burden, which is essential for the practical deployment of large charging-point networks. At the implementation level, local information processing and storage improve privacy protection for both EV users and grid operators, and the computational burden can be maintained within a time scale suitable for real-time or near-real-time applications.

Overall, the internet of smart charging points provides a promising pathway for moving from isolated charging facilities toward coordinated, networked, and grid-supportive EV infrastructures. Its significance lies not only in improving charging efficiency, but also in enabling EVs to participate as flexible resources in active distribution systems. Looking ahead, further development may extend this framework toward broader market participation, stronger cyber-security mechanisms, richer user-interaction models, and larger-scale multi-energy coordination. These directions are important for advancing practical V2G deployment and for supporting the deeper integration of EVs into low-carbon power systems.

References

[1] Chen W, Li J, Bao Q, Gao Z, Cheng T, Yu Y. Evaluation of straw open burning prohibition effect on provincial air quality during October and November 2018 in jilin province. Atmosphere July 2019;10(7):375.

[2] Zhang X, Wang Q, Qin W, Guo L. Sustainable policy evaluation of vehicle exhaust control—Empirical data from China's air pollution control. Sustainability 2019;12(1):1–8.
[3] Salah WA, Albreem MA, Alsayid B, Zneid BA, Alkhasawneh M, Al-Mofleh A, et al. Electric vehicle technology impacts on energy. Int J Power Electron Drive Syst March 1, 2019;10(1):1.
[4] Williams G. Rare earths outlook 2019: EV production to drive demand, vol 25; 2019. p. 2–6.
[5] Zheng Y, Shao Z, Zhang Y, Jian L. A systematic methodology for mid-and-long term electric vehicle charging load forecasting: The case study of Shenzhen, China. Sustain Cities Soc May 1, 2020;56:102084.
[6] Nikoobakht A, Aghaei J, Khatami R, Mahboubi-Moghaddam E, Parvania M. Stochastic flexible transmission operation for coordinated integration of plug-in electric vehicles and renewable energy sources. Appl energy March 15, 2019;238:225–38.
[7] Liu C, Chau KT, Wu D, Gao S. Opportunities and challenges of vehicle-to-home, vehicle-to-vehicle, and vehicle-to-grid technologies. Proc IEEE 2013;101(11):2409–27.
[8] García-Villalobos J, Zamora I, Knezović K, Marinelli M. Multi-objective optimization control of plug-in electric vehicles in low voltage distribution networks. Appl Energy October 15, 2016;180:155–68.
[9] Zecchino A, Prostejovsky AM, Ziras C, Marinelli M. Large-scale provision of frequency control via V2G: The Bornholm power system case. Elec Power Syst Res May 1, 2019;170:25–34.
[10] Zhao J, Wan C, Xu Z, Wong KP. Spinning reserve requirement optimization considering integration of plug-in electric vehicles. IEEE Trans Smart Grid August 2, 2016;8 (4):2009–21.
[11] Cao Y, Kaiwartya O, Wang R, Jiang T, Cao Y, Aslam N, et al. Toward efficient, scalable, and coordinated on-the-move EV charging management. IEEE Wireless Commun April 24, 2017;24(2):66–73.
[12] Rassaei F, Soh WS, Chua KC. Distributed scalable autonomous market-based demand response via residential plug-in electric vehicles in smart grids. IEEE Trans Smart Grid November 12, 2016;9(4):3281–90.
[13] Fathabadi H. Novel stand-alone, completely autonomous and renewable energy based charging station for charging plug-in hybrid electric vehicles (PHEVs). Appl Energy February 15, 2020;260:114194.
[14] Zhang T, Pota H, Chu CC, Gadh R. Real-time renewable energy incentive system for electric vehicles using prioritization and cryptocurrency. Appl Energy September 15, 2018;226:82–594.
[15] Zheng Y, Shang Y, Shao Z, Jian L. A novel real-time scheduling strategy with near-linear complexity for integrating large-scale electric vehicles into smart grid. Appl Energy May 1, 2018;217:1–13.
[16] Cardona JE, López JC, Rider MJ. Decentralized electric vehicles charging coordination using only local voltage magnitude measurements. Elec Power Syst Res August 2018;161:139–51.
[17] An D, Yang Q, Yu W, Li D, Zhao W. LoPrO: Location privacy-preserving online auction scheme for electric vehicles joint bidding and charging. Future Gener Comput Syst June 1, 2020;107:394–407.
[18] Yucel F, Akkaya K, Bulut E. Efficient and privacy preserving supplier matching for electric vehicle charging. Ad Hoc Netw July 1, 2019;90:101730.

[19] Jian L, Zheng Y, Xiao X, Chan CC. Optimal scheduling for vehicle-to-grid operation with stochastic connection of plug-in electric vehicles to smart grid. Appl Energy May 15, 2015;146:150–61.

[20] Dong X, Mu Y, Xu X, Jia H, Wu J, Yu X, et al. A charging pricing strategy of electric vehicle fast charging stations for the voltage control of electricity distribution networks. Appl Energy September 1, 2018;225:857–68.

[21] Shang Y, Zheng Y, Jian L. A user independent choice charging strategy for electric vehicle in household smart micro-grid. IEEE Int Conf Inf Autom (ICIA) July 2017;18:925–30.

[22] Yang Z, Li K, Foley A. Computational scheduling methods for integrating plug-in electric vehicles with power systems: A review. Renew Sustain Energy Rev November 1, 2015;51:396–416.

[23] Wang Y, Wang X, Shao C, Gong N. Distributed energy trading for an integrated energy system and electric vehicle charging stations: A Nash bargaining game approach. Renew Energy March 6, 2020:254.

[24] Du J, Ouyang D. Progress of Chinese electric vehicles industrialization in 2015: A review. Appl Energy February 15, 2017;188:529–46.

[25] Zheng Y, Niu S, Shang Y, Shao Z, Jian L. Integrating plug-in electric vehicles into power grids: A comprehensive review on power interaction mode, scheduling methodology and mathematical foundation. Renew Sustain Energy Rev September 1, 2019;112:424–39.

[26] Fachrizal R, Shepero M, van der Meer D, Munkhammar J, Widén J. Smart charging of electric vehicles considering photovoltaic power production and electricity consumption: a review. eTransp March 24, 2020:100056.

[27] Hannan MA, Tan SY, Al-Shetwi AQ, Jern KP, Begum RA. Optimised controller for renewable energy sources integration into microgrid: Functions, constraints and suggestions. J Clean Prod February 4, 2020:120419.

[28] Mozafar MR, Moradi MH, Amini MH. A simultaneous approach for optimal allocation of renewable energy sources and electric vehicle charging stations in smart grids based on improved GA-PSO algorithm. Sustain Cities Soc July 1, 2017;32:627–37.

[29] Wu H, Pang GK, Choy KL, Lam HY. A charging-scheme decision model for electric vehicle battery swapping station using varied population evolutionary algorithms. Appl Soft Comput December 1, 2017;61:905–20.

[30] Hafez O, Bhattacharya K. Integrating EV charging stations as smart loads for demand response provisions in distribution systems. IEEE Trans Smart Grid June 6, 2016;9 (2):1096–106.

[31] Tang Y, Yang J, Yan J, He H. Intelligent load frequency controller using GrADP for island smart grid with electric vehicles and renewable resources. Neurocomputing December 25, 2015;170:406–16.

[32] Liu T, Hu X. A bi-level control for energy efficiency improvement of a hybrid tracked vehicle. IEEE Trans Ind Inf January 24, 2018;14(4):1616–25.

[33] Shi W, Cao J, Zhang Q, Li Y, Xu L. Edge computing: Vision and challenges. IEEE Internet Things J June 9, 2016;3(5):637–46.

[34] Zhou Z, Tan L, Xu G. Blockchain and edge computing based vehicle-to-grid energy trading in energy internet. In: 2018 2nd IEEE Conference on Energy Internet and Energy System Integration (EI2); 2018. p. 1–5 [Beijing].

[35] Xu X, Xue Y, Qi L, Yuan Y, Zhang X, Umer T, et al. An edge computing-enabled computation offloading method with privacy preservation for internet of connected vehicles. Future Gener Comput Syst 2019;96:89–100.

[36] Tan KM, Ramachandaramurthy VK, Yong JY, Tariq M. Experimental verification of a flexible vehicle-to-grid charger for power grid load variance reduction. Energy April 2021;7:120560.

[37] Watts DJ, Strogatz SH. Collective dynamics of small-world networks. Nature 1998;393 (6684):440–2.

[38] Shang Y, Liu M, Shao Z, Jian L. Internet of smart charging points with photovoltaic Integration: A high-efficiency scheme enabling optimal dispatching between electric vehicles and power grids. Appl Energy 2020;278:115640.

[39] Solcast. Global solar irradiance data and PV system power output data. URL, https://solcast.com/; 2019.

Index

'*Note:* Page numbers followed by "f" indicate figures and "t" indicate tables.'